A First Course in Differential Geometry

Surfaces in Euclidean Space

Differential geometry is the study of curved spaces using the techniques of calculus. It is a mainstay of undergraduate mathematics education and a cornerstone of modern geometry. It is also the language used by Einstein to express general relativity, and so is an essential tool for astronomers and theoretical physicists.

This introductory textbook originates from a popular course given to third-year students at Durham University for over 20 years, first by the late Lyndon Woodward and then by John Bolton (and others). It provides a thorough introduction by focussing on the beginnings of the subject as studied by Gauss: curves and surfaces in Euclidean space. While the main topics are the classics of differential geometry – the definition and geometric meaning of Gaussian curvature, the Theorema Egregium of Gauss, geodesics, and the Gauss–Bonnet Theorem – the treatment is modern and student-friendly, taking direct routes to explain, prove, and apply the main results. It includes many exercises to test students' understanding of the material, and ends with a supplementary chapter on minimal surfaces that could be used as an extension towards advanced courses or as a source of student projects.

John Bolton earned his Ph.D. at the University of Liverpool and joined Durham University in 1970, where he was joined in 1971 by **Lyndon Woodward,** who obtained his D.Phil. from the University of Oxford. They embarked on a long and fruitful collaboration, co-authoring over 30 research papers in differential geometry, particularly on generalisations of "soap film" surfaces. Between them, they have over 70 years' teaching experience, being well regarded as enthusiastic, clear, and popular lecturers. Lyndon Woodward passed away in 2000.

A First Course in Differential Geometry

Surfaces in Euclidean Space

L. M. WOODWARD
University of Durham

J. BOLTON
University of Durham

CAMBRIDGE
UNIVERSITY PRESS

CAMBRIDGE
UNIVERSITY PRESS

University Printing House, Cambridge CB2 8BS, United Kingdom

One Liberty Plaza, 20th Floor, New York, NY 10006, USA

477 Williamstown Road, Port Melbourne, VIC 3207, Australia

314–321, 3rd Floor, Plot 3, Splendor Forum, Jasola District Centre, New Delhi – 110025, India

79 Anson Road, #06–04/06, Singapore 079906

Cambridge University Press is part of the University of Cambridge.

It furthers the University's mission by disseminating knowledge in the pursuit of education, learning, and research at the highest international levels of excellence.

www.cambridge.org
Information on this title: www.cambridge.org/9781108424936
DOI: 10.1017/9781108348072

First published 2019
Reprinted 2020

Printed in the United Kingdom by TJ International Ltd. Padstow, Cornwall 2019

A catalogue record for this publication is available from the British Library.

Library of Congress Cataloging-in-Publication Data
Names: Bolton, John (Mathematics professor), author. | Woodward, L. M. (Lyndon M.), author.
Title: A first course in differential geometry surfaces in Euclidean space / John Bolton (University of Durham), L.M. Woodward.
Description: Cambridge ; New York, NY : Cambridge University Press, 2019. | Includes index.
Identifiers: LCCN 2018037033 | ISBN 9781108424936
Subjects: LCSH: Geometry, Differential – Textbooks.
Classification: LCC QA641 .B625925 2019 | DDC 516.3/6–dc23
LC record available at https://lccn.loc.gov/2018037033

ISBN 978-1-108-42493-6 Hardback
ISBN 978-1-108-44102-5 Paperback

Additional resources for this publication at www.cambridge.org/Woodward&Bolton

Brief Contents

Contents

Sections marked with † denote optional sections

Preface

We believe that the differential geometry of surfaces in Euclidean space is an ideal topic to present at advanced undergraduate level. It allows a mix of calculational work (both routine and advanced) with more theoretical material. Moreover, one may draw pictures of surfaces in Euclidean 3-space, so that the results can actually be visualised. This helps to develop geometrical intuition, and at the same time builds confidence in mathematical methods. One of our aims is to convey our enthusiasm for, and enjoyment of, this subject.

The book covers material presented for many years to advanced undergraduate Mathematics and Natural Sciences students at Durham University in a module entitled "Differential Geometry". This module constitutes one sixth of the academic content of their third year. The two main prerequisites are basic linear algebra and many-variable calculus.

We have three main targets.

(i) Gaussian curvature: we seek to explain this important function, and illustrate the geometrical information it carries. We further demonstrate its importance when we discuss the Theorema Egregium of Gauss.

(ii) Geodesics: these are the most important and interesting curves on a surface. They are the analogues for surfaces of straight lines in a plane.

(iii) The Gauss–Bonnet Theorem: among other things, this theorem shows that Gaussian curvature (which is defined using local properties of a surface) influences the global overall properties of that surface.

The Theorema Egregium and the Gauss–Bonnet Theorem are both very surprising, but readily understood and appreciated. They are also very important and influential from a historical perspective, having had a profound effect on the development of differential geometry as a whole.

We have tried to present the material needed to attain these targets using the minimum amount of theory, and have, for the most part, resisted the temptation to include extra material (but this resistance has crumbled spectacularly in Chapter 9!). This means that we have been rather selective in our choice of applications and results. However, each chapter contains some optional material, clearly signposted by a dagger symbol †, to provide flexibility in the module and to add interest and mental stimulation to the more committed student. The optional material also provides opportunities for additional reading as the module progresses.

There should be time to cover at least some of the optional material, and choices may be made between the technical, the slightly more advanced, and some interesting topics which are not specifically needed to attain our three targets mentioned above. There is also some optional material on surfaces in higher dimensional Euclidean spaces (and on general abstract surfaces), which is designed to whet the appetite of the students, and help the transition to more advanced topics.

In a forty lecture module, we would suggest that the material in the first four chapters should be covered in the first half of the module (and perhaps a start made on Chapter 5), with between four and six lectures on each of the first three chapters, and perhaps four lectures on the material in Chapter 4.

The pace picks up in the second half of the module. We suggest seven lectures for Chapter 5 and three for Chapter 6. Five lectures could be allowed for Chapter 7, and four for Chapter 8. However, this may only be achievable if students are asked to read for themselves the proofs of some of the results.

This may leave a couple of lectures to briefly discuss the contents of the optional Chapter 9 (on minimal and CMC surfaces). Although the material in this chapter is more advanced, it is included because the mathematics is so beautiful, and is suitable for self-study by an interested student. It could also form the starting point of a student project at senior undergraduate or beginning postgraduate level.

Our aim throughout is to make the material appealing and understandable, while at the same time building up confidence and geometrical intuition. Topics are presented in bite-sized sections, and concrete criteria or formulae are clearly stated for the various objects under discussion. We give as many worked examples as possible, given the time constraints imposed by the module, and have also included many exercises at the end of each of the chapters (and provided brief hints or solutions to some of them). On-line solutions to all the exercises are available to instructors on application to the publishers.

We have been heavily influenced by the excellent text *Differential Geometry of Curves and Surfaces* by Manfredo Do Carmo (Dover Books on Mathematics). However, we have omitted many of the more advanced topics found in that book, and at the same time have further elucidated, where we thought appropriate, the material we believe may be reasonably covered in our forty lecture module.

Finally, our sincere thanks to Roger Astley and his team at Cambridge University Press, who have been encouraging and patient throughout the rather long gestation period of the book.

Please enjoy the book.

Internal referencing

There are inevitably very many definitions which have to be included in a book of this nature. Rather than numbering these and referring back to them each time they are used, we thought it best to italicise the terms being defined and then include all these terms in the index.

Results and Examples are numbered in a single sequence within each section. A typical internal reference might be, for instance, Theorem 3 of §2.5. If no section reference is given, the result or example is in the current section. Equations to be referred to later in the book are numbered consecutively within each chapter (so, for instance, equation (3.7) is the seventh numbered equation in Chapter 3).

1 Curves in $\mathbb{R}^n$

This book provides an account of the differential geometry of surfaces, principally (but not exclusively) in Euclidean 3-space. We shall be studying their metric geometry; both internal, or *intrinsic* geometry, and their external, or *extrinsic* geometry.

As a preliminary, in this chapter we study curves in the vector space $\mathbb{R}^n$ with its standard inner product. For the most part n will be 2 or 3 since we wish to emphasize the geometrical aspects in a way which can be easily visualized. The crucial properties of the curves we study are that they are 1-dimensional and may be approximated up to first order near any point by a straight line, the *tangent line* at that point. The intrinsic geometry of these curves is somewhat simple, consisting of the *arc length* along the curve between any two points on the curve, while the most important measure of the extrinsic geometry is the *curvature*, the rate at which the curve bends away from its tangent line.

The ideas in this chapter are important for what follows in the rest of the book for several reasons. Firstly, many of the ideas extend in a natural way to surfaces (and to the more general study of n-dimensional objects called *differentiable manifolds*), and so a number of important concepts are introduced here in the simplest possible situation. Secondly, the intrinsic and extrinsic geometry of a surface are most easily and intuitively studied by using curves on the surface. For instance, the geometry of a surface may be studied by means of its *geodesics*, which are the analogues for surfaces of straight lines in the plane. Finally, curves on a surface may often be regarded in a natural way as curves in the plane where this latter is now endowed with a non-standard metric, and many of the ideas we develop in this chapter may be extended to study this new situation.

There is a large and interesting body of work concerned with the local and global theory of curves in Euclidean space, but we have been rather ruthless in our selection of material. Other than the material on involutes and evolutes in §1.4 (some or all of which may be omitted if desired, since the material is not used directly in the rest of the book), we have restricted ourselves to those aspects of the theory that have most relevance to our study of surfaces.

The layout of the chapter is as follows. After some preliminary definitions and examples we consider the local theory of plane curves, where the notion of curvature is introduced. We then seek to give some familiarity with the ideas in the optional section on involutes and evolutes. Finally, we consider the local theory of space curves, where the behaviour is governed by two invariants, namely the curvature and the torsion.

1.1 Basic definitions

For each positive integer n, let $\mathbb{R}^n$ denote the n-dimensional vector space of n-tuples of real numbers, with vector addition and multiplication by a scalar λ carried out component-wise. Specifically,

$$(x_1, \dots, x_n) + (y_1, \dots, y_n) = (x_1 + y_1, \dots, x_n + y_n)$$

and

$$\lambda(x_1, \dots, x_n) = (\lambda x_1, \dots, \lambda x_n) .$$

A *smooth parametrised curve* (henceforth called a *smooth curve*) in $\mathbb{R}^n$ is a smooth map $\boldsymbol{\alpha} : I \to \mathbb{R}^n$, where I is a possibly infinite open interval of real numbers. Thus $\boldsymbol{\alpha}(u) = (x_1(u), \dots, x_n(u))$, where $x_1, \dots, x_n : I \to \mathbb{R}$, are infinitely differentiable functions of u. The variable u is called the *parameter* and the image $\boldsymbol{\alpha}(I) \subset \mathbb{R}^n$ is called the *trace* of $\boldsymbol{\alpha}$. Intuitively, we are thinking of a curve as the path traced out by a point moving in $\mathbb{R}^n$.

The metric properties of such a curve (or indeed a surface) are derived from the metric properties of the containing Euclidean space $\mathbb{R}^n$. These are determined by the *inner product* (also called the *scalar* or *dot* product) on $\mathbb{R}^n$ which assigns to each pair of vectors $\boldsymbol{v} = (v_1, \dots, v_n)$, $\boldsymbol{w} = (w_1, \dots, w_n)$ the scalar $\boldsymbol{v}.\boldsymbol{w}$ given by

$$\boldsymbol{v}.\boldsymbol{w} = v_1 w_1 + \cdots + v_n w_n .$$

The *length* $|\boldsymbol{v}|$ of a vector $\boldsymbol{v}$ in $\mathbb{R}^n$ is defined by $|\boldsymbol{v}| = \sqrt{\boldsymbol{v}.\boldsymbol{v}}$, and the *angle* θ between two non-zero vectors $\boldsymbol{v}$, $\boldsymbol{w}$ is given by

$$\boldsymbol{v}.\boldsymbol{w} = |\boldsymbol{v}|\,|\boldsymbol{w}| \cos\theta , \quad 0 \le \theta \le \pi .$$

We let $x'(u)$ denote the derivative of a function $x(u)$. Then the *tangent vector* to a smooth curve $\boldsymbol{\alpha}$ at u is given by $\boldsymbol{\alpha}'(u) = \big(x_1'(u), \dots, x_n'(u)\big)$. As mentioned at the start of the chapter, the crucial property of the curves we wish to study is that they may be approximated up to first order near any point by a straight line, the *tangent line*. For this reason, we shall for the most part consider *regular curves*; these are smooth curves for which $\boldsymbol{\alpha}'(u)$ is never zero. The tangent line is then the line though $\boldsymbol{\alpha}(u)$ in direction $\boldsymbol{\alpha}'(u)$, and the *unit tangent vector* $\boldsymbol{t}$ to $\boldsymbol{\alpha}$ (Figure 1.1) is given by

$$\boldsymbol{t} = \frac{\boldsymbol{\alpha}'}{|\boldsymbol{\alpha}'|} .$$

In the above, and elsewhere when no confusion should arise, we omit specific reference to the parameter u.

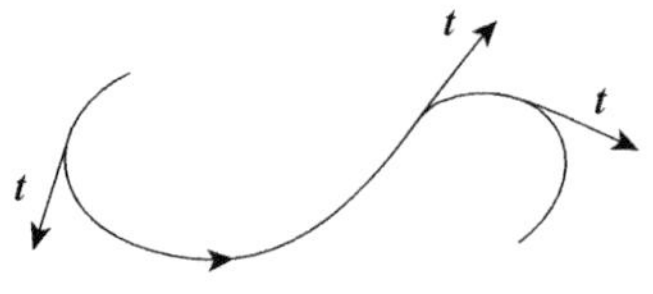

Figure 1.1 The trace of a regular curve

We shall often think of u as a time parameter, in which case $|\boldsymbol{\alpha}'|$ gives the *speed*, and $\boldsymbol{t}$ the *direction* of travel along $\boldsymbol{\alpha}$.

Example 1 (Ellipse) Let $\boldsymbol{\alpha} : \mathbb{R} \to \mathbb{R}^2$ be defined by

$$\boldsymbol{\alpha}(u) = (a \cos u, b \sin u), \quad u \in \mathbb{R},$$

where a and b are distinct positive real numbers. Then

$$\boldsymbol{\alpha}' = (-a \sin u, b \cos u) \neq \mathbf{0},$$

so that

$$\boldsymbol{t} = \frac{(-a \sin u, b \cos u)}{(a^2 \sin^2 u + b^2 \cos^2 u)^{1/2}},$$

and we see that $\boldsymbol{\alpha}$ is a regular curve whose trace is the ellipse defined by the equation

$$\frac{x^2}{a^2} + \frac{y^2}{b^2} = 1.$$

A point at which a smooth curve has vanishing derivative will be called a *singular point*.

Example 2 (Cusp point) Let $\boldsymbol{\alpha} : \mathbb{R} :\to \mathbb{R}^2$ be defined by

$$\boldsymbol{\alpha}(u) = (u^3, u^2), \quad u \in \mathbb{R}.$$

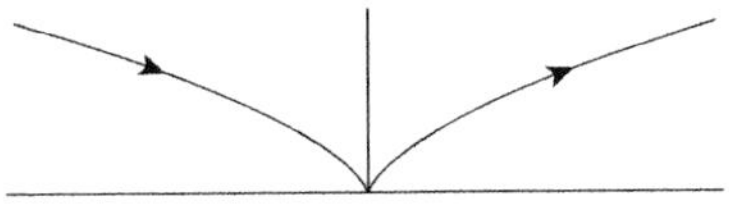

Figure 1.2 Cusp at $\boldsymbol{\alpha}(0)$

Then $\boldsymbol{\alpha}$ is smooth but not regular since $\boldsymbol{\alpha}'(0) = \mathbf{0}$. The trace of $\boldsymbol{\alpha}$ (Figure 1.2) is the curve $y^3 = x^2$ which has a cusp at $\boldsymbol{\alpha}(0)$. This is an example of the type of behaviour we exclude when we consider regular curves.

Of course, the restriction of the curve $\boldsymbol{\alpha}$ in Example 2 to $(0, \infty)$ and to $(-\infty, 0)$ are both regular curves, as is the restriction of any regular curve to an open subinterval of its domain of definition.

Example 3 (Helix) Let $\boldsymbol{\alpha} : \mathbb{R} :\to \mathbb{R}^3$ be defined by

$$\boldsymbol{\alpha}(u) = (a \cos u, a \sin u, bu), \quad u \in \mathbb{R},$$

where $a > 0$ and $b \neq 0$. Then

$$\boldsymbol{\alpha}' = (-a \sin u, a \cos u, b) \neq \mathbf{0},$$

so that

$$\boldsymbol{t} = \frac{(-a \sin u, a \cos u, b)}{(a^2 + b^2)^{1/2}},$$

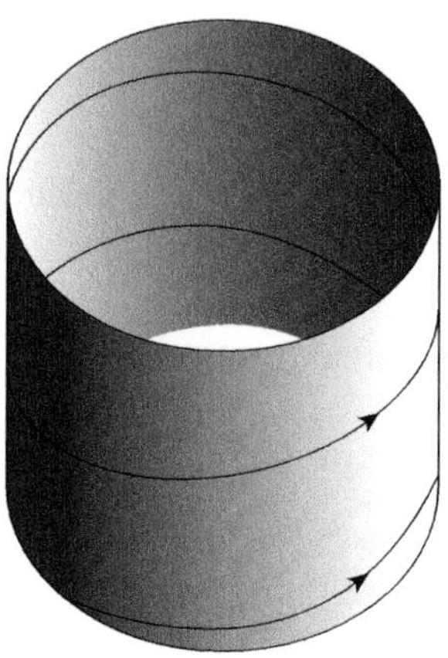

Figure 1.3 Helix on a cylinder

and we see that $\boldsymbol{\alpha}$ is a regular curve, called a *helix* (Figure 1.3); its trace lies on the cylinder $x^2 + y^2 = a^2$ in $\mathbb{R}^3$.

The *pitch* of the helix is $2\pi b$; this is the vertical distance between the points $\boldsymbol{\alpha}(u)$ and $\boldsymbol{\alpha}(u + 2\pi)$, one point being obtained from the other after one complete revolution of the helix round the cylinder. We note that $|\boldsymbol{\alpha}'|$ is constant, so with this parametrisation we travel along the curve with constant speed.

Example 4 (Graph of a function) Let $g : I \to \mathbb{R}$ be a smooth function defined on an open interval I of real numbers. The *graph* $\Gamma(g)$ of g is the trace of the regular curve in $\mathbb{R}^2$ given by

$$\boldsymbol{\alpha}(u) = (u, g(u)) \ , \quad u \in I \ .$$

For example, the graph of $g(u) = u^2$ gives the parabola $y = x^2$.

The trace of the graph of a function g has equation $y - g(x) = 0$. It may be expected that a wealth of other examples may be written down using equations of the form $f(x, y) = c$, where c is constant and $f(x, y)$ is a smooth function of x and y. In fact, an equation of this type does not always give the trace of a regular curve (for instance $x^2 + y^2 = 0$, or, as we have seen, $y^3 = x^2$), and even when it does, we do not have a natural associated parameter. For these reasons, we discuss sets of points satisfying equations in the next chapter in the context of surfaces in $\mathbb{R}^3$.

We conclude this section with a slight extension of our treatment of curves. A smooth (*resp.* regular) curve on a **closed** interval $[a, b]$ is one which may be extended to a smooth (*resp.* regular) curve on an open interval containing $[a, b]$. A *closed* curve $\boldsymbol{\alpha} : [a, b] \to \mathbb{R}^n$ is a regular curve such that $\boldsymbol{\alpha}$ and all its derivatives agree at the end points of the interval; that is,

$$\boldsymbol{\alpha}(a) = \boldsymbol{\alpha}(b) \ , \quad \boldsymbol{\alpha}'(a) = \boldsymbol{\alpha}'(b) \ , \quad \boldsymbol{\alpha}''(a) = \boldsymbol{\alpha}''(b), \ \ldots \ .$$

For example, the restriction to $[-\pi, \pi]$ of the curve $\boldsymbol{\alpha}$ in Example 1 is a closed curve – it travels once round the ellipse, starting and ending at $(-a, 0)$.

1.2 Arc length

It is important to note that, as far as geometry is concerned, it is the **trace** (or **image**) of a smooth curve which is of interest; the parametrisation is just a convenient device for describing and studying this. A good choice of parametrisation is often helpful, however, as this can lead to a great simplification of a given problem. In this section we describe an intrinsic parametrisation for any regular curve; it is defined by taking the arc length in the direction of travel measured from some given point on the curve. This parametrisation is of fundamental importance in the general theory of regular curves but, as we shall indicate, finding such a parametrisation is impracticable for most examples and so is usually best avoided in explicit calculations.

Let $\boldsymbol{\alpha} : (a, b) \to \mathbb{R}^n$ be a smooth curve and let $u_0 \in (a, b)$. We define $s : (a, b) \to \mathbb{R}$ by integrating the speed of travel between $\boldsymbol{\alpha}(u_0)$ and $\boldsymbol{\alpha}(u)$. Thus

$$s(u) = \int_{u_0}^{u} |\boldsymbol{\alpha}'(v)| dv \tag{1.1}$$

is the *arc length* along $\boldsymbol{\alpha}$ measured from $\boldsymbol{\alpha}(u_0)$. Note that $s(u)$ is positive for $u > u_0$, and negative for $u < u_0$.

Example 1 (Ellipse) Let a and b be distinct positive real numbers and let $\boldsymbol{\alpha} : \mathbb{R} \to \mathbb{R}^2$ be the ellipse

$$\boldsymbol{\alpha}(u) = (a \cos u, b \sin u)\,, \quad u \in \mathbb{R}\,.$$

Then

$$\boldsymbol{\alpha}' = (-a \sin u, b \cos u)\,,$$

and so, if $s(u)$ denotes arc length measured from $\boldsymbol{\alpha}(0)$, then

$$s(u) = \int_0^u \sqrt{a^2 \sin^2 v + b^2 \cos^2 v}\, dv\,.$$

This integral cannot be expressed in terms of elementary functions such as trigonometric functions, and serves to define a special class of functions called *elliptic functions*.

As the above example indicates, it may be difficult to write down explicit expressions in closed form (that is to say, in terms of standard functions) for functions describing the geometry, even in quite simple cases. In the following example, however, the calculations are all fairly straightforward.

Example 2 (Cycloid) This is the curve in the plane traced out by a point on a circle which rolls without slipping along a line (Figure 1.4).

Assuming that the radius of the circle is 1 and the circle rolls on the x-axis in $\mathbb{R}^2$, the curve may be parametrised by $\boldsymbol{\alpha} : \mathbb{R} \to \mathbb{R}^2$ where

$$\boldsymbol{\alpha}(u) = (u - \sin u, 1 - \cos u)\,.$$

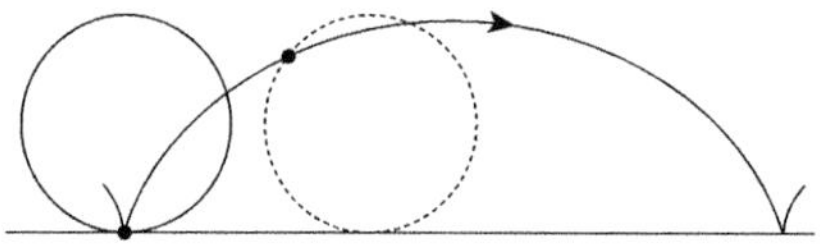

Figure 1.4 Cycloid

Then

$$\begin{aligned}\boldsymbol{\alpha}' &= (1 - \cos u, \sin u)\\ &= \left(2\sin^2\frac{u}{2}, 2\sin\frac{u}{2}\cos\frac{u}{2}\right)\\ &= 2\sin\frac{u}{2}\left(\sin\frac{u}{2}, \cos\frac{u}{2}\right),\end{aligned}$$

so that $\boldsymbol{\alpha}$ has singular points when $u = 2n\pi$, where n is an integer. These singular points correspond to the points where the cycloid touches the x-axis; at these points the cycloid has the characteristic cusp shape pointed out in Example 2 of §1.1.

Furthermore,

$$\begin{aligned}|\boldsymbol{\alpha}'| &= \left|2\sin\frac{u}{2}\right|\\ &= 2\sin\frac{u}{2}, \qquad \text{for } 0 \leq u \leq 2\pi.\end{aligned}$$

Thus, for $0 \leq u \leq 2\pi$, if $s(u)$ denotes arc length measured from $\boldsymbol{\alpha}(0)$, then

$$\begin{aligned}s(u) &= \int_0^u 2\sin\frac{v}{2}dv\\ &= 4(1 - \cos\frac{u}{2}).\end{aligned}$$

In particular, the length of a single arch of the cycloid is 8.

We now show that we may use arc length s to parametrise a regular curve, and describe some consequences of doing so. The most useful results we obtain are equation (1.4) and its immediate consequence that when we parametrise a regular curve by arc length we travel along it at unit speed.

We begin by noting that the arc length $s(u)$ along a regular curve $\boldsymbol{\alpha}(u)$ in $\mathbb{R}^n$ is a smooth function and, from (1.1),

$$\frac{ds}{du} = |\boldsymbol{\alpha}'| > 0. \tag{1.2}$$

Hence s is an increasing function of u, and we may use arc length to parametrise the trace of the curve in the same direction of travel. The chain rule for differentiation then tells us that

$$\frac{d}{du} = \frac{ds}{du}\frac{d}{ds}. \tag{1.3}$$

We now give a brief explanation of why (1.3) holds; this paragraph may be omitted by those who are happy with the chain rule as stated in (1.3). Let $\boldsymbol{\alpha}(u)$ be a regular curve, and parametrise it by arc length by letting $\tilde{\boldsymbol{\alpha}}(s)$ be the point on the trace of $\boldsymbol{\alpha}$ having arc length s from a chosen base point $\boldsymbol{\alpha}(u_0)$. Then $\boldsymbol{\alpha}(u) = \tilde{\boldsymbol{\alpha}}(s(u))$. More generally, given a function $\tilde{f}(s)$, we let $f(u) = \tilde{f}(s(u))$. Then, since the derivative of a composite is the product of the derivatives,

$$\left.\frac{df}{du}\right|_u = \left.\frac{d\tilde{f}}{ds}\right|_{s(u)} \left.\frac{ds}{du}\right|_u .$$

Following commonly used convention, we do not usually mention the points at which the differentiation takes place, and also, when there is no danger of confusion, we omit the $\tilde{}$ and simply write

$$\frac{df}{du} = \frac{df}{ds}\frac{ds}{du} ,$$

which gives the operator equation (1.3). This completes the optional paragraph of explanation of (1.3).

Returning to our account of the parametrisation of a regular curve using its arc length s, the chain rule (1.3), together with (1.2), shows that

$$\frac{d}{ds} = \frac{1}{|\boldsymbol{\alpha}'|}\frac{d}{du} , \tag{1.4}$$

and, in particular,

$$\frac{d\boldsymbol{\alpha}}{ds} = \frac{1}{|\boldsymbol{\alpha}'|}\boldsymbol{\alpha}' = \boldsymbol{t} , \tag{1.5}$$

so that when we parametrise a regular curve by arc length we travel along it at unit speed. With such a parametrisation, the arc length along $\boldsymbol{\alpha}$ from $\boldsymbol{\alpha}(s_0)$ to $\boldsymbol{\alpha}(s_1)$ is equal to $s_1 - s_0$.

Note that when, as above, we are considering two different parametrisations with the same trace, the notation $'$ for derivative must be used with care in order to avoid confusion between d/du and d/ds. We shall always use $'$ to denote d/du, the derivative with respect to the given parameter u of the curve, and we shall use d/ds to denote differentiation with respect to the arc length parameter.

We summarise the content of this section in the following theorem.

Theorem 3 *Let $\boldsymbol{\alpha}(u)$ be a regular curve in $\mathbb{R}^n$. Then we may parametrise the trace of $\boldsymbol{\alpha}$ using arc length s from a point $\boldsymbol{\alpha}(u_0)$ on $\boldsymbol{\alpha}$. If we do this, then $d\boldsymbol{\alpha}/ds$ is the unit tangent vector $\boldsymbol{t}$ to $\boldsymbol{\alpha}$ in the direction of travel. In particular, $\boldsymbol{t}$ is smoothly defined along $\boldsymbol{\alpha}$, and, when using arc length as parameter, we travel along $\boldsymbol{\alpha}$ at unit speed. The arc length along $\boldsymbol{\alpha}$ from $\boldsymbol{\alpha}(s_0)$ to $\boldsymbol{\alpha}(s_1)$ is equal to $s_1 - s_0$.*

It is important to note that if a curve is not regular then it cannot usually be parametrised by arc length past a singular point. For instance, the unit tangent vector in the direction of travel of the cycloid has discontinuities (and so is not smooth) at the singular points. A similar comment holds for the cusp curve in Example 2 of §1.1.

As mentioned at the start of this section, and as we shall see later, the existence of the arc length parameter is very important for theoretical work. However, arc length is not usually a good choice of parameter to use in calculations since in general it is difficult to find explicitly, as illustrated by Example 1.

1.3 The local theory of plane curves

In this section we introduce the signed curvature κ of a regular curve in the plane $\mathbb{R}^2$, which describes the way in which the curve is bending in the plane. We then discuss the fundamental theorem of the local theory of plane curves, which shows that a regular plane curve is determined essentially uniquely by its curvature as a function of arc length. In Chapter 6, we shall discuss Bonnet's Theorem, which is the analogous result for surfaces in $\mathbb{R}^3$.

The main goals of the first half of this section are to explain the moving frame equations (1.6) and (1.7), and to give examples of their use.

Let $\boldsymbol{\alpha} : I \to \mathbb{R}^2$ be a regular curve defined on an open interval I, and, as usual, let d/ds denote differentiation with respect to arc length along $\boldsymbol{\alpha}$. As we have seen, the unit tangent vector is given by $\boldsymbol{t} = d\boldsymbol{\alpha}/ds$, and we let $\boldsymbol{n}$ be the unit vector obtained by rotating $\boldsymbol{t}$ anticlockwise through $\pi/2$. Thus, if $\boldsymbol{t} = (a, b)$ then $\boldsymbol{n} = (-b, a)$. Then $\{\boldsymbol{t}, \boldsymbol{n}\}$ is an *adapted orthonormal moving frame* along $\boldsymbol{\alpha}$ (Figure 1.5).

Since $\boldsymbol{t}.\boldsymbol{t} = 1$, we may use the product rule for differentiation to deduce that $\dfrac{d\boldsymbol{t}}{ds}.\boldsymbol{t} = 0$. Hence

$$\frac{d\boldsymbol{t}}{ds} = \kappa \boldsymbol{n} \tag{1.6}$$

for a uniquely determined smooth function κ called the *signed curvature* (or simply the *curvature*) of $\boldsymbol{\alpha}$. Similarly, $\dfrac{d\boldsymbol{n}}{ds}.\boldsymbol{n} = 0$, so that $\dfrac{d\boldsymbol{n}}{ds}$ is a scalar multiple of $\boldsymbol{t}$. Differentiating the expression $\boldsymbol{t}.\boldsymbol{n} = 0$ and applying (1.6) we see that

$$\frac{d\boldsymbol{n}}{ds} = -\kappa \boldsymbol{t}\ . \tag{1.7}$$

As we shall see, κ measures the rate of rotation of $\boldsymbol{t}$ (and $\boldsymbol{n}$) in an anticlockwise direction as we travel along the curve at unit speed.

Curvature is a measure of acceleration, and hence plays a big part in all our lives. For instance, it shows itself as the sideways force we, and our coffee cups(!), feel as we go round a bend on a railway train. When travelling at a given speed, the more the track bends, the quicker the coffee cup slides (or falls over, if the curvature is really big). When we are facing the direction of travel, the cup slides to our right if the curvature is positive, and to our left if it is negative.

Equations (1.6) and (1.7), which give the rate of change of each element of the moving frame $\{\boldsymbol{t}, \boldsymbol{n}\}$ in terms of the frame itself, are called the *moving frame equations*.

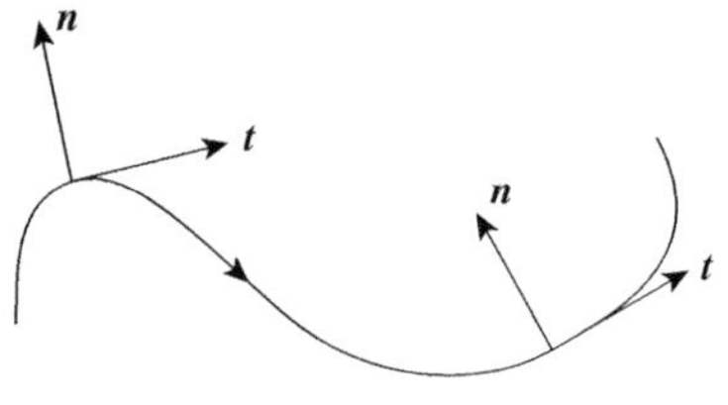

Figure 1.5 A moving frame

Example 1 (Circle of radius r) The circle with centre $\mathbf{a}$ and radius $r > 0$ traversed in an anticlockwise direction has constant curvature $\kappa = 1/r$. For, parametrising the circle by arc length, we have

$$\boldsymbol{\alpha}(s) = \mathbf{a} + r\left(\cos\frac{s}{r}, \sin\frac{s}{r}\right),$$

so that

$$\boldsymbol{t} = \frac{d\boldsymbol{\alpha}}{ds} = \left(-\sin\frac{s}{r}, \cos\frac{s}{r}\right),$$

and

$$\boldsymbol{n} = -\left(\cos\frac{s}{r}, \sin\frac{s}{r}\right).$$

Then

$$\frac{d\boldsymbol{t}}{ds} = -\frac{1}{r}\left(\cos\frac{s}{r}, \sin\frac{s}{r}u\right) = \frac{1}{r}\boldsymbol{n},$$

so that $\boldsymbol{\alpha}$ has curvature $1/r$. If the circle is traversed in a clockwise direction then it has curvature $-1/r$.

We now give an example to show how we may find the curvature of a regular curve $\boldsymbol{\alpha}$ which is not parametrised by arc length. In this, and much of the following, we repeatedly use equation (1.4). This equation will also be very useful in the following sections.

Example 2 (Cycloid) Recall from Example 2 in §1.2 that the cycloid may be parametrised as

$$\boldsymbol{\alpha}(u) = (u - \sin u, 1 - \cos u),$$

and that, using $'$ for d/du as usual,

$$\boldsymbol{\alpha}' = 2\sin\frac{u}{2}\left(\sin\frac{u}{2}, \cos\frac{u}{2}\right).$$

Hence, for $0 < u < 2\pi$,

$$\begin{aligned}
\boldsymbol{t} &= \left(\sin\frac{u}{2}, \cos\frac{u}{2}\right), \\
\boldsymbol{n} &= \left(-\cos\frac{u}{2}, \sin\frac{u}{2}\right), \\
|\boldsymbol{\alpha}'| &= 2\sin\frac{u}{2}.
\end{aligned}$$

Thus, using (1.4),

$$\begin{aligned}
\frac{d\boldsymbol{t}}{ds} &= \frac{1}{|\boldsymbol{\alpha}'|}\boldsymbol{t}' \\
&= \frac{1}{2\sin(u/2)}\frac{1}{2}\left(\cos\frac{u}{2}, -\sin\frac{u}{2}\right) \\
&= -\frac{1}{4\sin(u/2)}\boldsymbol{n}.
\end{aligned}$$

The curvature, for $0 < u < 2\pi$, is therefore given by

$$\kappa = -\frac{1}{4\sin(u/2)}, \qquad 0 < u < 2\pi .$$

In fact, for all values of the parameter u,

$$\kappa = -\frac{1}{4|\sin(u/2)|} .$$

Notice that the minimum of the absolute value $|\kappa|$ of the curvature for $0 < u < 2\pi$ is $1/4$ at $u = \pi$, and that the curvature approaches $-\infty$ as u approaches 0 and 2π. Indeed, the absolute value of the curvature decreases from ∞ to $1/4$ as u increases from 0 to π and then increases from $1/4$ to ∞ as u increases from π to 2π. This can be seen in the diagram of the curve in Figure 1.4, as can the clockwise direction of rotation of the unit tangent vector $\boldsymbol{t}$ (which is why the curvature is negative).

Now that we have obtained the moving frame equations and given examples of their use, in the remainder of this section we give a geometrical interpretation of the curvature κ, and then state and prove a basic existence and uniqueness theorem for regular curves in the plane.

As may be seen from (1.6), the curvature κ is a measure of how quickly the trace of the curve is bending away from its tangent line when the trace is traversed at unit speed. This is reflected in the following result.

Lemma 3 *The curvature κ of a regular plane curve $\boldsymbol{\alpha}$ is identically zero if and only if $\boldsymbol{\alpha}$ is a straight line.*

Proof If $\kappa = 0$ at each point of $\boldsymbol{\alpha}$ then (1.6) shows that $\boldsymbol{t} = \boldsymbol{c}$, a constant unit vector. In this case, $d\boldsymbol{\alpha}/ds = \boldsymbol{c}$, so $\boldsymbol{\alpha}(s) = \boldsymbol{b} + s\boldsymbol{c}$, for some constant vector $\boldsymbol{b}$. Thus $\boldsymbol{\alpha}$ is the straight line through $\boldsymbol{b}$ in direction $\boldsymbol{c}$. Conversely, a line may be parametrised by arc length as $\boldsymbol{\alpha}(s) = \boldsymbol{b} + s\boldsymbol{c}$, where $\boldsymbol{b}$ is a point on the line and $\boldsymbol{c}$ is a unit vector in the direction of the line. That $\kappa = 0$ at each point of $\boldsymbol{\alpha}$ is now easily checked. □

As mentioned earlier, we may interpret κ as the rate of rotation in the anticlockwise direction of the unit tangent vector $\boldsymbol{t}$, or equivalently of the unit normal vector $\boldsymbol{n}$, as we travel along the curve at unit speed. Here is the proof.

Lemma 4 *Let $\boldsymbol{e}_1, \boldsymbol{e}_2$ denote the standard basis vectors* $(1,0), (0,1)$ *respectively in $\mathbb{R}^2$. If θ is the angle from $\boldsymbol{e}_1$ to $\boldsymbol{t}$ measured in an anticlockwise direction (or equivalently, the angle from $\boldsymbol{e}_2$ to $\boldsymbol{n}$), then*

$$\kappa = \frac{d\theta}{ds} .$$

Proof The unit tangent vector $\boldsymbol{t}$ is given by (Figure 1.6)

$$\boldsymbol{t} = (\cos\theta, \sin\theta), \quad s \in I ,$$

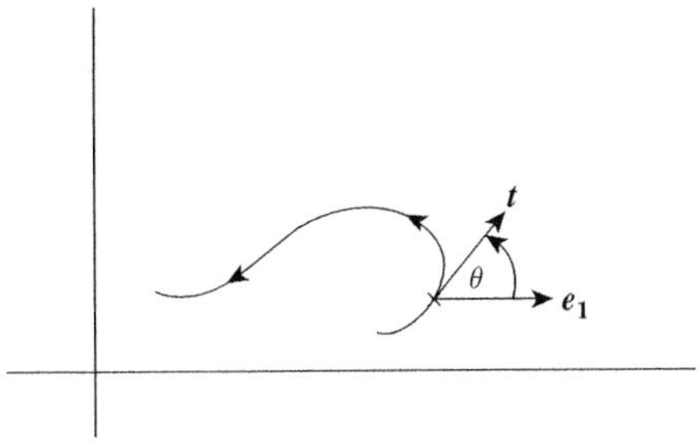

Figure 1.6 For the proof of Lemma 4

and so, using the chain rule,

$$\frac{dt}{ds} = (-\sin\theta, \cos\theta)\frac{d\theta}{ds} \, .$$

Since $\boldsymbol{n} = (-\sin\theta, \cos\theta)$, we see that

$$\kappa = \frac{d\theta}{ds} \, ,$$

which completes the proof. □

Remark 5 Using $'$ for d/du as usual, we may use (1.4) and (1.5) to show that

$$\boldsymbol{n}' = |\boldsymbol{\alpha}'|\frac{d\boldsymbol{n}}{ds} = -\kappa|\boldsymbol{\alpha}'|\boldsymbol{t} = -\kappa\boldsymbol{\alpha}' \, ,$$

so that $|\kappa|$ may be interpreted as the ratio of the speed of travel along the curve $\boldsymbol{n}$ to the speed of travel along $\boldsymbol{\alpha}$. In Section 5.12 we shall see a similar interpretation of the Gaussian curvature of a surface in $\mathbb{R}^3$.

Remark 6 For a regular plane curve $\boldsymbol{\alpha}(u)$, not necessarily parametrised by arc length,

$$\frac{d\boldsymbol{t}}{ds} = \frac{1}{|\boldsymbol{\alpha}'|}\left(\frac{1}{|\boldsymbol{\alpha}'|}\boldsymbol{\alpha}'\right)' ,$$

and, using this, one can show (see Exercise 1.8) that if $\boldsymbol{\alpha}(u) = (x(u), y(u))$ then

$$\kappa = \frac{x'y'' - x''y'}{(x'^2 + y'^2)^{3/2}} \, . \tag{1.8}$$

For example, it is now straightforward to use the parametrisation of the cycloid given in Example 2 to confirm that the curvature of the cycloid is $(-4|\sin(u/2)|)^{-1}$, but we prefer to use the calculation of the curvature along the lines indicated in Example 2 rather than using formula (1.8), since the calculations given there illustrate the theory (and similar calculations will be needed later on).

We now show that a regular plane curve is determined up to rigid motions of $\mathbb{R}^2$ by its curvature as a function of arc length. We can see this intuitively if we think of taking a straight piece of wire which is to be bent in order to fit a given curve in the plane. In order to do this it suffices to specify the amount by which the wire has to be bent at each point; that is to say to specify the signed curvature.

Theorem 7 (The Fundamental Theorem of the Local Theory of Plane Curves) *Let $\kappa : I \to \mathbb{R}$ be a smooth function defined on an open interval I. Then there is a regular curve $\boldsymbol{\alpha} : I \to \mathbb{R}^2$ parametrised by arc length s with curvature κ. Moreover, $\boldsymbol{\alpha}$ is unique up to rigid motions of $\mathbb{R}^2$.*

Proof We use the ideas introduced in the statement and proof of Lemma 4.

We first prove existence. Let $\theta : I \to \mathbb{R}$ be an indefinite integral of κ (so that θ is a smooth function with $\theta' = \kappa$), and let x_1, x_2 be indefinite integrals of $\cos\theta$, $\sin\theta$ respectively. If we let

$$\boldsymbol{\alpha}(s) = (x_1(s), x_2(s)) \ , \quad s \in I \ ,$$

then $\boldsymbol{\alpha}$ is a smooth curve, and

$$\boldsymbol{\alpha}' = (\cos\theta, \sin\theta) \ .$$

Hence

$$\boldsymbol{\alpha}'' = \theta'(-\sin\theta, \cos\theta) \ ,$$

so that $\boldsymbol{\alpha}$ is parametrised by arc length and has curvature $\theta' = \kappa$.

We now prove the statement concerning uniqueness. So, let $\boldsymbol{\alpha}_1(s)$ and $\boldsymbol{\alpha}_2(s)$ be parametrised by arc length, both having the same curvature κ. We let $\boldsymbol{t}_1 = (\cos\theta_1, \sin\theta_1)$ and $\boldsymbol{t}_2 = (\cos\theta_2, \sin\theta_2)$ be the unit tangent vectors to $\boldsymbol{\alpha}_1$ and $\boldsymbol{\alpha}_2$ respectively. Picking a base-point $s_0 \in I$ we may assume, by applying a suitable rigid motion of $\mathbb{R}^2$, that

$$\boldsymbol{\alpha}_1(s_0) = \boldsymbol{\alpha}_2(s_0) \ , \quad \boldsymbol{t}_1(s_0) = \boldsymbol{t}_2(s_0) \ .$$

Using Lemma 4, we see that $d\theta_1/ds = d\theta_2/ds$, so that $\theta_1 - \theta_2$ is constant and hence, since $\theta_1(s_0) = \theta_2(s_0)$ by assumption, we see that $\theta_1 = \theta_2$, so that $\boldsymbol{t}_1 = \boldsymbol{t}_2$. But then, $d\boldsymbol{\alpha}_1/ds = d\boldsymbol{\alpha}_2/ds$, so a similar argument shows that $\boldsymbol{\alpha}_1 = \boldsymbol{\alpha}_2$, and the uniqueness statement is proved. □

We illustrate the existence part of the above proof by constructing directly all plane curves with constant non-zero curvature.

Example 8 (Curves of constant curvature) If κ is a positive constant, we set $r = 1/\kappa$. Then, in the notation of the previous proof, $d\theta/ds = 1/r$ so that

$$\theta(s) = \frac{s}{r} + c \ , \quad c \text{ constant.}$$

Thus

$$\begin{aligned}\boldsymbol{\alpha}(s) &= \left(\int \cos\left(\frac{s}{r} + c\right) ds, \int \sin\left(\frac{s}{r} + c\right) ds\right) \\ &= \left(r\sin\left(\frac{s}{r} + c\right), -r\cos\left(\frac{s}{r} + c\right)\right) + \boldsymbol{b} \ , \quad \boldsymbol{b} \text{ constant,}\end{aligned}$$

and $\boldsymbol{\alpha}(s)$ is the circle with centre $\boldsymbol{b}$ and radius r parametrised by arc length in an anti-clockwise direction. If we assume that κ is a negative constant, then the circle will be parametrised in a clockwise direction.

Unfortunately, for a given non-constant κ it is usually very difficult to determine $\boldsymbol{\alpha}$ explicitly, as you will find if you try the case where $\kappa(s) = s$. In fact, this seemingly simple example leads to a so-called *Fresnel integral.* Such integrals can't be evaluated in terms of standard functions.

1.4 Involutes and evolutes of plane curves †

As indicated by the † symbol, the material in this section is not needed for the rest of the book. However, it is included because of its historical and intrinsic interest, and to provide practice at the type of local calculations which are useful in the study of differential geometry. It should also help to build geometrical intuition, and we would recommend covering at least the material up to and including Example 1, the calculation of the involute of the cycloid.

We shall be considering two curves $\boldsymbol{\alpha}$ and $\boldsymbol{\beta}$ in this section. To avoid confusion, we denote objects corresponding to each curve by the appropriate suffix.

Let $\boldsymbol{\alpha} : I \to \mathbb{R}^2$ be a regular curve and let $\boldsymbol{\beta} : I \to \mathbb{R}^2$ be defined by

$$\boldsymbol{\beta}(u) = \boldsymbol{\alpha}(u) - s_{\alpha}(u)\boldsymbol{t}_{\alpha}(u)\,, \tag{1.9}$$

where s_α denotes arc length along $\boldsymbol{\alpha}$ measured from some point $\boldsymbol{\alpha}(u_0)$. Then (Figure 1.7) $\boldsymbol{\beta}$ is a smooth curve, and, if the curvature κ_α of $\boldsymbol{\alpha}$ is never zero, the only singular point of $\boldsymbol{\beta}$ is at $u = u_0$ (see Exercise 1.10). The curve $\boldsymbol{\beta}$ is called an *involute* of $\boldsymbol{\alpha}$. One physical interpretation of $\boldsymbol{\beta}$ is that $\boldsymbol{\beta}$ is the path described by the end of a piece of string as it is "unwound" from $\boldsymbol{\alpha}$ starting at $\boldsymbol{\alpha}(u_0)$.

Example 1 (Cycloid) We shall consider the reflection in the x-axis of the cycloid considered in Example 2 of §1.2. We parametrise this by

$$\boldsymbol{\alpha}(u) = (u - \sin u, \cos u - 1)\,,$$

and we shall find the involute which starts at the lowest point $(\pi, -2)$ (corresponding to $u = \pi$) of the cycloid.

In terms of our physical interpretation of the involute; if we imagine a pendulum made of a bob at the end of a piece of string of length 4 whose top end is supported at $(2\pi, 0)$

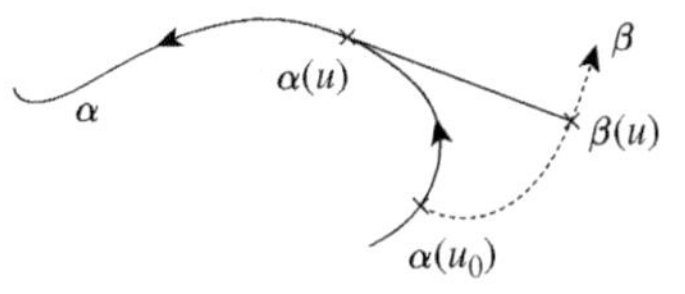

Figure 1.7 Involute of $\boldsymbol{\alpha}$

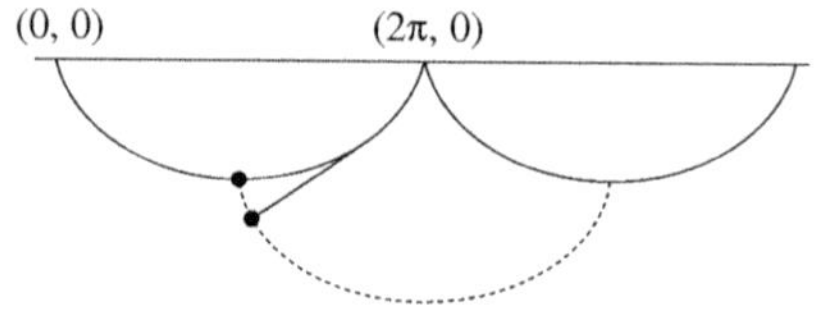

Figure 1.8 Cycloidal pendulum

and which is wound around the cycloid so that the bob is at the lowest point $(\pi, -2)$, then we are finding the path traced out by the bob as the pendulum is left to swing under gravity (Figure 1.8).

The calculation is as follows. Since

$$\boldsymbol{\alpha}' = 2\sin\frac{u}{2}\left(\sin\frac{u}{2}, -\cos\frac{u}{2}\right),$$

it follows that for $u \in (\pi, 2\pi)$,

$$\boldsymbol{t}_\alpha = \left(\sin\frac{u}{2}, -\cos\frac{u}{2}\right)$$

and

$$\begin{aligned} s_\alpha &= \int_\pi^u 2\sin\frac{v}{2}\,dv \\ &= -4\cos\frac{u}{2}. \end{aligned}$$

Hence

$$\begin{aligned} \boldsymbol{\beta}(u) &= \boldsymbol{\alpha}(u) + 4\cos\frac{u}{2}\left(\sin\frac{u}{2}, -\cos\frac{u}{2}\right) \\ &= (u - \sin u, \cos u - 1) + (2\sin u, -2\cos u - 2) \\ &= (u + \sin u, -3 - \cos u) \\ &= \Big((u-\pi) - \sin(u-\pi), \cos(u-\pi) - 1\Big) + (\pi, -2). \end{aligned}$$

Thus $\boldsymbol{\beta}$ is also a cycloid, obtained by translating the original one.

This example is of historical importance since it enabled Huyghens in the seventeenth century to construct a pendulum, called the cycloidal pendulum, whose bob traces out a cycloid. It is known that (neglecting friction) a particle moving under gravity on a cycloid performs simple harmonic motion, so the period of a cycloidal pendulum is independent of the amplitude of swing. Examples of clocks with this type of pendulum may be seen in the British Museum.

This concludes the minimum amount of material that we suggested you cover from this section. If you would like to continue, we shall now use the techniques given earlier to find the relation between the geometrical quantities $\boldsymbol{t}_\beta, \boldsymbol{n}_\beta, \kappa_\beta$ of an involute $\boldsymbol{\beta}$ of a regular curve $\boldsymbol{\alpha}$ and the geometrical quantities $\boldsymbol{t}_\alpha, \boldsymbol{n}_\alpha, \kappa_\alpha$ of $\boldsymbol{\alpha}$. In these calculations, which are quite intricate, we often make use of equations (1.2) to (1.7).

Lemma 2 *Let $\epsilon = s_\alpha \kappa_\alpha / |s_\alpha \kappa_\alpha|$ (so that $\epsilon = 1$ if $s_\alpha \kappa_\alpha > 0$ and $\epsilon = -1$ if $s_\alpha \kappa_\alpha < 0$). Then*

$$\boldsymbol{t}_\beta = -\epsilon\, \boldsymbol{n}_\alpha\,, \tag{1.10}$$

$$\boldsymbol{n}_\beta = \epsilon\, \boldsymbol{t}_\alpha\,, \tag{1.11}$$

$$\kappa_\beta = \frac{\epsilon}{s_\alpha}\,. \tag{1.12}$$

Proof Following the method explained in Example 2 of §1.3, we first differentiate (1.9) with respect to the given parameter u and obtain

$$\begin{aligned}
\boldsymbol{\beta}' &= \boldsymbol{\alpha}' - {s_\alpha}' \boldsymbol{t}_\alpha - s_\alpha {\boldsymbol{t}_\alpha}' \\
&= \boldsymbol{\alpha}' - |\boldsymbol{\alpha}'| \boldsymbol{t}_\alpha - s_\alpha {\boldsymbol{t}_\alpha}' \\
&= -s_\alpha {\boldsymbol{t}_\alpha}' \\
&= -s_\alpha |\boldsymbol{\alpha}'| \frac{d\boldsymbol{t}_\alpha}{ds_\alpha} \\
&= -s_\alpha \kappa_\alpha |\boldsymbol{\alpha}'|\, \boldsymbol{n}_\alpha\,,
\end{aligned} \tag{1.13}$$

so that (1.10) and (1.11) now follow.

To find κ_β, we continue to follow our method for finding curvature. We first note from (1.13) that

$$|\boldsymbol{\beta}'| = |s_\alpha \kappa_\alpha|\, |\boldsymbol{\alpha}'|\,, \tag{1.14}$$

so, differentiating (1.10) with respect to s_β, and using (1.11), we find that

$$\begin{aligned}
\frac{d\boldsymbol{t}_\beta}{ds_\beta} &= -\frac{\epsilon^2}{s_\alpha \kappa_\alpha |\boldsymbol{\alpha}'|} {\boldsymbol{n}_\alpha}' \\
&= -\frac{1}{s_\alpha \kappa_\alpha} \frac{d\boldsymbol{n}_\alpha}{ds_\alpha} \\
&= \frac{1}{s_\alpha} \boldsymbol{t}_\alpha \\
&= \frac{\epsilon}{s_\alpha} \boldsymbol{n}_\beta\,,
\end{aligned}$$

which gives our required expression (1.12) for κ_β. □

The definition of involute given in (1.9) defines $\boldsymbol{\beta}$ in terms of geometrical quantities associated with $\boldsymbol{\alpha}$. We now obtain an expression for $\boldsymbol{\alpha}$ in terms of geometrical quantities associated with $\boldsymbol{\beta}$.

Lemma 3

$$\boldsymbol{\alpha} = \boldsymbol{\beta} + \frac{1}{\kappa_\beta} \boldsymbol{n}_\beta\,. \tag{1.15}$$

Proof The definition (1.9) of $\boldsymbol{\beta}$ in terms of $\boldsymbol{\alpha}$ gives that

$$\boldsymbol{\alpha} = \boldsymbol{\beta} + s_\alpha \boldsymbol{t}_\alpha\,. \tag{1.16}$$

However, (1.11) and (1.12) show that

$$s_\alpha \boldsymbol{t}_\alpha = \frac{1}{\kappa_\beta}\boldsymbol{n}_\beta ,$$

and the result follows. □

For each parameter value u, the quantity $1/|\kappa_\beta(u)|$ is called the *radius of curvature* of $\boldsymbol{\beta}$ at u, and $\boldsymbol{\beta}(u) + \frac{1}{\kappa_\beta}\boldsymbol{n}_\beta(u)$ is the *centre of curvature* of $\boldsymbol{\beta}$ at u. The circle with centre at the centre of curvature of $\boldsymbol{\beta}$ at u and with radius $1/|\kappa_\beta(u)|$ has second order contact with $\boldsymbol{\beta}$ at $\boldsymbol{\beta}(u)$.

The locus of the centres of curvature of a regular plane curve is called the *evolute* of that curve. So, the evolute of a regular curve $\boldsymbol{\beta}(u)$ is the curve $\boldsymbol{\alpha}$ given in Lemma 3. If we imagine the curve $\boldsymbol{\beta}$ to be a light filament, then its evolute would be the curve in the plane of maximum illumination. The evolute is often called the *caustic*.

Since the evolute of a curve is an important object associated with the curve, we restate Lemma 3 as a proposition.

Proposition 4 *Let $\boldsymbol{\beta}$ be an involute of a regular curve $\boldsymbol{\alpha}$ in $\mathbb{R}^2$. Then $\boldsymbol{\alpha}$ is the evolute of $\boldsymbol{\beta}$.*

Example 5 (Cycloid) Example 1 would lead us to expect that the evolute of a cycloid is a translate of that cycloid, and we shall verify this directly. As in Example 1, we parametrise the cycloid as

$$\boldsymbol{\alpha}(u) = (u - \sin u, \cos u - 1) ,$$

and, as we have seen, for $0 < u < 2\pi$,

$$|\boldsymbol{\alpha}'| = 2\sin\frac{u}{2} , \quad \boldsymbol{t}_\alpha = \left(\sin\frac{u}{2}, -\cos\frac{u}{2}\right) .$$

Hence

$$\boldsymbol{n}_\alpha = \left(\cos\frac{u}{2}, \sin\frac{u}{2}\right) ,$$

and it quickly follows that

$$\kappa_\alpha = \frac{1}{4\sin\frac{u}{2}} .$$

The evolute $\boldsymbol{\beta}$ of $\boldsymbol{\alpha}$ is thus given by

$$\begin{aligned}
\boldsymbol{\beta} &= \boldsymbol{\alpha} + \frac{1}{\kappa_\alpha}\boldsymbol{n}_\alpha \\
&= (u - \sin u, \cos u - 1) + 4\sin\frac{u}{2}\left(\cos\frac{u}{2}, \sin\frac{u}{2}\right) \\
&= (u - \sin u, \cos u - 1) + (2\sin u, 2 - 2\cos u) \\
&= (u + \sin u, 1 - \cos u) \\
&= ((u - \pi) - \sin(u - \pi), \cos(u - \pi) - 1) + (\pi, 2) ,
\end{aligned}$$

so that, as we anticipated, the evolute of the cycloid is another cycloid which is a translate of the first (Figure 1.9).

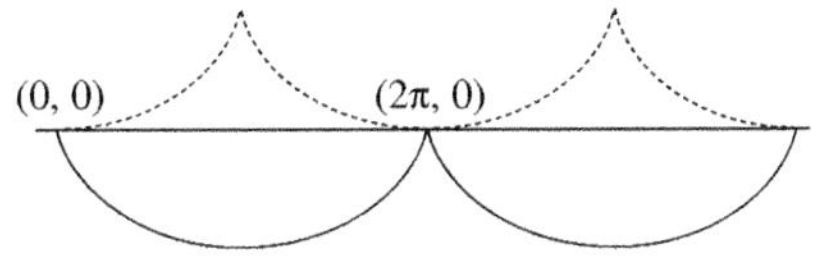

Figure 1.9 Evolute of a cycloid

Returning to our physical interpretation of the involute and the unwinding string, equation (1.10) shows that the direction of travel of the involute is everywhere orthogonal to the direction of the unwinding string; put another way, the involute is an orthogonal trajectory of the pencil of lines formed by the tangents to the original curve. We shall have more to say about orthogonal trajectories in §3.5.

1.5 The local theory of space curves

In §1.3, we showed that a plane curve is essentially uniquely determined by one scalar invariant, the curvature κ. We did this by constructing an adapted orthonormal moving frame $\{\boldsymbol{t}, \boldsymbol{n}\}$ along the curve, and using the moving frame equations (1.6) and (1.7).

In this section, we carry out a similar process for a regular curve $\boldsymbol{\alpha}$ in $\mathbb{R}^3$. This time, we need two scalar invariants, the curvature and the torsion, to describe the curve. The main results of this section are the Serret–Frenet formulae (1.20), and the basic existence and uniqueness theorem for regular curves in Euclidean 3-space given in Theorem 4.

Let $\boldsymbol{\alpha} : I \to \mathbb{R}^3$ be a regular curve defined on an open interval I, and, as usual, let d/ds denote differentiation with respect to arc length along $\boldsymbol{\alpha}$. Then the unit tangent vector is given by $\boldsymbol{t} = d\boldsymbol{\alpha}/ds$. Since $\boldsymbol{t}.\boldsymbol{t} = 1$ we have $\dfrac{d\boldsymbol{t}}{ds}.\boldsymbol{t} = 0$, so that $\dfrac{d\boldsymbol{t}}{ds}$ is orthogonal to $\boldsymbol{t}$. We define the *curvature* κ of $\boldsymbol{\alpha}$ by

$$\kappa = \left|\frac{d\boldsymbol{t}}{ds}\right| .$$

Note that, in contrast with the case of plane curves, the definition of curvature κ of a space curve implies that $\kappa \geq 0$. This is because the notions of "clockwise" and "anticlockwise" rotations (which we used to define the signed curvature of a plane curve) do not apply in $\mathbb{R}^3$.

At points where $\kappa \neq 0$ we define the *principal normal* $\boldsymbol{n}$ of $\boldsymbol{\alpha}$ by setting

$$\frac{d\boldsymbol{t}}{ds} = \kappa \boldsymbol{n} , \tag{1.17}$$

and the *binormal* $\boldsymbol{b}$ of $\boldsymbol{\alpha}$ by

$$\boldsymbol{b} = \boldsymbol{t} \times \boldsymbol{n} , \tag{1.18}$$

where we have used the vector cross product in $\mathbb{R}^3$ on the right hand side. Then $\{\boldsymbol{t}, \boldsymbol{n}, \boldsymbol{b}\}$ is the *adapted orthonormal moving frame* along $\boldsymbol{\alpha}$. Note that there is no natural choice of principal normal or binormal to $\boldsymbol{\alpha}$ at those points where the curvature is zero.

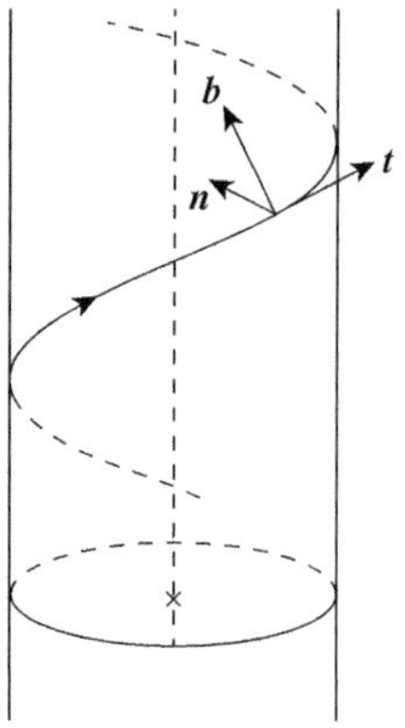

Figure 1.10 The moving frame along a helix

Figure 1.10 shows this frame at a typical point of a helix. As will be clear from Example 2, the principal normal $\boldsymbol{n}$ is a horizontal unit vector pointing towards the z-axis. Hence (anticipating some material from Chapter 3, but clear from intuition), $\boldsymbol{n}$ is orthogonal to the cylinder on which the helix lies, so that $\boldsymbol{t}$ and $\boldsymbol{b}$ are both tangential to the cylinder.

We now find the *moving frame equations*, of which (1.17) is the first, which describe the rate of change of each element of the moving frame $\{\boldsymbol{t}, \boldsymbol{n}, \boldsymbol{b}\}$ in terms of the frame itself.

We first differentiate (1.18) and use (1.17) to find that

$$\frac{d\boldsymbol{b}}{ds} = \left(\frac{d\boldsymbol{t}}{ds} \times \boldsymbol{n}\right) + \left(\boldsymbol{t} \times \frac{d\boldsymbol{n}}{ds}\right) = \boldsymbol{t} \times \frac{d\boldsymbol{n}}{ds},$$

and in particular

$$\frac{d\boldsymbol{b}}{ds}.\boldsymbol{t} = 0.$$

Also

$$\boldsymbol{b}.\boldsymbol{b} = 1 \quad \text{so that} \quad \frac{d\boldsymbol{b}}{ds}.\boldsymbol{b} = 0.$$

Thus

$$\frac{d\boldsymbol{b}}{ds} = \tau\boldsymbol{n} \tag{1.19}$$

for some function τ called the *torsion* of $\boldsymbol{\alpha}$. (Please be aware that some authors use $-\tau$ in place of τ.) Since

$$\boldsymbol{n} = \boldsymbol{b} \times \boldsymbol{t} = -\boldsymbol{t} \times \boldsymbol{b},$$

we have, using (1.17) and (1.19),

$$\begin{aligned}\frac{d\boldsymbol{n}}{ds} &= -\left(\frac{d\boldsymbol{t}}{ds} \times \boldsymbol{b}\right) - \left(\boldsymbol{t} \times \frac{d\boldsymbol{b}}{ds}\right) \\ &= -\kappa\boldsymbol{t} - \tau\boldsymbol{b}.\end{aligned}$$

Thus we have our required moving frame equations, called the *Serret–Frenet formulae*. We write them down grouped together for convenience.

$$\begin{aligned} \frac{dt}{ds} &= \kappa n\,, \\ \frac{dn}{ds} &= -\kappa t - \tau b\,, \\ \frac{db}{ds} &= \tau n\,. \end{aligned} \tag{1.20}$$

We now discuss the geometry of these equations. The line through $\alpha(u)$ in direction $t(u)$ is the tangent line to α at $\alpha(u)$. As mentioned earlier, this is the line having first order contact with α, and κ measures the rate at which the trace of the curve is bending away from this line when the trace is traversed at unit speed. The plane through $\alpha(u)$ spanned by $t(u), n(u)$ is called the *osculating plane* (from *'osculans'*, Latin for 'kissing') to α at $\alpha(u)$. As is clear from the first Serret–Frenet formula, this is the plane with which α has second order contact at $\alpha(u)$, in the sense that the curve touches the plane there, and α' and α'' are both tangential to the plane (Figure 1.11).

Since b is the unit normal to the osculating plane, db/ds measures the rate of change of the osculating plane. The third Serret–Frenet formula shows that the osculating plane is rotating about the tangent vector t at each point, and τ measures this rate of rotation. This is the rate at which the curve is twisting away from its osculating plane. Finally, the *normal plane* at $\alpha(u)$ is the plane spanned by $n(u), b(u)$; as we move along the curve, the normal plane rotates about the binormal, the rate of rotation being measured by κ.

The above comments on the osculating plane and its rate of change would lead us to suppose that a curve should have everywhere zero torsion if and only if the curve is contained in a plane, and we now demonstrate this.

Lemma 1 *The torsion τ of a regular space curve α is identically zero if and only if α is contained in a plane.*

Proof First suppose that α lies in a plane with unit normal b_0, say. Then, for some real constant c,

$$\alpha . b_0 = c\,.$$

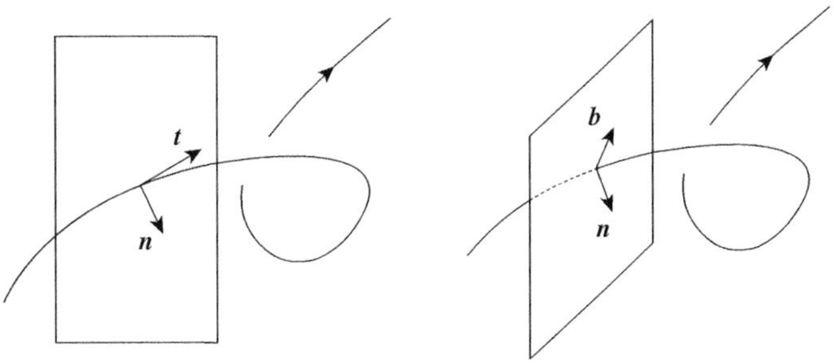

Figure 1.11 Osculating plane and normal plane

Differentiating with respect to arc length s along $\boldsymbol{\alpha}$, we find that

$$\frac{d\boldsymbol{\alpha}}{ds}.\boldsymbol{b}_0 = 0 \quad \text{and} \quad \frac{d^2\boldsymbol{\alpha}}{ds^2}.\boldsymbol{b}_0 = 0\,.$$

Thus, if $\boldsymbol{\alpha}$ has nowhere vanishing curvature,

$$\boldsymbol{t}.\boldsymbol{b}_0 = 0 \quad \text{and} \quad \boldsymbol{n}.\boldsymbol{b}_0 = 0\,,$$

so that $\boldsymbol{b} = \pm\boldsymbol{b}_0$ and $\tau = 0$. All the osculating planes of $\boldsymbol{\alpha}$ coincide with the plane in which $\boldsymbol{\alpha}$ lies.

Conversely, if $\boldsymbol{\alpha}$ has nowhere vanishing curvature and if $\boldsymbol{b}$ is constant, say $\boldsymbol{b} = \boldsymbol{b}_0$, then

$$\frac{d\boldsymbol{\alpha}}{ds}.\boldsymbol{b}_0 = \boldsymbol{t}.\boldsymbol{b}_0 = 0\,,$$

so that $\boldsymbol{\alpha}.\boldsymbol{b}_0 = c$ and $\boldsymbol{\alpha}$ lies in a plane. □

We now give an example to illustrate a method of finding the curvature and torsion of a regular space curve which is not parametrised by arc length. As with the corresponding method for plane curves, it is not usually a good idea to attempt to re-parametrise the curve by arc length (although, in this example it is rather easy). Rather, one should exploit the chain rule by using equation (1.4).

Example 2 (Helix) Recall from Example 3 of §1.1 that this space curve may be parametrised by

$$\boldsymbol{\alpha}(u) = (a\cos u, a\sin u, bu)\,, \qquad u \in \mathbb{R}\,,$$

where $a > 0$ and $b \neq 0$. Then

$$\boldsymbol{\alpha}' = (-a\sin u, a\cos u, b)\,,$$

so that

$$|\boldsymbol{\alpha}'| = (a^2 + b^2)^{1/2}$$

and

$$\boldsymbol{t} = \frac{(-a\sin u, a\cos u, b)}{(a^2+b^2)^{1/2}}\,.$$

Also, using (1.4), we have

$$\frac{d}{ds} = \frac{1}{(a^2+b^2)^{1/2}}\frac{d}{du}\,.$$

Hence

$$\kappa\boldsymbol{n} = \frac{d\boldsymbol{t}}{ds} = \frac{1}{a^2+b^2}(-a\cos u, -a\sin u, 0)\,,$$

so that

$$\kappa = \frac{a}{a^2+b^2} \quad \text{and} \quad \boldsymbol{n} = (-\cos u, -\sin u, 0)\,.$$

But then

$$\boldsymbol{b} = \boldsymbol{t} \times \boldsymbol{n} = \frac{1}{(a^2 + b^2)^{1/2}}(b \sin u, -b \cos u, a)\,,$$

so that

$$\frac{d\boldsymbol{b}}{ds} = \frac{b}{a^2 + b^2}(\cos u, \sin u, 0)\,.$$

Hence

$$\tau = -\frac{b}{a^2 + b^2}\,.$$

The helix thus has constant curvature and constant torsion. If $\tau < 0$ then the helix is right-handed; if $\tau > 0$ it is left-handed.

Remark 3 For a regular space curve $\boldsymbol{\alpha}(u)$, not necessarily parametrised by arc length, we may find expressions for κ and τ directly in terms of $\boldsymbol{\alpha}'$, $\boldsymbol{\alpha}''$ and $\boldsymbol{\alpha}'''$. In fact, in Exercise 1.15 you are asked to show that

$$\kappa = \frac{|\boldsymbol{\alpha}' \times \boldsymbol{\alpha}''|}{|\boldsymbol{\alpha}'|^3}\,, \qquad \tau = -\frac{(\boldsymbol{\alpha}' \times \boldsymbol{\alpha}'')\,.\,\boldsymbol{\alpha}'''}{|\boldsymbol{\alpha}' \times \boldsymbol{\alpha}''|^2}\,,$$

where, as usual, $\times$ is vector cross product in $\mathbb{R}^3$. However, for the reasons given in Remark 6 of §1.3, we prefer to use the calculation of the curvature and torsion along the lines indicated in Example 2 rather than using the above formulae.

The importance of the curvature and torsion of a space curve is that they determine the curve up to rigid motions of $\mathbb{R}^3$. As with the case of plane curves we can see this intuitively if we think of taking a straight piece of wire which is to be manipulated in order to fit a given curve in $\mathbb{R}^3$. In order to do this it suffices to specify the amount by which the wire has to be bent and twisted at each point; that is to say to specify the curvature and torsion of the given curve.

Theorem 4 (The Fundamental Theorem of the Local Theory of Space Curves) *Let $\kappa : I \to \mathbb{R}$, $\tau : I \to \mathbb{R}$ be smooth functions defined on an open interval I, and assume that $\kappa > 0$. Then there is a regular curve $\boldsymbol{\alpha} : I \to \mathbb{R}^3$ parametrised by arc length s with curvature κ and torsion τ. Moreover, $\boldsymbol{\alpha}$ is unique up to rigid motions of $\mathbb{R}^3$.*

Proof The proof depends on the existence and uniqueness theorem for linear systems of ordinary differential equations. We shall refer to this as the *ODE theorem*.

For a given κ and τ, the Serret–Frenet formulae form a linear system of three first order ordinary differential equations for the $\mathbb{R}^3$-valued functions $\boldsymbol{t}$, $\boldsymbol{n}$ and $\boldsymbol{b}$, and the ODE theorem tells us that such a system has a unique solution $\{\boldsymbol{t}, \boldsymbol{n}, \boldsymbol{b}\}$ on I for any set of initial conditions $\{\boldsymbol{t}(s_0), \boldsymbol{n}(s_0), \boldsymbol{b}(s_0)\}$. Then the six quantities $\boldsymbol{t}.\boldsymbol{t}$, $\boldsymbol{n}.\boldsymbol{n}$, $\boldsymbol{b}.\boldsymbol{b}$, $\boldsymbol{t}.\boldsymbol{n}$, $\boldsymbol{t}.\boldsymbol{b}$, $\boldsymbol{n}.\boldsymbol{b}$ satisfy a linear system of six first order ordinary differential equations (one of which, for instance, is $\frac{d}{ds}(\boldsymbol{t}.\boldsymbol{b}) = \kappa\boldsymbol{n}.\boldsymbol{b} + \tau\boldsymbol{t}.\boldsymbol{n}$) for which $\boldsymbol{t}.\boldsymbol{t} = \boldsymbol{n}.\boldsymbol{n} = \boldsymbol{b}.\boldsymbol{b} = 1$, $\boldsymbol{t}.\boldsymbol{n} = \boldsymbol{t}.\boldsymbol{b} = \boldsymbol{n}.\boldsymbol{b} = 0$ is easily seen to be a solution. Thus, using the ODE theorem again, we see that any solution of the Serret–Frenet formulae with initial trihedron being right-handed orthonormal will stay right-handed orthonormal.

We now prove the existence part of the theorem. For given functions κ and τ, let $\{\boldsymbol{t}, \boldsymbol{n}, \boldsymbol{b}\}$ be a right-handed orthonormal solution of the Serret–Frenet formulae, and let $\boldsymbol{\alpha} : I \to \mathbb{R}^3$ be an indefinite integral of $\boldsymbol{t}$. Then $\boldsymbol{\alpha}$ is a smooth curve with $d\boldsymbol{\alpha}/ds = \boldsymbol{t}$ and $d^2\boldsymbol{\alpha}/ds^2 = d\boldsymbol{t}/ds = \kappa\boldsymbol{n}$. It follows that $\boldsymbol{\alpha}$ is parametrised by arc length, that $\boldsymbol{t}$ is the unit tangent vector, that $\boldsymbol{n}$ is the principal normal vector and κ is the curvature. Thus $\boldsymbol{b}$ is the binormal, from which it follows that $\boldsymbol{\alpha}$ has torsion τ.

This completes the proof of existence, and we now prove uniqueness. Let $\boldsymbol{\alpha}_1$ and $\boldsymbol{\alpha}_2$ be smooth curves parametrised by arc length, both having the same curvature κ and torsion τ, and let $\{\boldsymbol{t}_1, \boldsymbol{n}_1, \boldsymbol{b}_1\}$, $\{\boldsymbol{t}_2, \boldsymbol{n}_2, \boldsymbol{b}_2\}$ be the corresponding unit tangent vectors, principal normals and binormals. Picking a base point $s_0 \in I$ we may assume, by applying a suitable rigid motion of $\mathbb{R}^3$, that

$$\boldsymbol{\alpha}_1(s_0) = \boldsymbol{\alpha}_2(s_0)\,, \quad \boldsymbol{t}_1(s_0) = \boldsymbol{t}_2(s_0)\,, \quad \boldsymbol{n}_1(s_0) = \boldsymbol{n}_2(s_0)\,, \quad \boldsymbol{b}_1(s_0) = \boldsymbol{b}_2(s_0)\,,$$

and the uniqueness part of the ODE theorem now shows that $\{\boldsymbol{t}_1, \boldsymbol{n}_1, \boldsymbol{b}_1\} = \{\boldsymbol{t}_2, \boldsymbol{n}_2, \boldsymbol{b}_2\}$. In particular, $\boldsymbol{t}_1 = \boldsymbol{t}_2$ so that $d\boldsymbol{\alpha}_1/ds = d\boldsymbol{\alpha}_2/ds$, and it follows that $\boldsymbol{\alpha}_1 - \boldsymbol{\alpha}_2$ is constant. Since we applied a rigid motion so that $\boldsymbol{\alpha}_1(s_0) = \boldsymbol{\alpha}_2(s_0)$ we see that $\boldsymbol{\alpha}_1 = \boldsymbol{\alpha}_2$, and uniqueness is proved. □

It follows from Theorem 4 that helices may be characterised as those curves having non-zero constant curvature and non-zero constant torsion. However, it is not hard to give a direct proof of this (see Exercise 1.16).

This concludes our treatment of the local theory of plane and space curves.

Exercises

1.1 The subset of the plane satisfying $x^{2/3} + y^{2/3} = 1$ is called the *astroid*. Show that $\boldsymbol{\alpha}(u) = (\cos^3 u, \sin^3 u)$, $u \in \mathbb{R}$, is a parametrisation of the astroid. Show that the parametrisation is regular except when u is an integer multiple of $\pi/2$. Sketch the astroid and mark the singular points of the parametrisation. Find the length of the astroid between parameter values $u = 0$ and $u = \pi/2$.

1.2 For each positive constant r, the smooth curve given by

$$\boldsymbol{\alpha}(u) = (2r \sin u - r \sin 2u,\ 2r \cos u - r \cos 2u)\,, \quad u \in \mathbb{R}\,,$$

is called an *epicycloid*. It is the curve traced out by a point on the circumference of a circle of radius r which rolls without slipping on a circle of the same radius. Sketch the trace of the curve, and find the length of $\boldsymbol{\alpha}$ between the singular points corresponding to $u = 0$ and $u = 2\pi$.

1.3 For each positive constant r, the smooth curve given by

$$\boldsymbol{\alpha}(u) = \frac{r}{\cosh u}(u \cosh u - \sinh u,\ 1)\,,$$

is called a *tractrix*.

Taking $r = 1$, show that, for $u > 0$, $\boldsymbol{\alpha}(u)$ is the curve traced out by a stone starting at (0,1) on the end of a piece of rope of length 1 when the tractor on the other end of the piece of rope drives along the positive x-axis starting at $(0, 0)$. In more mathematical terms, show that $\boldsymbol{\alpha}(u) + \boldsymbol{t}(u)$ is on the positive x-axis for $u > 0$ (and that $\boldsymbol{\alpha}(0) = (0, 1)$). Sketch the trace of the curve for all real values of u.

1.4 Let $g : I \to \mathbb{R}$ be a smooth function, and parametrise its graph by $\boldsymbol{\alpha}(u) = (u, g(u))$. Use the method of Example 2 of §1.3 to show that the curvature κ of $\boldsymbol{\alpha}$ is given by

$$\kappa = g'' \left(1 + (g')^2\right)^{-3/2} .$$

Now check this by using the formula given in Remark 6 of §1.3 (and also given in Exercise 1.8).

1.5 Show that, for $u > 0$, the curvature of the tractrix parametrised as in Exercise 1.3 (taking $r = 1$ for simplicity) is given by $\kappa = \operatorname{cosech} u$.

1.6 For each positive constant k, the smooth curve given by

$$\boldsymbol{\alpha}(u) = \left(u, k \cosh \frac{u}{k}\right)$$

is called a *catenary*. The trace of a catenary is the shape taken by a uniform chain hanging under the action of gravity. Use the same set of axes to sketch the catenary given by various values of k. Find the curvature of the catenary $\boldsymbol{\alpha}(u) = (u, \cosh u)$.

1.7 Let $\boldsymbol{\alpha}$ be a regular plane curve, and let ℓ be a real number. The corresponding *parallel* curve to $\boldsymbol{\alpha}$ is given by $\boldsymbol{\alpha}_\ell = \boldsymbol{\alpha} + \ell \boldsymbol{n}$. Show that the curvature κ_ℓ of $\boldsymbol{\alpha}_\ell$ is given by

$$\kappa_\ell = \frac{\kappa}{|1 - \kappa \ell|} .$$

1.8 Show that if $\boldsymbol{\alpha}(u) = (x(u), y(u))$ is a regular plane curve, then its curvature κ is given by

$$\kappa = \frac{x'y'' - x''y'}{(x'^2 + y'^2)^{3/2}} .$$

1.9 *(This exercise uses material in the optional §1.4.)* Let $\boldsymbol{\alpha}(u) = (u, \cosh u)$ be the parametrisation of a catenary discussed in Exercise 1.6. Show that:

(i) the involute of $\boldsymbol{\alpha}$ starting from $(0, 1)$ is the tractrix given in Exercise 1.3 (with $r = 1$);

(ii) the evolute of $\boldsymbol{\alpha}$ is the curve given by

$$\boldsymbol{\beta}(u) = (u - \sinh u \cosh u, 2 \cosh u) .$$

Find the singular points of $\boldsymbol{\beta}$ and sketch its trace.

1.10 *(This exercise uses material in the optional §1.4.)* Let $\boldsymbol{\alpha}(u)$ be a regular plane curve with nowhere vanishing curvature. Show that the involute of $\boldsymbol{\alpha}$ starting from $\boldsymbol{\alpha}(u_0)$ is a smooth curve whose only singular point is at $u = u_0$.

1.11 *(This exercise uses material in the optional §1.4.)* For ease of calculation, in this exercise you might prefer to consider the special case in which $u_0 < u_1 < u$ and $\boldsymbol{\alpha}$ has positive curvature.

Let $\boldsymbol{\alpha}(u)$ be a regular plane curve with nowhere vanishing curvature, and let u_0, u_1 be real numbers in the domain of $\boldsymbol{\alpha}$. Let $\boldsymbol{\beta}_0$ and $\boldsymbol{\beta}_1$ be the involutes of $\boldsymbol{\alpha}$ starting at $\boldsymbol{\alpha}(u_0)$ and $\boldsymbol{\alpha}(u_1)$ respectively. Use equation (1.12) to write down the curvature κ_0 of $\boldsymbol{\beta}_0$ and κ_1 of $\boldsymbol{\beta}_1$ in terms of arc length s_0, s_1 along $\boldsymbol{\alpha}$ measured from $\boldsymbol{\alpha}(u_0)$, $\boldsymbol{\alpha}(u_1)$, respectively.

Show that $\boldsymbol{\beta}_1$ is a parallel curve to $\boldsymbol{\beta}_0$ (as in Exercise 1.7), and check that the expressions for κ_0 and κ_1 you have just written down satisfy the formula for the curvature of parallel curves given in Exercise 1.7.

1.12 Let $\boldsymbol{\alpha}(u)$ be the curve in $\mathbb{R}^3$ parametrised by

$$\boldsymbol{\alpha}(u) = e^u(\cos u, \sin u, 1)\,, \quad u \in \mathbb{R}\,.$$

Sketch the trace of the curve.

If $0 < \lambda_0 < \lambda_1$, find the length of the segment of $\boldsymbol{\alpha}$ which lies between the planes $z = \lambda_0$ and $z = \lambda_1$. Show also that the curvature and torsion of $\boldsymbol{\alpha}$ are both inversely proportional to e^u.

1.13 Let $\boldsymbol{\alpha}(u)$ be the curve in $\mathbb{R}^3$ parametrised by

$$\boldsymbol{\alpha}(u) = (\cosh u, \sinh u, u)\,, \quad u \in \mathbb{R}\,.$$

Show that the curvature and torsion of $\boldsymbol{\alpha}$ are given by

$$\kappa = \frac{1}{2\cosh^2 u}\,, \qquad \tau = -\frac{1}{2\cosh^2 u}\,.$$

1.14 Find all regular curves in $\mathbb{R}^3$ with everywhere zero curvature.

1.15 Let $\boldsymbol{\alpha}(u)$ be a regular curve in $\mathbb{R}^3$. Show that the curvature κ and the torsion τ of $\boldsymbol{\alpha}$ are given by

$$\kappa = \frac{|\boldsymbol{\alpha}' \times \boldsymbol{\alpha}''|}{|\boldsymbol{\alpha}'|^3}\,, \qquad \tau = -\frac{(\boldsymbol{\alpha}' \times \boldsymbol{\alpha}'')\,.\,\boldsymbol{\alpha}'''}{|\boldsymbol{\alpha}' \times \boldsymbol{\alpha}''|^2}\,,$$

where, as usual, $'$ denotes differentiation with respect to u, and $\times$ is vector cross product in $\mathbb{R}^3$.

1.16 The cylinder with centre-line ℓ and radius $a > 0$ consists of those points in $\mathbb{R}^3$ at perpendicular distance a from the line ℓ. The *generating lines* or *rulings* on the cylinder are those lines on the cylinder parallel to the centre-line. A *helix* on the cylinder is a regular curve with non-zero torsion whose trace lies on the cylinder and whose unit tangent vector $\boldsymbol{t}$ makes a constant angle with the generating lines.

(i) Let $\boldsymbol{\alpha}$ be a regular curve on the cylinder $x^2 + y^2 = a^2$ $(a > 0)$ which has a parametrisation of the form

$$\boldsymbol{\alpha}(v) = (a\cos\theta(v), a\sin\theta(v), v + c)\,, \quad v \in \mathbb{R}\,,$$

where $\theta(v)$ is a smooth function of v and c is a constant (a parametrisation of this form exists on any open interval for which $\boldsymbol{\alpha}$ is nowhere perpendicular to the generating lines of the cylinder). If $\boldsymbol{\alpha}(0) = (a, 0, 0)$ show that the trace of $\boldsymbol{\alpha}$ is a helix if and only if $\boldsymbol{\alpha}$ may be parametrised as in Example 3 of §1.1; that is to say, in the form

$$\boldsymbol{\alpha}(u) = (a \cos u, a \sin u, bu), \quad u \in \mathbb{R},$$

for some non-zero constant b.

(ii) We saw in Example 2 of §1.5 that a helix has constant non-zero curvature and constant non-zero torsion. Conversely, without using the Fundamental Theorem of the Local Theory of Space Curves, show that if $\boldsymbol{\alpha}$ is a regular curve in $\mathbb{R}^3$ with constant non-zero curvature κ and constant non-zero torsion τ, then $\boldsymbol{\alpha}$ is a helix on a cylinder of radius $a = \kappa/(\kappa^2 + \tau^2)$. To simplify calculations, you may assume, without loss of generality, that $\boldsymbol{\alpha}$ is parametrised by arc length s. (*Hint:* first show that $\tau \boldsymbol{t} - \kappa \boldsymbol{b}$ is a constant vector, $\boldsymbol{X}_0$, say, and then show that the curve $\boldsymbol{\alpha} + a\boldsymbol{n}$ is a straight line in direction $\boldsymbol{X}_0$.)

1.17 Let $\boldsymbol{\alpha}$ be a regular curve in $\mathbb{R}^3$ with non-zero curvature and torsion. Prove that the tangent lines to $\boldsymbol{\alpha}$ make a constant angle with a fixed direction in $\mathbb{R}^3$ if and only if κ/τ is constant. Such a curve is called a *generalised helix*. (*Hint:* if $\kappa/\tau = k$ for some constant k, consider the vector $\boldsymbol{t} - k\boldsymbol{b}$.)

1.18 Let $\boldsymbol{\alpha}(u)$ be a regular curve in $\mathbb{R}^3$ and assume that there is a point $p \in \mathbb{R}^3$ such that, for each parameter value u, the line through $\boldsymbol{\alpha}(u)$ in direction $\boldsymbol{n}(u)$ passes through p. Prove that $\boldsymbol{\alpha}$ is (part of) a circle. *(In this exercise you could assume, without loss of generality, that $\boldsymbol{\alpha}$ is parametrised by arc length. However, it doesn't make much difference in the solution.)*

1.19 Regular curves $\boldsymbol{\alpha}(u)$, $\boldsymbol{\beta}(u)$ in $\mathbb{R}^3$ are said to be *Bertrand mates* if, for each parameter value u, the line through $\boldsymbol{\alpha}(u)$ in direction $\boldsymbol{n}_{\alpha}(u)$ is equal to the line through $\boldsymbol{\beta}(u)$ in direction $\boldsymbol{n}_{\beta}(u)$. Prove that, if $\boldsymbol{\alpha}(u)$ and $\boldsymbol{\beta}(u)$ are Bertrand mates then:

(i) the angle between $\boldsymbol{t}_{\alpha}(u)$ and $\boldsymbol{t}_{\beta}(u)$ is independent of u; and
(ii) $\boldsymbol{\beta}(u) = \boldsymbol{\alpha}(u) + r\boldsymbol{n}_{\alpha}(u)$ for some constant real number r.

1.20 In this exercise, we extend the idea of involutes to space curves. Let $\boldsymbol{\alpha}(u)$ be a regular curve in $\mathbb{R}^3$, and let s_{α} denote arc length along $\boldsymbol{\alpha}$ measured from some point $\boldsymbol{\alpha}(u_0)$. Then the curve $\boldsymbol{\beta}(u)$ defined by

$$\boldsymbol{\beta}(u) = \boldsymbol{\alpha}(u) - s_{\alpha}(u)\boldsymbol{t}_{\alpha}(u)$$

is the *involute* of $\boldsymbol{\alpha}$ starting from $\boldsymbol{\alpha}(u_0)$. Let $\boldsymbol{\alpha}$ be the helix parametrised in the usual way as

$$\boldsymbol{\alpha}(u) = (a \cos u, a \sin u, bu), \quad a > 0, \quad b \neq 0,$$

and let $\boldsymbol{\beta}$ be the involute of $\boldsymbol{\alpha}$ starting from $\boldsymbol{\alpha}(0)$. Show that $\boldsymbol{\beta}$ lies in the plane $z = 0$ and is the involute starting from $(a, 0, 0)$ of the circle of intersection of the plane $z = 0$ with the cylinder $x^2 + y^2 = a^2$.

1.21 Let $\boldsymbol{\alpha}(s)$ be a regular curve in $\mathbb{R}^3$ parametrised by arc length with nowhere vanishing curvature κ and torsion τ. Show that $\boldsymbol{\alpha}$ lies on a sphere if and only if

$$\frac{\tau}{\kappa} = \frac{d}{ds}\left(\frac{1}{\tau\kappa^2}\frac{d\kappa}{ds}\right).$$

2 Surfaces in $\mathbb{R}^n$

In this chapter we introduce the main objects of study in the book, namely surfaces in $\mathbb{R}^n$. The most easily visualised situation is that in which $n = 3$, so we give emphasis to this. Indeed, if preferred, the reader may take $n = 3$ throughout this chapter, in which case the second half of §2.4 should be omitted. However, many interesting and challenging ideas emerge in the study of surfaces in higher dimensional Euclidean spaces, so we have included material on this for those who are interested.

For us, surfaces are subsets of $\mathbb{R}^n$ which, locally at least, can be smoothly identified with open subsets of the plane $\mathbb{R}^2$. The crucial properties of the surfaces we study are that they are 2-dimensional and may be approximated up to first order near any point by a flat plane, the *tangent plane* at that point. In this book we shall study the metric intrinsic and extrinsic geometry of surfaces, and the inter-relation between them. The intrinsic properties we study include the lengths of curves on surfaces, their angles of intersection, and the area of suitable regions. For the extrinsic geometry, we shall study various measures of the way in which a surface is bending away from its tangent plane. As one would expect, and as we shall see, these two aspects of the geometry of a surface are related in many interesting ways.

We first give our definition of a surface in $\mathbb{R}^n$, and then spend most of the rest of the chapter discussing methods of constructing and recognising surfaces. This will enable us to build up a large number of examples. Although we postpone the formal definition of tangent plane until the next chapter, the existence of these planes at all points of a surface should be intuitively clear in the examples.

2.1 Definition of a surface

As mentioned above, a surface in $\mathbb{R}^n$ is a subset of $\mathbb{R}^n$ which locally looks like an open subset of $\mathbb{R}^2$ which has been smoothly deformed. Globally, however, a surface can be very different from an open subset of $\mathbb{R}^2$.

We begin with the basic definitions. Let U be an open subset of $\mathbb{R}^m$ and let

$$f = (f_1, \dots, f_n) : U \to \mathbb{R}^n$$

be a map. The real-valued functions $f_1, \dots, f_n$ are the *coordinate functions* of f, and f is *smooth* if all partial derivatives of all orders of each f_i exist at each point and are continuous. The *image* $f(U)$ of U under f is given by

$$f(U) = \{f(u) \in \mathbb{R}^n : u \in U\} .$$

Example 1 Let

$$x(u, v) = (\cos v \cos u, \cos v \sin u, \sin v) .$$

Then x is a smooth map from $\mathbb{R}^2$ to $\mathbb{R}^3$, and, denoting partial derivatives by the appropriate subscript,

$$x_u = (-\cos v \sin u, \cos v \cos u, 0)$$

and

$$x_v = (-\sin v \cos u, -\sin v \sin u, \cos v) .$$

The image of x is the unit sphere $S^2(1) = \{(x, y, z) \in \mathbb{R}^3 : x^2 + y^2 + z^2 = 1\}$.

We now define the basic objects we study. For the purposes of this book, a non-empty subset S of $\mathbb{R}^n$ is a *surface* if for every point $p \in S$ there is an open subset U of $\mathbb{R}^2$ and a smooth map $x : U \to \mathbb{R}^n$ such that $p \in x(U)$ and (Figure 2.1)

(S1) $x(U) \subseteq S$,

(S2) there is an open subset W of $\mathbb{R}^n$ with $W \cap S = x(U)$ and a smooth map $F : W \to \mathbb{R}^2$ such that

$$Fx(u, v) = (u, v) , \quad \forall (u, v) \in U .$$

The above definition has been chosen to accord with our intuition of something which is "2-dimensional". Intuitively speaking, although the image of a smooth map may have dimension less than that of the domain (for example, $(x, y) \mapsto (x, 0, 0)$), the image cannot have higher dimension. Thus, the image of x has dimension at most two, but (S2) ensures that it can't be 1-dimensional since F must map the image of x back on to the whole of the 2-dimensional set U. Thus (S1) and (S2) ensure that surfaces do look "2-dimensional" at all points. The idea is that x takes the open subset U of $\mathbb{R}^2$ and moulds it smoothly onto part of S in a way that may be smoothly reversed (this being the role of F). For example, as we shall see, the sphere $S^2(r)$ of radius $r > 0$ in $\mathbb{R}^3$ is a surface, but neither the cone nor the "folded sheet" (Figure 2.2) are surfaces (although they **are** surfaces if the vertex is removed from the cone and the fold line is removed from the folded sheet).

The condition $W \cap S = x(U)$ in (S2) implies that, in our definition, a surface cannot have self-intersections. In fact, we could include the latter situation (Figure 2.3), but it leads to technicalities which we prefer to avoid.

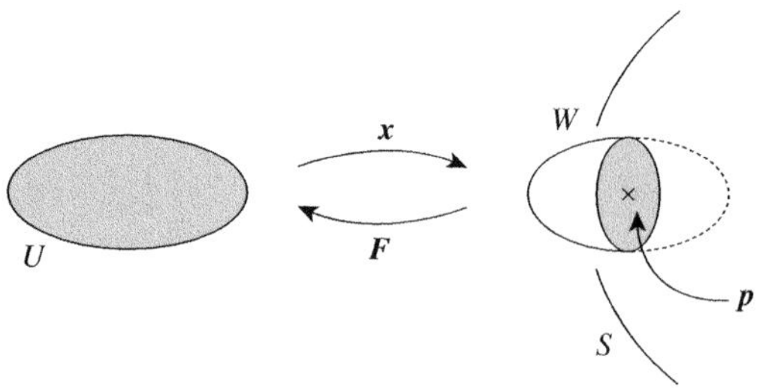

Figure 2.1 For the definition of surface

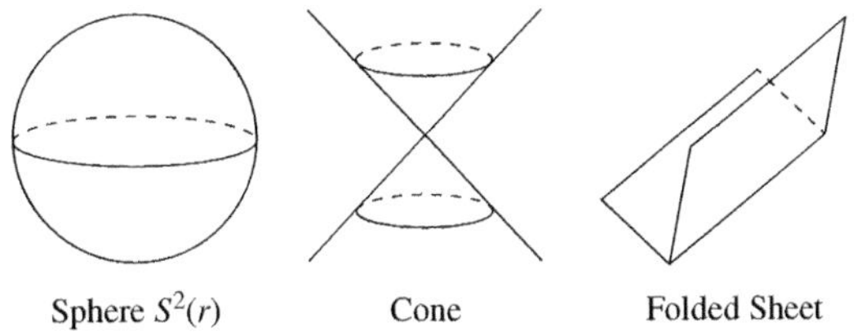

Figure 2.2 The cone and the folded sheet are not surfaces

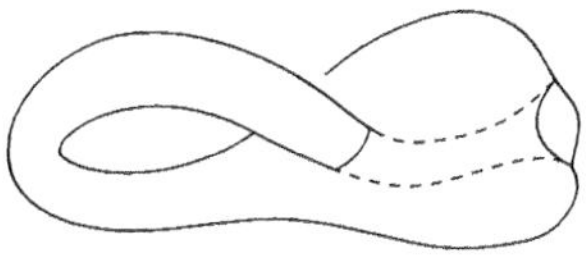

Figure 2.3 A Klein bottle: a "surface" with self-intersections

Example 2 (Hyperboloid) Let

$$S = \{(x, y, z) \in \mathbb{R}^3 : x^2 + y^2 = z^2 - 1,\ z > 0\}\,,$$

so that S is the upper sheet of a hyperboloid of two sheets (Figure 2.4).

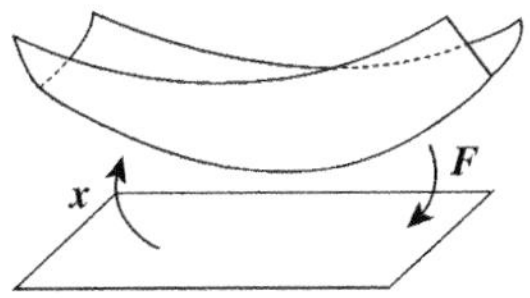

Figure 2.4 Upper sheet of a hyperboloid of two sheets

If we let U be the whole of $\mathbb{R}^2$, and define $\boldsymbol{x} : U \to \mathbb{R}^3$ by

$$\boldsymbol{x}(u, v) = \left(u, v, \sqrt{u^2 + v^2 + 1}\right), \quad u, v \in \mathbb{R}\,,$$

then $\boldsymbol{x}$ may be thought of as pushing the (horizontal) (x, y)-plane vertically upwards onto S. The way to reverse this process is to squash things flat again, so we take $W = \mathbb{R}^3$ and $\boldsymbol{F} : \mathbb{R}^3 \to \mathbb{R}^2$ to be given by $\boldsymbol{F}(x, y, z) = (x, y)$. It is now very easy to check that (S1) and (S2) both hold, and since every point of S is in the image of $\boldsymbol{x}$, this shows that S is a surface in $\mathbb{R}^3$. In fact, this surface is the graph of the function $g : \mathbb{R}^2 \to \mathbb{R}$ given by $g(u, v) = \sqrt{u^2 + v^2 + 1}$, and we shall see in §2.2 that this example may be easily generalised to show that the graph of any smooth real-valued function defined on an open set of $\mathbb{R}^2$ is a surface in $\mathbb{R}^3$.

A smooth map $\boldsymbol{x} : U \to S \subseteq \mathbb{R}^n$ with the properties described in (S1) and (S2) is called a *local parametrisation* of S.

Lemma 3 *A local parametrisation $\boldsymbol{x}$ of a surface S is injective. That is to say, if $\boldsymbol{x}(u_1, v_1) = \boldsymbol{x}(u_2, v_2)$ then $(u_1, v_1) = (u_2, v_2)$.*

Proof Taking notation from (S1) and (S2), if $\boldsymbol{x}(u_1, v_1) = \boldsymbol{x}(u_2, v_2)$ then, applying $\boldsymbol{F}$, we see that $(u_1, v_1) = (u_2, v_2)$. □

The image $\boldsymbol{x}(U)$ of a local parametrisation $\boldsymbol{x}$, which may be regarded as being differentiably equivalent to the open subset U of the plane, is called a *coordinate neighbourhood* on S. Thus a surface is a subset of $\mathbb{R}^n$ which may be covered by (possibly overlapping) coordinate neighbourhoods. However, it is important to note that, as far as geometry is concerned, it is the surface S as a subset of $\mathbb{R}^n$ which is of interest; the role of the local parametrisations is to help to describe and study the surface. (You may recall a similar remark being made in Chapter 1 for regular curves.)

Example 2 is rather simple since the whole of S may be covered using just one coordinate neighbourhood.

Example 4 (Sphere) We shall show that the sphere

$$S^2(r) = \{(x, y, z) \in \mathbb{R}^3 : x^2 + y^2 + z^2 = r^2\}$$

with radius $r > 0$ and centre at the origin of $\mathbb{R}^3$ is a surface. We begin by finding a local parametrisation whose image covers the northern hemisphere, which consists of those points of $S^2(r)$ with $z > 0$. To do this, we may proceed in a similar way to Example 2 and regard the northern hemisphere as the graph of the function $g : U \to \mathbb{R}$ defined on the open disc $U = \{(u, v) \in \mathbb{R}^2 : u^2 + v^2 < r^2\}$ given by $g(u, v) = \sqrt{r^2 - u^2 - v^2}$.

Specifically, we take

$$\boldsymbol{x}(u, v) = \left(u, v, \sqrt{r^2 - u^2 - v^2}\right), \quad u^2 + v^2 < r^2 .$$

Then, taking

$$W = \{(x, y, z) \in \mathbb{R}^3 : z > 0\},$$

so that $W \cap S^2(r)$ is the northern hemisphere, and defining $\boldsymbol{F} : W \to \mathbb{R}^2$ by

$$\boldsymbol{F}(x, y, z) = (x, y),$$

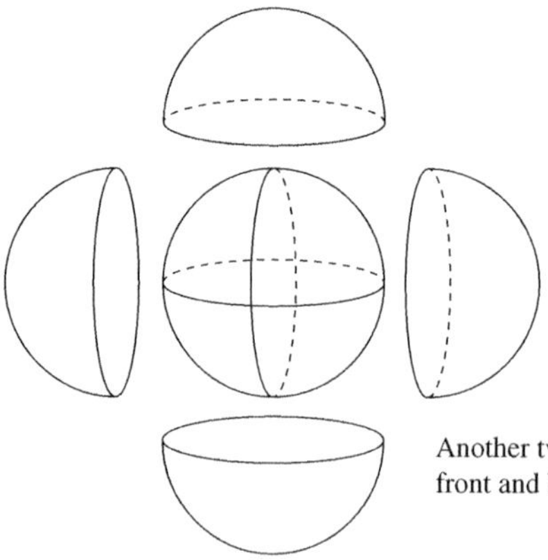

Figure 2.5 Covering a sphere with six coordinate neighbourhoods

we see that $\boldsymbol{x}$ and $\boldsymbol{F}$ are both smooth, $\boldsymbol{x}(U) = W \cap S^2(r)$ and $\boldsymbol{F}\boldsymbol{x}(u, v) = (u, v)$ for all $(u, v) \in U$. Thus conditions (S1) and (S2) hold for all points of the northern hemisphere.

To show by this line of argument that $S^2(r)$ is a surface, we have to show that every point of $S^2(r)$ is in some coordinate neighbourhood. This may be done by using six local parametrisations of the above type, each of which covers a hemisphere (Figure 2.5).

A surface may be covered by coordinate neighbourhoods in many different ways. For instance, we shall see later that the sphere $S^2(r)$ may be covered using just two coordinate neighbourhoods, namely $S^2(r)\backslash\{(0, 0, r)\}$ and $S^2(r)\backslash\{(0, 0, -r)\}$. In fact, it is often possible to choose particularly nice local parametrisations for a surface, and this is an important skill to acquire for both theoretical and calculational work. The interesting question of the minimum number of coordinate neighbourhoods needed to cover a surface is related to a topological invariant called the *Lusternik–Schnirelmann category*.

As mentioned in the introduction to this chapter, one crucial property of a surface is that it may be approximated up to first order near any point by a flat plane, the *tangent plane*. We shall see that this is the plane spanned by the partial derivatives $\boldsymbol{x}_u$ and $\boldsymbol{x}_v$ of a local parametrisation $\boldsymbol{x}$. For this to work, we first need to know that $\boldsymbol{x}_u$ and $\boldsymbol{x}_v$ do indeed span a plane.

Proposition 5 *Let $\boldsymbol{x}$ be a local parametrisation of a surface S. Then $\boldsymbol{x}_u$ and $\boldsymbol{x}_v$ are linearly independent at each point, and so span a plane.*

Proof We first recall that two vectors are linearly independent if they are both non-zero, and one is not a scalar multiple of the other. We shall prove the proposition for surfaces in $\mathbb{R}^3$, since the generalisation to surfaces in $\mathbb{R}^n$ is clear but the notation in the latter case is more cumbersome.

So, taking our notation from the definition of surface, and letting $\boldsymbol{x}(u, v) = (x(u, v), y(u, v), z(u, v))$, we have from (S2) that

$$\boldsymbol{F}\,(x(u, v), y(u, v), z(u, v)) = (u, v)\,.$$

Differentiating with respect to u and using the chain rule, we find

$$x_u \boldsymbol{F}_x + y_u \boldsymbol{F}_y + z_u \boldsymbol{F}_z = (1, 0)\,,$$

while differentiating with respect to v gives

$$x_v \boldsymbol{F}_x + y_v \boldsymbol{F}_y + z_v \boldsymbol{F}_z = (0, 1)\,.$$

It follows that both $\boldsymbol{x}_u = (x_u, y_u, z_u)$ and $\boldsymbol{x}_v = (x_v, y_v, z_v)$ are never zero, and, for instance, $(x_u, y_u, z_u) = \lambda(x_v, y_v, z_v)$ would give the contradiction that $(1, 0) = (0, \lambda)$. □

We conclude this section by noting that a suitable piece of a surface is itself a surface. A subset X of a surface S in $\mathbb{R}^n$ is said to be an *open subset* of S if X is the intersection of S with an open subset of $\mathbb{R}^n$. In Exercise 2.2 you are asked to prove the following lemma.

Lemma 6 *A non-empty open subset of a surface in $\mathbb{R}^n$ is itself a surface in $\mathbb{R}^n$.*

In each of the next three sections we discuss a particular method of constructing surfaces. This will enable us to build up a large collection of interesting examples.

2.2 Graphs of functions

We begin by recalling that if $g : U \to \mathbb{R}$ is a smooth function defined on a subset U of $\mathbb{R}^2$ then the *graph* $\Gamma(g)$ of g is the subset of $\mathbb{R}^3$ given by

$$\Gamma(g) = \left\{(u, v, g(u, v)) \in \mathbb{R}^3 : (u, v) \in U\right\} .$$

Proposition 1 *Let U be an open subset of $\mathbb{R}^2$, and let $g : U \to \mathbb{R}$ be a smooth function. Then the graph $\Gamma(g)$ of g is a surface in $\mathbb{R}^3$.*

Proof We generalise the method of Example 2 in §2.1 as follows. Take $W = \mathbb{R}^3$, and let $\boldsymbol{x} : U \to \mathbb{R}^3$, $\boldsymbol{F} : \mathbb{R}^3 \to \mathbb{R}^2$ be the smooth maps defined by

$$\boldsymbol{x}(u, v) = (u, v, g(u, v)) \ , \quad \boldsymbol{F}(x, y, z) = (x, y) \, . \tag{2.1}$$

Then $\boldsymbol{x}(U) = W \cap \Gamma(g) = \Gamma(g)$, and $\boldsymbol{F}\boldsymbol{x}(u, v) = (u, v)$ for all $(u, v) \in U$. Since every point of $\Gamma(g)$ lies in the image of $\boldsymbol{x}$, it follows from the definition that $\Gamma(g)$ is a surface in $\mathbb{R}^3$. □

Graphs provide quite simple examples of surfaces since they may be covered by a single coordinate neighbourhood; this is not the case for the examples in the next section.

As an illustration of Proposition 5 of §2.1, we note that if $\boldsymbol{x}$ is as in (2.1) then

$$\boldsymbol{x}_u = (1, 0, g_u) \, , \quad \boldsymbol{x}_v = (0, 1, g_v) \, ,$$

which are clearly linearly independent at all points.

Example 2 (Elliptic paraboloid) Let a, b be positive real numbers and let $g : \mathbb{R}^2 \to \mathbb{R}$ be given by

$$g(u, v) = \frac{u^2}{a^2} + \frac{v^2}{b^2} \, , \quad u, v \in \mathbb{R} \, .$$

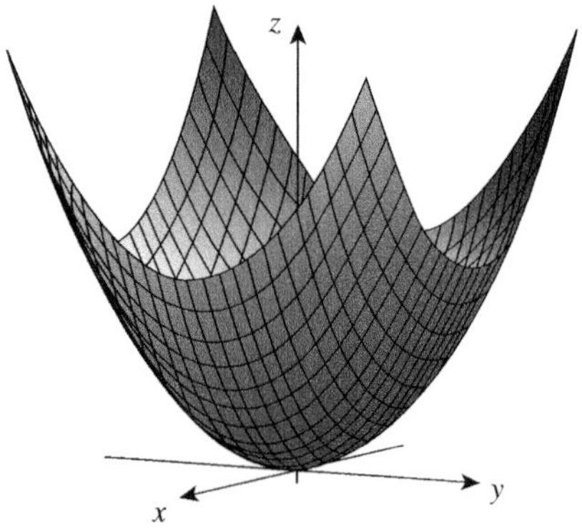

Figure 2.6 Elliptic paraboloid

The graph $\Gamma(g)$ of g is an elliptic paraboloid (Figure 2.6),

$$\Gamma(g) = \left\{ \left(u, v, \frac{u^2}{a^2} + \frac{v^2}{b^2} \right) : u, v \in \mathbb{R} \right\}.$$

Example 3 (Hyperbolic paraboloid) Let a, b be positive real numbers and let $g : \mathbb{R}^2 \to \mathbb{R}$ be given by

$$g(u, v) = \frac{u^2}{a^2} - \frac{v^2}{b^2}, \quad u, v \in \mathbb{R}.$$

The graph of g is a hyperbolic paraboloid, which is a saddle-shaped surface (Figure 2.7).

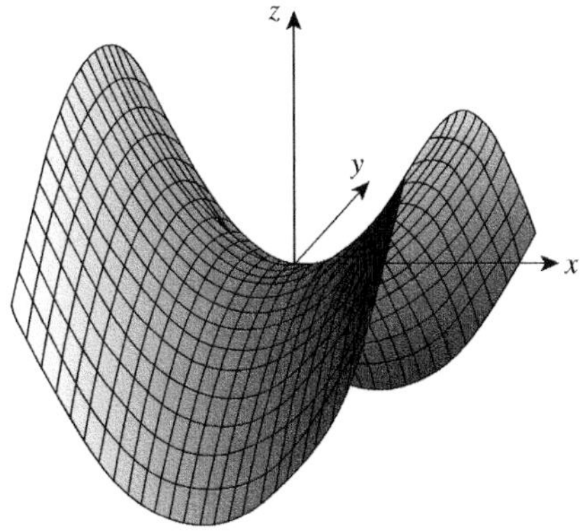

Figure 2.7 Hyperbolic paraboloid

2.3 Surfaces of revolution

A *surface of revolution* in $\mathbb{R}^3$ is a surface S which is setwise invariant under rotations of $\mathbb{R}^3$ about a line – the *axis of rotation* of S.

Let I be an open interval and let $\boldsymbol{\alpha}(v) = (f(v), 0, g(v))$, $v \in I$, be a regular curve without self-intersections in the plane $y = 0$ in $\mathbb{R}^3$. We assume that $f(v) > 0$ for all $v \in I$, so, in particular, $\boldsymbol{\alpha}$ does not meet the z-axis (when $f(v)$ would be zero). The subset S of $\mathbb{R}^3$ swept out as we rotate $\boldsymbol{\alpha}$ about the z-axis is given by (Figure 2.8)

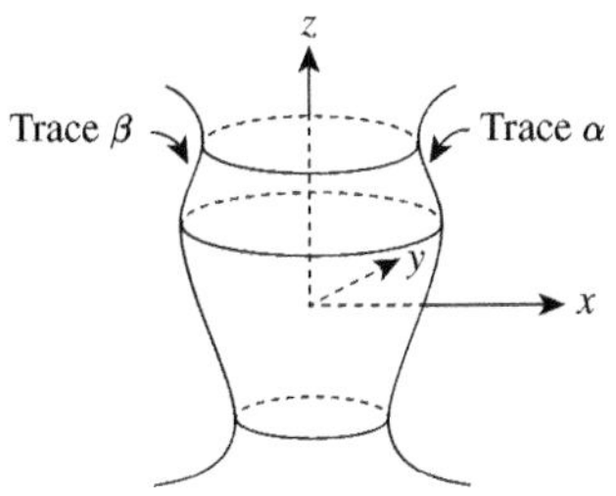

Figure 2.8 Surface of revolution

$$S = \{(f(v)\cos u,\ f(v)\sin u, g(v)) : u \in \mathbb{R}, v \in I\}, \tag{2.2}$$

which is clearly setwise invariant under rotations about the z-axis. We shall show in Proposition 5 that S is a surface by covering S using just two coordinate neighbourhoods, namely $S \setminus \text{trace}\, \boldsymbol{\alpha}$ and $S \setminus \text{trace}\, \boldsymbol{\beta}$, where $\boldsymbol{\beta}$ is the reflection of $\boldsymbol{\alpha}$ in the z-axis. Before doing this, however, we give some examples.

Example 1 (Catenoid) This is obtained by rotating the catenary

$$\boldsymbol{\alpha}(v) = \left(k\cosh\frac{v}{k}, 0, v\right), \quad v \in \mathbb{R},$$

about the z-axis (Figure 2.9). Here, k is any positive constant.

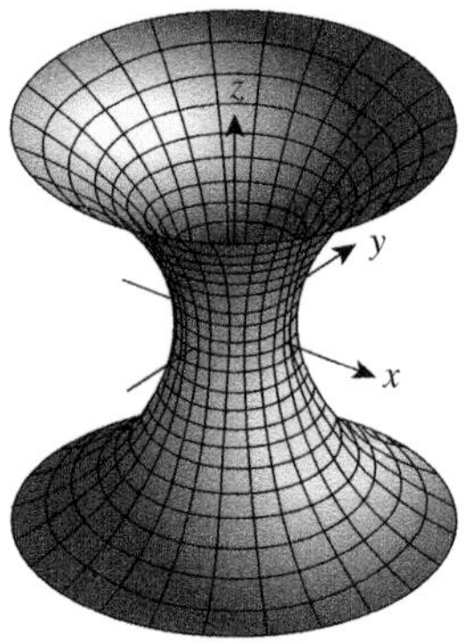

Figure 2.9 A catenoid

Example 2 (Pseudosphere) This is obtained by rotating the tractrix

$$\boldsymbol{\alpha}(v) = (\text{sech}\, v, 0, v - \tanh v), \quad v > 0,$$

about the z-axis (Figure 2.10). Motivation for the name "pseudosphere" will be given when we study Gaussian curvature in Chapter 5.

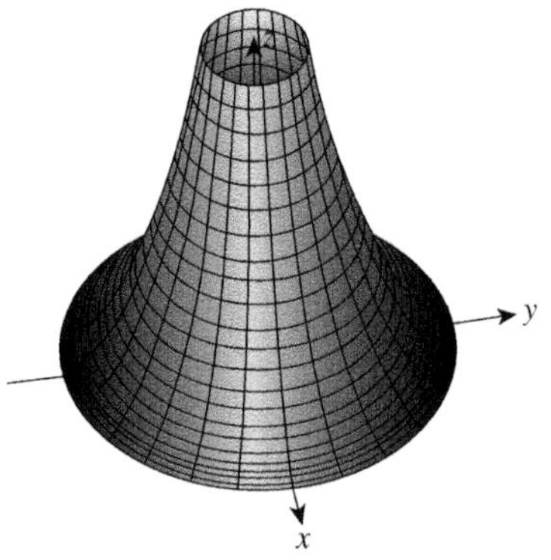

Figure 2.10 The pseudosphere

If we rotate a regular curve in the xz-plane which intersects the z-axis, then we may get singular points there; for instance, if we rotate the line $z = x$ then we obtain a cone. However, sometimes we do obtain a surface, as we see in the following example.

Example 3 (Sphere) The sphere $S^2(r)$ of radius $r > 0$ and centre the origin is obtained by rotating the semicircle

$$\boldsymbol{\alpha}(v) = (r\cos v, 0, r\sin v)\,, \qquad -\frac{\pi}{2} \le v \le \frac{\pi}{2}\,,$$

about the z-axis (Figure 2.11).

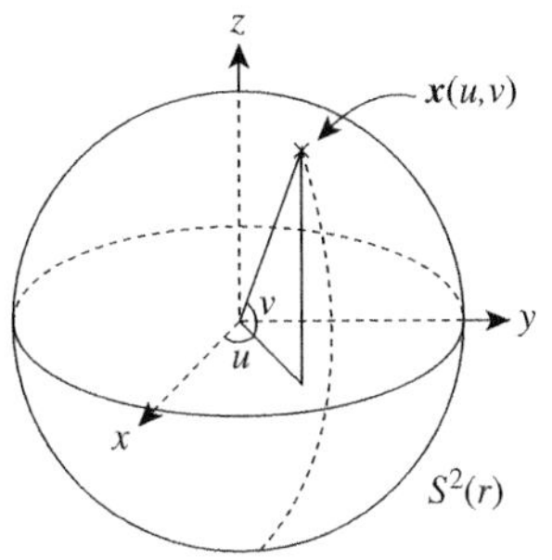

Figure 2.11 Sphere parametrised as a surface of revolution

Note that, in this example, we have rotated a curve defined on a closed interval whose end points are on the axis of rotation. Points at which a surface of revolution intersects its axis are called *poles*. We shall see in Chapter 4 that, in order for the surface to be smooth at these points, the curve must intersect the axis orthogonally.

Example 4 (Torus of revolution) Let a, b be positive real numbers with $a > b$. Then the torus of revolution $T_{a,b}$ is obtained by rotating, about the z-axis, the circle of radius b and centre $(a, 0, 0)$ in the xz-plane (Figure 2.12).

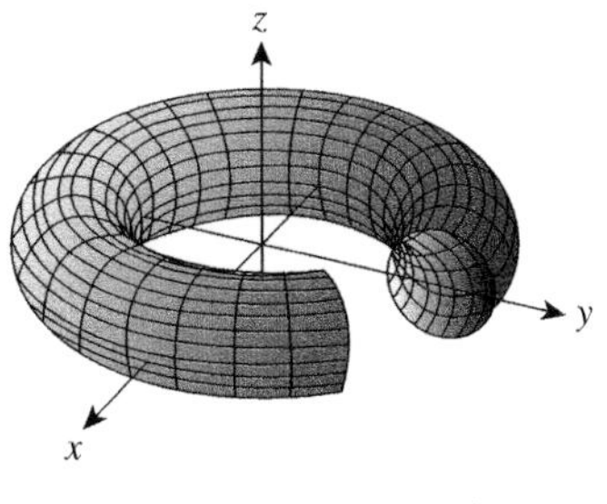

Figure 2.12 A torus of revolution

Note that this example is also slightly different from the preceeding ones in that we have rotated a closed curve rather than one defined on an open interval.

We now show that surfaces of revolution are indeed surfaces according to our definition.

Proposition 5 *Let I be an open interval of real numbers and let $\boldsymbol{\alpha}(v) = (f(v), 0, g(v))$, $v \in I$, be a regular curve in the xz-plane without self-intersections and with $f(v) > 0$ for all $v \in I$. Then the set*

$$S = \{(f(v)\cos u,\ f(v)\sin u, g(v)) : u \in \mathbb{R}, v \in I\}$$

is a surface in $\mathbb{R}^3$.

Proof In the course of the proof, and elsewhere in the book, we shall use the notation $A \times B$ for the *cartesian product* of two sets A and B; specifically,

$$A \times B = \{(a,b) : a \in A, b \in B\}\,.$$

Our first guess at a local parametrisation of S might be to define $\boldsymbol{x} : \mathbb{R} \times I \to \mathbb{R}^3$ by

$$\boldsymbol{x}(u,v) = (f(v)\cos u,\ f(v)\sin u, g(v))\,, \quad u \in \mathbb{R}\,,\ v \in I\,.$$

The image of this map certainly covers S, but Lemma 3 of §2.1 shows that $\boldsymbol{x}$ can't be a local parametrisation since $\boldsymbol{x}(u + 2\pi, v) = \boldsymbol{x}(u, v)$. We overcome this difficulty by restricting the domain of $\boldsymbol{x}$; we let $U = (-\pi, \pi) \times I$ and define $\boldsymbol{x} : U \to \mathbb{R}^3$ by

$$\boldsymbol{x}(u,v) = (f(v)\cos u,\ f(v)\sin u, g(v))\,, \quad -\pi < u < \pi\,,\ v \in I\,. \tag{2.3}$$

We now show how to find an open subset W of $\mathbb{R}^3$ and a smooth map $\boldsymbol{F} : W \to \mathbb{R}^2$ as in condition (S2) for a surface. We take $W = \mathbb{R}^3 \setminus \Pi$, where Π is the half-plane on which $y = 0$ and $x \leq 0$, and, in order to simplify the rest of the proof, we shall restrict ourselves to finding a map $\boldsymbol{F} : W \to \mathbb{R}^2$ as in condition (S2) for a surface in the following two special cases:

Case 1: $\boldsymbol{\alpha}(v) = (f(v), 0, v)$,
Case 2: $\boldsymbol{\alpha}(v) = (v, 0, g(v))$ (so, in particular, we assume $v > 0$).

The finding of $\boldsymbol{F}$ in these special cases contains the essential ideas, but avoids a technical difficulty (which may be overcome by use of an important result, the *Inverse Function Theorem*, taken from the theory of differential calculus of functions between Euclidean spaces – more about this later).

Case 1: Here we let $\boldsymbol{F} : W \to \mathbb{R}^2$ be given by

$$\boldsymbol{F}(x,y,z) = (\mathrm{Arg}\,(x + iy), z)\,,$$

where $-\pi < \mathrm{Arg} < \pi$ denotes the principal argument of a complex number. Then $\boldsymbol{x}$ and $\boldsymbol{F}$ are smooth, and, since we assume that $f(v) > 0$ for all $v \in I$, it follows that $W \cap S = \boldsymbol{x}(U)$. Also,

$$\boldsymbol{F}\boldsymbol{x}(u,v) = \boldsymbol{F}\,(f(v)\cos u,\ f(v)\sin u, v) = (u,v)\,, \quad (u,v) \in U\,.$$

It now follows that (S2) holds, so that $\boldsymbol{x}$ is a local parametrisation of S.

Case 2: This time we take $\boldsymbol{F} : W \to \mathbb{R}^2$ to be

$$\boldsymbol{F}(x,y,z) = \left(\mathrm{Arg}\,(x + iy), \sqrt{x^2 + y^2}\right),$$

and a similar argument shows that (S2) holds, so that $\boldsymbol{x}$ is again a local parametrisation of S.

The price we pay for restricting the domain of $\boldsymbol{x}$ to $(-\pi,\pi)\times I$ is that the image no longer covers the whole of S (the trace of the reflection $\boldsymbol{\beta}$ of $\boldsymbol{\alpha}$ in the z-axis is omitted). In order to cover S with coordinate neighbourhoods we use another local parametrisation which is given by the same formula as $\boldsymbol{x}$ but has a different domain of definition, namely $\boldsymbol{y}:(0,2\pi)\times I\to\mathbb{R}^3$ given by

$$\boldsymbol{y}(u,v)=(f(v)\cos u,\ f(v)\sin u, g(v)),\quad 0<u<2\pi,\ v\in I. \tag{2.4}$$

The image of $\boldsymbol{y}$ omits the trace of $\boldsymbol{\alpha}$, but the two local parametrisations $\boldsymbol{x}$ and $\boldsymbol{y}$ given in (2.3) and (2.4) cover the whole of S, which shows that S is indeed a surface. □

For $\boldsymbol{x}(u,v)$ as in (2.3), we note that

$$\begin{aligned}\boldsymbol{x}_u &= (-f(v)\sin u,\ f(v)\cos u, 0)\,,\\ \boldsymbol{x}_v &= \left(f'(v)\cos u,\ f'(v)\sin u, g'(v)\right),\end{aligned}$$

which, in accordance with Proposition 5 of §2.1, are easily seen to be linearly independent at each point.

As already mentioned, the sphere discussed in Example 3 and the torus of revolution discussed in Example 4 are slightly different from the surfaces of revolution considered in Proposition 5. For the sphere, if we restrict the domain of definition of the curve $\boldsymbol{\alpha}$ given in Example 3 to $-\pi/2<v<\pi/2$ then the standard local parametrisations discussed above cover the whole of $S^2(r)$ except for the the two poles.

The torus of revolution discussed in Example 4 is obtained by rotating a closed curve, namely, a circle. This circle (with one point omitted) may be parametrised by

$$\boldsymbol{\alpha}(v)=(a+b\cos v, 0, b\sin v)\,,\quad -\pi<v<\pi\,,$$

and the torus may be covered using four coordinate neighbourhoods corresponding to the local parametrisations

$$\boldsymbol{x}(u,v)=((a+b\cos v)\cos u,\ (a+b\cos v)\sin u,\ b\sin v)\,,\quad (u,v)\in U\,,$$

where U is taken in turn to be

$$\begin{aligned}&(0,2\pi)\times(0,2\pi)\,,\quad (0,2\pi)\times(-\pi,\pi)\,,\\ &(-\pi,\pi)\times(0,2\pi)\,,\quad (-\pi,\pi)\times(-\pi,\pi)\,.\end{aligned}$$

Returning to the general situation of a surface of revolution, the curve $\boldsymbol{\alpha}(v)=(f(v),0,g(v))$ is the *generating curve* of the corresponding surface of revolution. The circle swept out by a point of $\boldsymbol{\alpha}$ is called a *parallel*, and the curve on S obtained by rotating $\boldsymbol{\alpha}$ through a fixed angle is called a *meridian*. Each parallel is the intersection of S with a horizontal plane $z=$ constant, while each meridian is the intersection of S with a half-plane with boundary the z-axis. In particular, the parallels and meridians of the sphere $S^2(r)$ described in Example 3 are the curves of latitude and longitude used in discussions of the geography of the Earth.

2.4 Surfaces defined by equations

Many surfaces in $\mathbb{R}^3$ can be defined by an equation. For example, the unit sphere has equation $x^2 + y^2 + z^2 = 1$, the ellipsoid has equation $(x/a)^2 + (y/b)^2 + (z/c)^2 = 1$, and the hyperboloid of two sheets has equation $(x/a)^2 + (y/b)^2 = (z/c)^2 - 1$.

However, not every equation defines a surface. For instance, the equation $xy = 0$ gives the union of two of the coordinate planes while $x^2 + y^2 = 0$ gives the z-axis.

In this section we look at conditions under which an equation defines a surface in $\mathbb{R}^3$ (and at the more general situation of surfaces in $\mathbb{R}^n$).

We first recall that, if $f : W \to \mathbb{R}$ is a smooth function defined on an open subset W of $\mathbb{R}^3$, then the *gradient* grad f of f is the vector-valued function given by grad $f = (f_x, f_y, f_z)$, where, as usual, partial differentiation is denoted by the appropriate suffix.

Theorem 1 *Let W be an open subset of $\mathbb{R}^3$ and let $f : W \to \mathbb{R}$ be a smooth function. Let k be a real number and let S be the subset of $\mathbb{R}^3$ defined by the equation $f(x, y, z) = k$. If S is non-empty and if grad f is never zero on S, then S is a surface in $\mathbb{R}^3$.*

The idea of the proof of the theorem is to show that (maybe after re-labelling the axes of $\mathbb{R}^3$) each sufficiently small piece of S is the graph of a function. The theorem (and its generalisation below) is the first of several results in this section and the next whose proofs use the Inverse Function Theorem. These proofs are not necessary for an appreciation of the material in the book, and they are also quite technical. For this reason, we present the proofs in an appendix of optional reading at the end of this chapter.

We note that if grad f does vanish at one or more points of the subset of $\mathbb{R}^3$ defined by the equation $f(x, y, z) = k$ then we cannot conclude that this subset is not a surface; the theorem simply gives us no information in that case (see, for example, Exercise 2.5).

Example 2 (Hyperboloids) Let a, b and c be positive real numbers and let $f(x, y, z) = \frac{x^2}{a^2} + \frac{y^2}{b^2} - \frac{z^2}{c^2}$. Then grad $f = \left(\frac{2x}{a^2}, \frac{2y}{b^2}, -\frac{2z}{c^2}\right)$, which is zero if and only if $x = y = z = 0$.

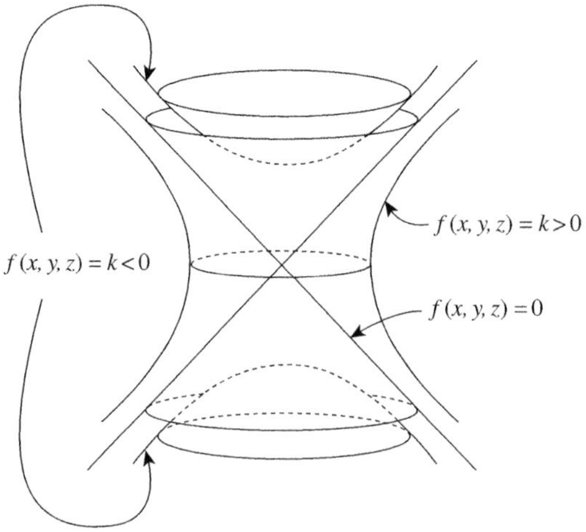

Figure 2.13 Hyperboloids and an elliptical cone

Since $f(0,0,0) = 0$ and since f takes all real values, we see that the equation $f(x, y, z) = k$ defines a surface S for all real $k \neq 0$. If $k > 0$ then S is a hyperboloid of one sheet, while if $k < 0$ then S is a hyperboloid of two sheets (Figure 2.13).

However $f(x, y, z) = 0$ is the equation of the elliptical cone formed by the lines passing through $(0, 0, 0)$ and the points on the ellipse in the plane $z = c$ given by

$$E = \{(x, y, z) \in \mathbb{R}^3 : \frac{x^2}{a^2} + \frac{y^2}{b^2} = 1,\ z = c\} .$$

This fails to be a surface at $(0, 0, 0)$, the point at which grad f is zero.

There is a standard notation which applies in this situation. Those points of the domain W of f at which grad f is zero are called *critical points* of f, and all other points of W are *regular points*. The inverse image of a real number k is the set of points $p \in W$ such that $f(p) = k$, and k is a *regular value* of f if its inverse image is non-empty and consists entirely of regular points. All other values of f are called *critical values*. Theorem 1 says that the inverse image of a regular value is a surface.

Many surfaces are examples of more than one of the three types of surface we have discussed in §2.2 to §2.4. For instance, if S is the graph of the smooth function $g : U \to \mathbb{R}$, where U is an open subset of $\mathbb{R}^2$, then S is also defined by the equation $z = g(x, y)$.

Another example is provided by surfaces of revolution obtained by rotating a regular curve of the form

$$\alpha(v) = (f(v), 0, v)\,, \quad f(v) > 0\ \forall v,$$

about the z-axis. Such a surface is defined by the equation

$$x^2 + y^2 = (f(z))^2 .$$

The methods in this section may be generalised to give a criterion for a system of equations to define a surface in $\mathbb{R}^n$. We now give a brief description of this, but **this material will not be used in an essential way in the book, so the rest of this section may be omitted if desired**.

In general, to describe a surface in $\mathbb{R}^n$ we need a suitable system of $n - 2$ equations.

Theorem 3[†] *Let W be an open subset of $\mathbb{R}^n$ and let $f = (f_1, \ldots, f_{n-2}) : W \to \mathbb{R}^{n-2}$ be a smooth map. Let $(k_1, \ldots, k_{n-2}) \in \mathbb{R}^{n-2}$ and let S be the subset of $\mathbb{R}^n$ defined by the equations*

$$f_i(x_1, \ldots, x_n) = k_i\,, \quad i = 1, \ldots, n-2 .$$

If S is non-empty and if grad $f_1, \ldots,$ grad f_{n-2} are linearly independent everywhere on S, then S is a surface in $\mathbb{R}^n$.

It should be clear how to define critical point, regular point, regular value and critical value of a function $f : W \to \mathbb{R}^{n-2}$ in such a way that Theorem 3 may be restated in a similar way to Theorem 1, namely that the inverse image of a regular value is a surface.

Example 4† **(A torus in $\mathbb{R}^4$)** Let r_1, r_2 be positive real numbers and let S be the subset of $\mathbb{R}^4$ defined by the equations

$$x_1{}^2 + x_2{}^2 = r_1{}^2 , \quad x_3{}^2 + x_4{}^2 = r_2{}^2 .$$

Then a straightforward application of Theorem 3 shows that S is a surface in $\mathbb{R}^4$. In fact, intuitively speaking, S is a product of two circles $S = S^1(r_1) \times S^1(r_2) \subset \mathbb{R}^2 \times \mathbb{R}^2$ and is, in a natural way, differentiably equivalent to the torus of revolution discussed in Example 4 of §2.3. However, as we shall see, the local intrinsic metric properties of the two surfaces are very different; the intrinsic geometry of the torus in $\mathbb{R}^4$ described here is locally like that of the plane (and so this surface is often called a *flat torus*), whereas that of the surface in Example 4 of §2.3 is not. To convince yourself of this try moulding a flat piece of paper onto a doughnut without crinkling the paper!

Example 5†**(Graphs of maps to $\mathbb{R}^{n-2}$)** Let U be an open subset of $\mathbb{R}^2$ and let $g = (g_1, \ldots, g_{n-2}) : U \to \mathbb{R}^{n-2}$ be a smooth map. Then the graph $\Gamma(g)$ of g is the subset of $\mathbb{R}^n$ given by

$$\Gamma(g) = \left\{ (u, v, g(u, v)) \in \mathbb{R}^n : (u, v) \in U \right\} .$$

It is quick to generalise the method used in §2.2 to show directly that $\Gamma(g)$ is a surface, but this may also be proved using Theorem 3 by noting that $\Gamma(g)$ is defined by the system of equations

$$x_{i+2} - g_i(x_1, x_2) = 0 , \quad i = 1, \ldots, n-2 .$$

2.5 Coordinate recognition

The surfaces discussed in §2.2 and §2.3 come equipped with natural local parametrisations but those of §2.4 do not. It is sometimes rather difficult to check criterion (S2) for local parametrisations because of problems associated with finding a function $\boldsymbol{F}$ with the properties required by that criterion, so we now describe three useful (but rather technical) results which help us to recognise local parametrisations without the need to verify that (S2) holds.

In order to provide some motivation and intuition, we first recall that one of the reasons for condition (S2) in the definition of surface is to ensure that the image $\boldsymbol{x}(U)$ of a local parametrisation is 2-dimensional. In particular, we saw in Lemma 3 of §2.1 that $\boldsymbol{x}$ is injective, while Proposition 5 of §2.1 showed that $\boldsymbol{x}_u$ and $\boldsymbol{x}_v$ are linearly independent at each point.

As we illustrate with some examples, it is often much easier to check the injectivity of $\boldsymbol{x}$ and the linear independence of $\boldsymbol{x}_u$ and $\boldsymbol{x}_v$ than to construct a map $\boldsymbol{F}$ satisfying condition (S2), and the motivation for the theorems we discuss in this section is that they will help us decide to what extent one or both of these easier conditions enable us to conclude that $\boldsymbol{x}$ is a local parametrisation of a surface. We shall be mostly interested in the case of surfaces in $\mathbb{R}^3$, and here the vector cross product often provides a convenient way of proving linear independence since the vector product of two vectors in $\mathbb{R}^3$ is zero if and only if they are linearly dependent.

The proofs of Theorems 1 and 3 in this section use the Inverse Function Theorem, and may be found in the appendix to this chapter. As already mentioned, it is a matter of time and taste as to whether the material in the appendix is included in a course of study.

Before stating the first theorem, we note that an *open neighbourhood* of a point $q \in \mathbb{R}^m$ is simply an open set in $\mathbb{R}^m$ containing q.

Theorem 1 *Let $\boldsymbol{x} : U \to \mathbb{R}^n$ be a smooth map defined on an open subset U of $\mathbb{R}^2$, and assume that $q \in U$ is such that the vectors $\boldsymbol{x}_u(q)$, $\boldsymbol{x}_v(q)$ are linearly independent (or, equivalently for a surface in $\mathbb{R}^3$, $(\boldsymbol{x}_u \times \boldsymbol{x}_v)(q) \neq 0$). Then U contains an open neighbourhood U_0 of q such that $\boldsymbol{x}(U_0)$ is a surface S in $\mathbb{R}^n$ and the restriction of $\boldsymbol{x}$ to U_0 is a parametrisation of the whole of S.*

Example 2 (Enneper's surface) This is an example of a minimal surface (the shape taken up by a soap film); these surfaces are studied in Chapter 9. Enneper's surface (for a picture, see Figure 9.2) is defined to be the image of

$$\boldsymbol{x}(u, v) = \left(u - \frac{u^3}{3} + uv^2,\ -v + \frac{v^3}{3} - u^2 v,\ u^2 - v^2\right), \quad u, v \in \mathbb{R}. \tag{2.5}$$

Here,

$$\boldsymbol{x}_u = (1 - u^2 + v^2,\ -2uv,\ 2u)\,, \quad \boldsymbol{x}_v = (2uv,\ -1 - u^2 + v^2,\ -2v)\,,$$

and a short calculation shows that

$$\boldsymbol{x}_u \times \boldsymbol{x}_v = \left(2u(1 + u^2 + v^2),\ 2v(1 + u^2 + v^2),\ (u^2 + v^2)^2 - 1\right).$$

This is clearly never zero, so that $\boldsymbol{x}_u$ and $\boldsymbol{x}_v$ are linearly independent at each point. Using Theorem 1, we see that sufficiently small pieces of Enneper's surface are indeed surfaces according to our definition. However, the finding of the corresponding map $\boldsymbol{F}$ would be daunting. We remark that the map $\boldsymbol{x}(u, v)$ given in (2.5) isn't injective (although it **is** injective when restricted to sufficiently small open sets in the plane); Enneper's surface has self-intersections and so is not a surface according to our definition.

We now discuss two further criteria for recognising local parametrisations. These are rather different from the one discussed above, since we need to know that we have a surface before we can apply the criteria. Once again, we leave the proof of the first of these results to the appendix to this chapter; the second result may be proved from the first by using topological arguments involving paracompactness.

Theorem 3 *Let S be a surface in $\mathbb{R}^n$ and let $\boldsymbol{x} : U \to \mathbb{R}^n$ be a smooth map defined on an open subset U of $\mathbb{R}^2$. If $\boldsymbol{x}(U) \subseteq S$ and if $q \in U$ is such that the vectors $\boldsymbol{x}_u(q)$, $\boldsymbol{x}_v(q)$ are linearly independent (or, equivalently for a surface in $\mathbb{R}^3$, $(\boldsymbol{x}_u \times \boldsymbol{x}_v)(q) \neq 0$), then U contains an open neighbourhood U_0 of q such that the restriction of $\boldsymbol{x}$ to U_0 is a local parametrisation of S.*

Theorem 4 (Parametrisation Recognition Theorem) *Let S be a surface in $\mathbb{R}^n$. A smooth map $\boldsymbol{x} : U \to \mathbb{R}^n$ defined on an open subset U of $\mathbb{R}^2$ is a local parametrisation of S if all of the following conditions hold:*

(1) $\boldsymbol{x}(U) \subseteq S$,
(2) $\boldsymbol{x}$ is injective,
(3) $\boldsymbol{x}_u$ and $\boldsymbol{x}_v$ are linearly independent at all points (or, equivalently for a surface in $\mathbb{R}^3$, $\boldsymbol{x}_u \times \boldsymbol{x}_v$ is never zero).

As mentioned earlier in this section, the converse to Theorem 4 is also true; if $\boldsymbol{x}(u, v)$ is a local parametrisation of a surface, then conditions (1), (2) and (3) all hold.

Example 5 (Hyperboloid of one sheet) We saw in Example 2 of §2.4 that if $f : \mathbb{R}^3 \to \mathbb{R}$ is given by

$$f(x, y, z) = \frac{x^2}{a^2} + \frac{y^2}{b^2} - \frac{z^2}{c^2},$$

then $f(x, y, z) = 1$ is the equation of a surface S (a hyperboloid of one sheet). This is not a surface of revolution (unless $a = b$), but our experience with these latter surfaces would encourage us to believe that the smooth map $\boldsymbol{x} : (-\pi, \pi) \times \mathbb{R} \to \mathbb{R}^3$ given by

$$\boldsymbol{x}(u, v) = (a \cosh v \cos u,\ b \cosh v \sin u,\ c \sinh v), \qquad -\pi < u < \pi, \quad v \in \mathbb{R},$$

would be a local parametrisation. The construction of a map $\boldsymbol{F}$ as in (S2) is not hard in this case, but we wish to illustrate the use of Theorem 4 by checking the three conditions required to apply that theorem.

First note that $f(\boldsymbol{x}(u, v)) = 1$ for all $(u, v) \in U$, so that the first condition of Theorem 4 holds. Also, $\boldsymbol{x}$ is injective since $\cos u$, $\sin u$ determine $u \in (-\pi, \pi)$ uniquely while $\sinh v$ determines v uniquely. Finally,

$$\boldsymbol{x}_u = (-a \cosh v \sin u,\ b \cosh v \cos u,\ 0),$$
$$\boldsymbol{x}_v = (a \sinh v \cos u,\ b \sinh v \sin u,\ c \cosh v),$$

and the vector product

$$\boldsymbol{x}_u \times \boldsymbol{x}_v = \cosh v\ (bc \cosh v \cos u,\ ca \cosh v \sin u,\ -ab \sinh v)$$

is clearly never zero.

The Parametrisation Recognition Theorem now shows that $\boldsymbol{x}$ is a local parametrisation of S. In order to cover S with coordinate neighbourhoods we could take, in addition to $\boldsymbol{x}$, the local parametrisation which is given by the same formula as $\boldsymbol{x}$ but has domain $(0, 2\pi) \times \mathbb{R}$.

2.6 Appendix: Proof of three theorems †

In this appendix we give a brief account of the Inverse Function Theorem, and show how it may be used to prove Theorem 1 of §2.4 and Theorems 1 and 3 of §2.5. As mentioned previously, this material is rather technical and could be omitted if desired.

We begin by recalling the theory of differentiation of functions between Euclidean spaces. So, let $f(u_1, \ldots, u_m)$ be a smooth $\mathbb{R}^n$-valued map defined on an open subset U of $\mathbb{R}^m$, and, as usual, write

$$f = (f_1, \ldots, f_n),$$

where $f_1, \ldots, f_n$ are smooth real-valued functions defined on U. The *derivative* of f at $p \in U$ is the linear map $df_p : \mathbb{R}^m \to \mathbb{R}^n$ given by matrix multiplication by the *Jacobian matrix* Jf_p of f at p. Specifically, using matrix multiplication on the right hand side of the equation,

$$df_p \begin{pmatrix} h_1 \\ \vdots \\ h_m \end{pmatrix} = Jf_p \begin{pmatrix} h_1 \\ \vdots \\ h_m \end{pmatrix},$$

where

$$Jf_p = \begin{pmatrix} \frac{\partial f_1}{\partial u_1}(p) & \cdots & \frac{\partial f_1}{\partial u_m}(p) \\ \vdots & & \vdots \\ \frac{\partial f_n}{\partial u_1}(p) & \cdots & \frac{\partial f_n}{\partial u_m}(p) \end{pmatrix}. \tag{2.6}$$

The derivative of f at $p \in U$ is very useful because it gives us a good linear approximation to f on an open neighbourhood of p in the following sense. If $h = (h_1, \ldots, h_m) \in \mathbb{R}^m$ and if $R(h)$ is the difference between $f(p + h)$ and $f(p) + df_p(h)$, that is to say, if

$$R(h) = f(p + h) - f(p) - df_p(h),$$

then $\lim_{h\to 0} R(h)/|h| = 0$.

Linear maps are usually easy to analyse, and the hope is that the behaviour of the linear map df_p will reflect the behaviour of f near p. The Inverse Function Theorem (Theorem 2) is a prime example of this hope being realised.

We now recall the chain rule, which says that the derivative of a composite is the composite of the derivatives.

Theorem 1 (Chain rule) *Let V be an open subset of $\mathbb{R}^q$ and U an open subset of $\mathbb{R}^m$. If $f : V \to \mathbb{R}^m$, $g : U \to \mathbb{R}^n$ are smooth maps then the composite map gf is smooth on*

$$f^{-1}(U) = \{p \in V : f(p) \in U\},$$

and if $p \in f^{-1}(U)$ then

$$d(gf)_p = dg_{f(p)} df_p.$$

Equivalently, in terms of Jacobian matrices and using matrix multiplication, $J(gf)_p = Jg_{f(p)} Jf_p$.

Let $f : U \to \mathbb{R}^n$ be a smooth map defined on an open subset U of $\mathbb{R}^m$. If $f(U)$ is an open subset of $\mathbb{R}^n$ and if there is a smooth map $g : f(U) \to U$ such that gf is the identity map on U and fg is the identity map on $f(U)$, then f is said to have a *smooth inverse*, and g is called the *inverse map*. It is often important to know when f has a smooth inverse g

because, if it has, then, in some sense, no information is lost when f is applied since this action can always be reversed by applying g.

However, the problem of deciding whether f has a smooth inverse is often very difficult. For instance, it is clear that if f has a smooth inverse then f is necessarily injective (not always easy to check), but this is not sufficient. The *Inverse Function Theorem* is a very powerful theorem because it shows that, if $p \in U$, then, on sufficiently small open neighbourhoods of p, the difficult problem of existence of a smooth inverse may be reduced to the easier problem of deciding whether the derivative df_p is a linear isomorphism, or, equivalently, whether the Jacobian matrix Jf_p is a square matrix with non-zero determinant.

Theorem 2 (Inverse Function Theorem) *Let U be an open subset of $\mathbb{R}^m$ and let $f : U \to \mathbb{R}^n$ be a smooth map. Suppose $p \in U$ and that df_p is a linear isomorphism (or, equivalently, $m = n$ and the Jacobian matrix Jf_p has non-zero determinant). Then U contains an open neighbourhood U_0 of p such that $f(U_0)$ is open in $\mathbb{R}^m$ and the restriction of f to U_0 has a smooth inverse map $g : f(U_0) \to U_0$.*

Using this theorem, we can now prove Theorem 1 of §2.4 and Theorems 1 and 3 of §2.5. In each case we take our notation from the statement of the relevant theorem.

Proof of Theorem 1 of §2.4 As already mentioned, the basic idea is to show that (maybe after re-labelling the axes of $\mathbb{R}^3$) each sufficiently small piece of S is the graph of a function. If $p \in S$ then the determinant of the Jacobian matrix at p of the map $\tilde{f}$ defined by $\tilde{f}(x, y, z) = (x, y, f(x, y, z))$ is equal to $(\partial f/\partial z)(p)$, which, by re-labelling the axes if necessary, we may assume is non-zero. The Inverse Function Theorem applied to $\tilde{f}$ now shows that W contains an open neighbourhood $\tilde{W}$ of p in $\mathbb{R}^3$ such that $\tilde{f}(\tilde{W})$ is open in $\mathbb{R}^3$ and $\tilde{f} : \tilde{W} \to \tilde{f}(\tilde{W})$ has a smooth inverse $\tilde{g} : \tilde{f}(\tilde{W}) \to \tilde{W}$, which in this case is necessarily of the form $\tilde{g}(u, v, w) = (u, v, g(u, v, w))$ for some smooth function $g(u, v, w)$. By taking $\tilde{W}$ smaller if necessary, we may assume that $\tilde{f}(\tilde{W})$ is of the form $U \times (k - \epsilon, k + \epsilon)$ for some open subset U of $\mathbb{R}^2$ and some $\epsilon > 0$ (Figure 2.14). We now show that $S \cap \tilde{W}$ is the graph of the function $(u, v) \mapsto g(u, v, k)$.

To do this, we define $\boldsymbol{x} : U \to \mathbb{R}^3$ by

$$\boldsymbol{x}(u, v) = \tilde{g}(u, v, k) = (u, v, g(u, v, k)). \tag{2.7}$$

Then $\tilde{f}\boldsymbol{x}(u, v) = \tilde{f}\tilde{g}(u, v, k) = (u, v, k)$, so that $f(\boldsymbol{x}(u, v)) = k$ from which it is clear that $\boldsymbol{x}$ satisfies (S1), and it follows immediately from (2.7) that (S2) is satisfied with $\boldsymbol{F} : \tilde{W} \to \mathbb{R}^2$ defined by $\boldsymbol{F}(x, y, z) = (x, y)$. □

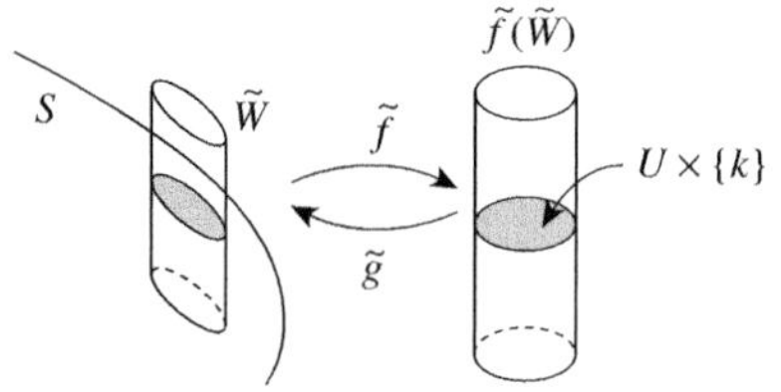

Figure 2.14 For the proof of Theorem 1 of §2.4

The above proof may be easily generalised to give a proof of Theorem 3 of §2.4.

Proof of Theorem 1 of §2.5 For simplicity we give the proof for the case $n = 3$, the general case being proved in an entirely similar manner. We also assume, without loss of generality, that the normal vector $\boldsymbol{x}_u(q) \times \boldsymbol{x}_v(q)$ to the plane spanned by $\boldsymbol{x}_u(q)$ and $\boldsymbol{x}_v(q)$ is not parallel to the xy-plane. Thus if

$$\boldsymbol{x}(u, v) = (x(u, v), y(u, v), z(u, v)),$$

we are assuming that $(x_u y_v - x_v y_u)(q) \neq 0$. Now let $\pi : \mathbb{R}^3 \to \mathbb{R}^2$ be defined by $\pi(x, y, z) = (x, y)$. Then the determinant of the Jacobian matrix of $\pi\boldsymbol{x}$ at q is $(x_u y_v - x_v y_u)(q)$, which is non-zero. So, by the Inverse Function Theorem, U contains an open neighbourhood U_0 of q such that $U_1 = \pi\boldsymbol{x}(U_0)$ is open in $\mathbb{R}^2$ and $\pi\boldsymbol{x} : U_0 \to U_1$ has a smooth inverse $g : U_1 \to U_0$ (Figure 2.15). Let $W = \pi^{-1}(U_1)$ and let $\boldsymbol{F} = g\pi : W \to U_0$. Then $\boldsymbol{F}$ is smooth and $\boldsymbol{F}\boldsymbol{x}(u, v) = g\pi\boldsymbol{x}(u, v) = (u, v)$ for all $(u, v) \in U_0$. The proof of the theorem now follows directly from the definition of a surface. □

Proof of Theorem 3 of §2.5 For each $q \in U$, we show that U contains an open neighbourhood U_0 of q such that there is a smooth map $\boldsymbol{F} : W \to \mathbb{R}^2$ defined on some open subset W of $\mathbb{R}^n$ with (S1) and (S2) holding for $\boldsymbol{x}|U_0$.

So, let $q \in U$, and, by taking U smaller if necessary, assume that $\boldsymbol{x}(U) = \tilde{W} \cap S$, where $\tilde{W}$ and $\tilde{\boldsymbol{F}}$ denote the set and map associated with a local parametrisation $\tilde{\boldsymbol{x}} : \tilde{U} \to S$ (Figure 2.16). Then, for each $r \in U$, there is a corresponding point $\tilde{r} \in \tilde{U}$ such that $\tilde{\boldsymbol{x}}(\tilde{r}) = \boldsymbol{x}(r)$. Then $\tilde{\boldsymbol{x}}\tilde{\boldsymbol{F}}\boldsymbol{x}(r) = \tilde{\boldsymbol{x}}\tilde{\boldsymbol{F}}\tilde{\boldsymbol{x}}(\tilde{r}) = \tilde{\boldsymbol{x}}(\tilde{r}) = \boldsymbol{x}(r)$, so that

$$\tilde{\boldsymbol{x}}\tilde{\boldsymbol{F}}\boldsymbol{x} = \boldsymbol{x}. \tag{2.8}$$

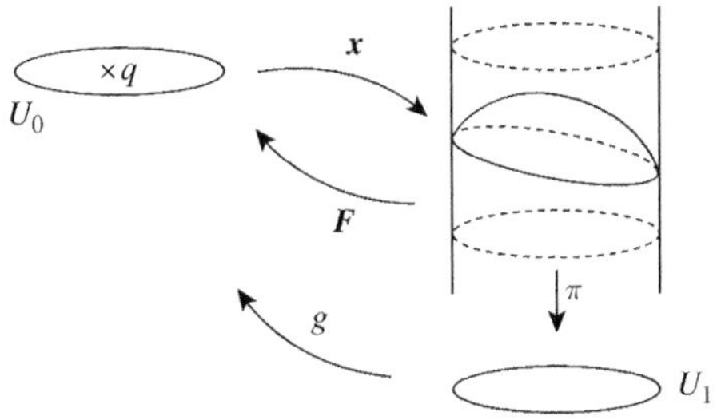

Figure 2.15 For the proof of Theorem 1 of §2.5

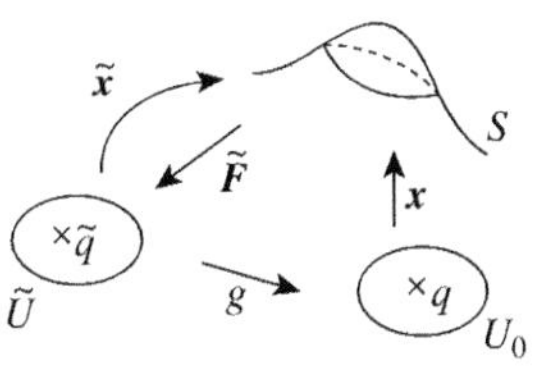

Figure 2.16 For the proof of Theorem 3 of §2.5

Using the chain rule (Theorem 1) and equation (2.8), we see that if $\tilde{\boldsymbol{F}}\boldsymbol{x}(q) = \tilde{q}$ then

$$J\tilde{\boldsymbol{x}}_{\tilde{q}}\, J(\tilde{\boldsymbol{F}}\boldsymbol{x})_q = J\boldsymbol{x}_q\,.$$

By assumption, the columns $\boldsymbol{x}_u$ and $\boldsymbol{x}_v$ of $J\boldsymbol{x}_q$ are linearly independent at q, so $J\boldsymbol{x}_q$ has rank two. It follows that $J(\tilde{\boldsymbol{F}}\boldsymbol{x})_q$ also has rank two, and hence has non-zero determinant. The Inverse Function Theorem now shows that U contains an open neighbourhood U_0 of q such that $\tilde{\boldsymbol{F}}\boldsymbol{x}|U_0$ has a smooth inverse g, say. Then $g\tilde{\boldsymbol{F}}\boldsymbol{x}$ is the identity map on U_0 so that (S2) holds for $\boldsymbol{x}|U_0$ with $\boldsymbol{F} = g\tilde{\boldsymbol{F}}$. Thus $\boldsymbol{x}|U_0$ is a local parametrisation as required. □

Exercises

None of these exercises require material in the Appendix.

2.1 Let $S^2(1)$ denote the unit sphere in $\mathbb{R}^3$, and, for $(u, v) \in \mathbb{R}^2$, let $\boldsymbol{x}(u, v)$ be the point of intersection with $S^2(1) \setminus \{(0, 0, 1)\}$ of the line in $\mathbb{R}^3$ through $(u, v, 0)$ and $(0, 0, 1)$. Show that

$$\boldsymbol{x}(u, v) = \frac{(2u, 2v, u^2 + v^2 - 1)}{u^2 + v^2 + 1}\,.$$

Let P be the plane in $\mathbb{R}^3$ with equation $z = 1$, and for each $(x, y, z) \in \mathbb{R}^3 \setminus P$, let $\boldsymbol{F}(x, y, z)$ be such that $(\boldsymbol{F}(x, y, z), 0)$ is the intersection with the xy-plane of the line through $(0, 0, 1)$ and (x, y, z). Show that

$$\boldsymbol{F}(x, y, z) = \frac{1}{1 - z}(x, y)\,.$$

Prove that $\boldsymbol{F}\boldsymbol{x}(u, v) = (u, v)$ for all $(u, v) \in \mathbb{R}^2$, and deduce that $\boldsymbol{x}$ is a local parametrisation of $S^2(1)$ which covers $S^2(1) \setminus \{(0, 0, 1)\}$.

2.2 A subset X of a surface S in $\mathbb{R}^n$ is said to be an *open subset* of S if X is the intersection of S with an open subset of $\mathbb{R}^n$. Prove that a non-empty open subset of a surface is itself a surface.

2.3 Consider the following two subsets of $\mathbb{R}^3$:

(a) the cylinder $x^2 + y^2 = 1$;
(b) the hyperboloid of two sheets $x^2 + y^2 = z^2 - 1$.

(i) Show directly from the definition that each subset is a surface.
(ii) Show that each subset is a surface of revolution.
(iii) Show that each subset is a surface S defined by an equation of the form $f(x, y, z) = 0$, where grad f does not vanish on S.

Sketch each surface, indicating the coordinate neighbourhoods you have used to parametrise the surface in your answers to (i). Indicate also the coordinate neighbourhoods arising from the standard parametrisations (as in (2.3) and (2.4)) as surfaces of revolution.

2.4 Let a be a positive real number, and let $f(x, y, z) = z^2 + \left(\sqrt{x^2 + y^2} - a\right)^2$. Find all points of $\mathbb{R}^3$ at which grad f vanishes. Show that if $0 < b < a$ then the equation $f(x, y, z) = b^2$ defines a torus of revolution. What happens if $0 < a < b$? Draw a picture of the set $f(x, y, z) = b^2$ in this case.

2.5 (a) Let $f(x, y, z) = (x + y + z - 1)^2$.

(i) Find all points at which grad f vanishes.

(ii) For which values of k does $f(x, y, z) = k$ define a surface?

(iii) Show that the set defined by $f(x, y, z) = 0$ is a surface.

(b) Repeat (i) and (ii) using the function $f(x, y, z) = xyz^2$. Is the set $xyz^2 = 0$ a surface?

2.6 **(The tangent surface of a curve)** Let $\boldsymbol{\alpha}(u)$ be a regular curve in $\mathbb{R}^3$ with curvature κ, and consider the map

$$\boldsymbol{x}(u, v) = \boldsymbol{\alpha}(u) + v\boldsymbol{\alpha}'(u)\,.$$

The image of $\boldsymbol{x}$ (Figure 2.17) is called the *tangent surface* of $\boldsymbol{\alpha}$ (although, as we shall see, it isn't actually a surface!). Show that $\boldsymbol{x}_u$ and $\boldsymbol{x}_v$ are linearly dependent at (u, v) if and only if either $\kappa(u) = 0$ or $v = 0$. Use Theorem 1 of §2.5 to deduce that at all other points each sufficiently small piece of the image of $\boldsymbol{x}$ is a surface. In fact, if κ is non-zero then the image of $\boldsymbol{x}$ gives two surfaces (possibly with self-intersections) with common boundary along the trace of $\boldsymbol{\alpha}$ (which corresponds to $v = 0$).

2.7 If $a, b, c > 0$ show that the ellipsoid S in $\mathbb{R}^3$ defined by

$$\frac{x^2}{a^2} + \frac{y^2}{b^2} + \frac{z^2}{c^2} = 1$$

is a surface. Show also that

$$\boldsymbol{x}(u, v) = (a \cos v \cos u,\ b \cos v \sin u,\ c \sin v)\,, \quad -\pi < u < \pi,\ -\pi/2 < v < \pi/2\,,$$

is a local parametrisation of S using each of the following methods.

(i) Find a suitable map $\boldsymbol{F}$ as in (S2) of the definition of surface.

(ii) Use Theorem 4 of §2.5.

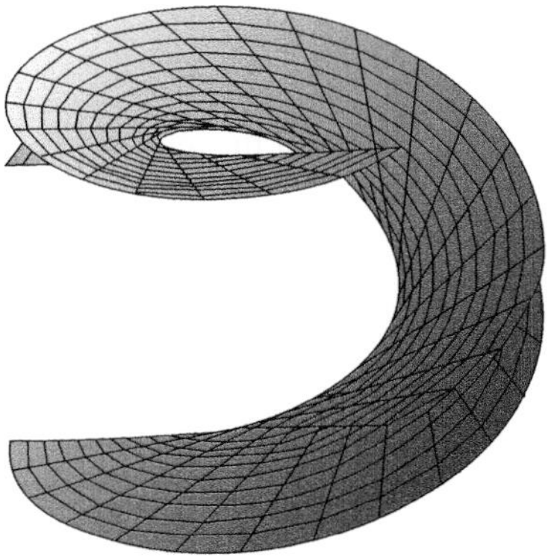

Figure 2.17 Tangent surface

2.8 (i) Show that the equation $z = x^2 - y^2$ defines a surface S in $\mathbb{R}^3$ (a hyperbolic paraboloid, see Figure 2.7). If we take

$$\boldsymbol{x}(u, v) = (v + \cosh u, v + \sinh u, 1 + 2v(\cosh u - \sinh u)), \quad u, v \in \mathbb{R},$$

show that $\boldsymbol{x}$ is a local parametrisation of S.

(ii) Show that the equation $xz + y^2 = 1$ defines a surface S in $\mathbb{R}^3$ (a hyperboloid of one sheet, see Figure 2.13). If we take

$$\boldsymbol{x}(u, v) = (\cos u + v(1 + \sin u), \sin u - v \cos u, \cos u - v(1 - \sin u)), \; 0 < u < 2\pi, \; v \in \mathbb{R},$$

show that $\boldsymbol{x}$ is a local parametrisation of S.

(The local parametrisations in (i) and (ii) may be written in the form $\boldsymbol{x}(u, v) = \boldsymbol{\alpha}(u) + v\boldsymbol{\beta}(u)$ with $\boldsymbol{\alpha}(u)$ being a regular curve and $\boldsymbol{\beta}(u)$ never equal to zero. This displays each surface as *ruled surface*, a surface swept out as a line is moved around in $\mathbb{R}^3$. There is more on ruled surfaces in §3.6.)

2.9 Show that the equation $x \sin z = y \cos z$ defines a surface S in $\mathbb{R}^3$. If

$$\boldsymbol{x}(u, v) = (v \cos u, v \sin u, u), \quad u, v \in \mathbb{R},$$

show that the image of $\boldsymbol{x}$ is the whole of S. Show also that $\boldsymbol{x}$ is a parametrisation of S using each of the following two methods:

(i) use the Parametrisation Recognition Theorem (Theorem 4 of §2.5);
(ii) find a map $\boldsymbol{F} : \mathbb{R}^3 \to \mathbb{R}^2$ such that $\boldsymbol{x}$ and $\boldsymbol{F}$ satisfy conditions (S1) and (S2) for a surface.

The surface S is a *helicoid*, a picture of which may be found in §3.6. Like the two surfaces in the previous question (and the surface in Exercise 2.6), the helicoid is a ruled surface.

2.10 **(Möbius band)** If we take a rectangle of rubber sheet which, for definiteness, we take to measure 4π by 2, and join the two ends of length 2 together after performing one twist, then (Figure 2.18) we have a model of a *Möbius band.* A mathematical model may be given as follows.

Let $\boldsymbol{\alpha}$ be the parametrisation of the unit circle in the (x, y)-plane given by

$$\boldsymbol{\alpha}(u) = (\sin u, \cos u, 0),$$

and let S be the image of the map $\boldsymbol{f} : \mathbb{R} \times (-1, 1) \to \mathbb{R}^3$ defined by

$$\boldsymbol{f}(u, v) = 2\boldsymbol{\alpha}(u) + v\boldsymbol{\beta}(u),$$

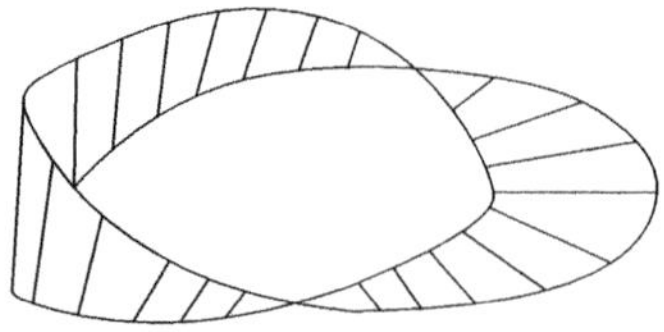

Figure 2.18 Möbius band

where

$$\boldsymbol{\beta}(u) = \cos\frac{u}{2}\boldsymbol{\alpha}(u) + \sin\frac{u}{2}(0,0,1)\,.$$

Assuming that S is a surface in $\mathbb{R}^3$, use the Parametrisation Recognition Theorem (Theorem 4 of §2.5) to show that S may be covered by two coordinate neighbourhoods. (*It may help to first show that* $|\boldsymbol{\beta}| = 1$ *and that* $\boldsymbol{\beta}$ *is orthogonal to* $\boldsymbol{\alpha}'$, *which enables you to deduce that* $\boldsymbol{f}_u$ *is orthogonal to* $\boldsymbol{f}_v$). Note that S is a ruled surface (see the previous two exercises); the rulings are indicated in Figure 2.18.

2.11 (*This exercise uses material in the optional second half of §2.4.*) For positive real numbers r_1 and r_2, let S be the flat torus in $\mathbb{R}^4$ (described in Example 4 of §2.4) defined by the equations

$$x_1^2 + x_2^2 = {r_1}^2\,, \quad x_3^2 + x_4^2 = {r_2}^2\,.$$

If $\boldsymbol{x}(u,v) = (r_1\cos u, r_1\sin u, r_2\cos v, r_2\sin v)$, $-\pi < u, v < \pi$, use the Parametrisation Recognition Theorem (Theorem 4 of §2.5) to show that $\boldsymbol{x}$ is a local parametrisation of S. Find a system of local parametrisations covering S.

3 Tangent planes and the first fundamental form

In the first section of this chapter we show that, at each point p of a surface S in $\mathbb{R}^n$, the set of vectors tangential to smooth curves on S through p form a 2-plane, the *tangent plane* of S at p. As already mentioned, the existence of this plane is crucial in the development of the geometry of surfaces.

We then discuss the measurement of length, angle and area on a surface in $\mathbb{R}^n$. These fundamental intrinsic properties are derived from the restriction of the inner product on $\mathbb{R}^n$ to the tangent planes of S. However, the ideas presented here extend readily to the more general study of abstract surfaces (which are not considered as subsets of $\mathbb{R}^n$) or even the study of length, angle and area on smooth manifolds of arbitrarily large dimension. (Manifolds are the generalisation of abstract surfaces to objects of higher dimension.) These metric properties are the defining features of a branch of mathematics called Riemannian geometry, named after B. Riemann (1826–1866), who may be regarded as the instigator of this study via the work in his doctoral thesis.

As in Chapter 2, the reader may continue to take $n = 3$ throughout this chapter.

3.1 The tangent plane

Let p be a point of a surface S in $\mathbb{R}^n$. A *tangent vector* (Figure 3.1) to S at p is a vector $\boldsymbol{X}$ such that $\boldsymbol{X} = \boldsymbol{\alpha}'(0)$ for some smooth curve $\boldsymbol{\alpha}(t)$ in $\mathbb{R}^n$ whose image lies on S and has $\boldsymbol{\alpha}(0) = p$. The main result in this section is Proposition 2, which says that the set of tangent vectors to S at p form a 2-plane. We then go on to show how to find this plane for each of the three types of surface discussed in Chapter 2, namely graphs of functions, surfaces of revolution and surfaces defined by equations.

We begin with a lemma which will be useful in many situations. If $\boldsymbol{x} : U \to \mathbb{R}^n$ is a local parametrisation of S, then the image under $\boldsymbol{x}$ of a smooth curve $\gamma(t) = (u(t), v(t))$ in U is the smooth curve $\boldsymbol{\alpha}(t) = \boldsymbol{x}\,(u(t), v(t))$ lying on S. We show that every smooth curve in $\mathbb{R}^n$ whose image lies on $\boldsymbol{x}(U)$ is of this form. This is an important result, because it shows that if $\boldsymbol{\alpha}(t)$ is a smooth curve whose image lies on S then sufficiently small pieces of $\boldsymbol{\alpha}$ may be described using local parametrisations.

Lemma 1 *Let $\boldsymbol{x} : U \to \mathbb{R}^n$ be a local parametrisation of S and let $\boldsymbol{\alpha}(t)$ be a smooth curve in $\mathbb{R}^n$ whose image lies on $\boldsymbol{x}(U)$. Then $\boldsymbol{\alpha}(t)$ is the image under $\boldsymbol{x}$ of a smooth curve in U, and hence may be written in the form*

$$\boldsymbol{\alpha}(t) = \boldsymbol{x}\,(u(t), v(t))$$

for smooth functions $u(t)$, $v(t)$.

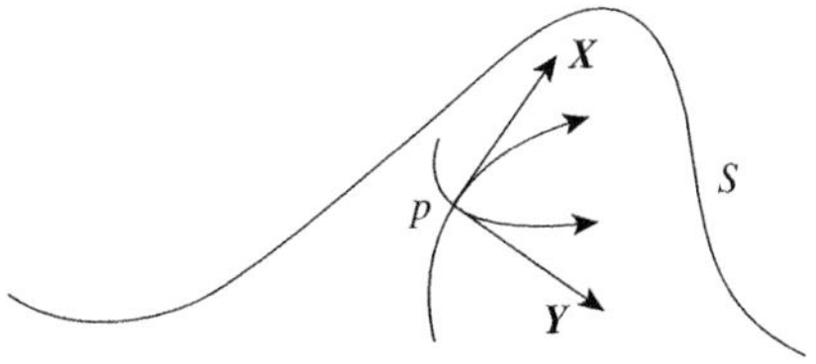

Figure 3.1 X and Y are tangent vectors to S at p

Proof We have seen (in Lemma 3 of §2.1) that a local parametrisation $\boldsymbol{x}$ is a bijective map onto $\boldsymbol{x}(U)$. Hence, for each t, there exists a unique point $\gamma(t) = (u(t), v(t)) \in U$ such that

$$\boldsymbol{x}\,(\gamma(t)) = \boldsymbol{\alpha}(t)\,. \tag{3.1}$$

It remains to show that $\gamma(t) = (u(t), v(t))$ is a smooth function of t. To do this, let $\boldsymbol{F} : W \to \mathbb{R}^2$ be a map as in (S2) of the definition of surface. If we apply $\boldsymbol{F}$ to both sides of (3.1), we obtain γ as the composite $\boldsymbol{F\alpha}$ of two smooth maps between open subsets of Euclidean spaces. It follows that γ is itself smooth. □

From now on, by a curve *on* a surface S we shall mean a smooth curve in $\mathbb{R}^n$ whose image lies on S. If $\boldsymbol{\alpha}(t)$ is such a curve and if $\boldsymbol{\alpha}(t_0)$ is in the image of a local parametrisation $\boldsymbol{x}(u, v)$ of S then continuity of $\boldsymbol{\alpha}$ implies that there exists $\epsilon > 0$ such that the image of the restriction of $\boldsymbol{\alpha}$ to $(t_0 - \epsilon, t_0 + \epsilon)$ is contained in the image of $\boldsymbol{x}$. We shall often encounter this sort of situation, and paraphrase by saying that if $\boldsymbol{\alpha}(t_0)$ lies in the image of a local parametrisation $\boldsymbol{x}(u, v)$ then, locally at least, the image of $\boldsymbol{\alpha}$ is contained in the image of $\boldsymbol{x}$.

The usefulness of Lemma 1 will be apparent in the proof of the following proposition, which, as already mentioned, is crucial in the study of the geometry of surfaces.

Proposition 2 *Let $\boldsymbol{x}(u, v)$ be a local parametrisation of S and assume that $\boldsymbol{x}(q) = p$. The set of tangent vectors to S at p is the plane spanned by $\boldsymbol{x}_u(q)$ and $\boldsymbol{x}_v(q)$.*

Proof We have already seen in Proposition 5 of §2.1 that $\boldsymbol{x}_u(q)$ and $\boldsymbol{x}_v(q)$ are linearly independent and hence do indeed span a plane. Now let $\boldsymbol{\alpha}(t)$ be a curve on S through p. Then, locally at least, we may write $\boldsymbol{\alpha}(t) = \boldsymbol{x}\,(u(t), v(t))$ as in Lemma 1. Thus, by the chain rule, $\boldsymbol{\alpha}' = u'\boldsymbol{x}_u + v'\boldsymbol{x}_v$, so that $\boldsymbol{\alpha}'$ is a linear combination of $\boldsymbol{x}_u$ and $\boldsymbol{x}_v$ (Figure 3.2).

Conversely, if λ and μ are real numbers, we let $\boldsymbol{\alpha}(t) = \boldsymbol{x}\,(\gamma(t))$, where $\gamma(t) = q + t(\lambda, \mu)$. Then, using the chain rule again,

$$\boldsymbol{\alpha}' = \lambda\boldsymbol{x}_u + \mu\boldsymbol{x}_v\,,$$

so that, by definition, $\lambda\boldsymbol{x}_u + \mu\boldsymbol{x}_v$ is a tangent vector. □

If $\boldsymbol{x}$ is a local parametrisation of S with $\boldsymbol{x}(q) = p$ then the vectors $\boldsymbol{x}_u(q)$ and $\boldsymbol{x}_v(q)$ are called the *coordinate vectors* at p, and Proposition 2 shows that we could have defined the set of tangent vectors to S at p to be the plane spanned by the coordinate vectors $\boldsymbol{x}_u(q)$ and $\boldsymbol{x}_v(q)$. However, as mentioned earlier, the surfaces are the objects of interest; the role

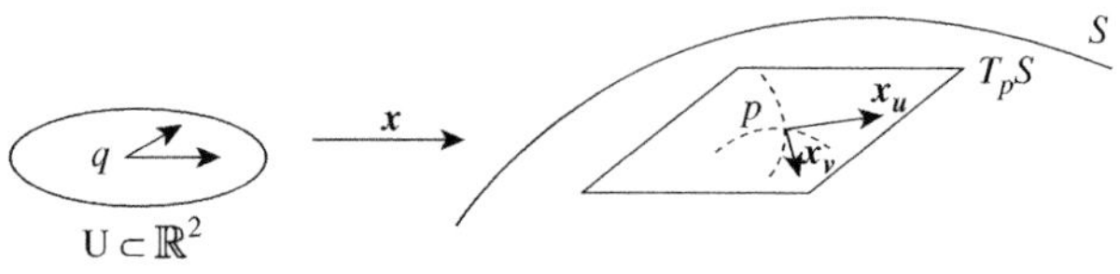

Figure 3.2 A tangent plane

of the local parametrisations is to help describe and study them. So, if we had defined the set of tangent vectors to S at p to be the plane spanned by $\boldsymbol{x}_u(q)$ and $\boldsymbol{x}_v(q)$, we would then have had to prove that this plane is independent of the choice of local parametrisation $\boldsymbol{x}$ (and we would still have to prove Lemma 1 in order to show that the tangent vector to any curve on S lies in the tangent plane). We have avoided this difficulty by choosing to define the set of tangent vectors without using a specific local parametrisation.

The set of tangent vectors to S at p is called the *tangent plane* of S at p, and denoted T_pS. Proposition 2 shows that T_pS is indeed a plane, and it is spanned by $\boldsymbol{x}_u(q)$ and $\boldsymbol{x}_v(q)$.

Example 3 (Tangent planes of graphs of functions) Let $g : U \to \mathbb{R}^{n-2}$ be a smooth function defined on an open subset U of $\mathbb{R}^2$. We saw in §2.2 that the graph

$$S = \{(u, v, g(u, v)) : (u, v) \in U\}$$

is a surface in $\mathbb{R}^n$ which may be parametrised by $\boldsymbol{x}(u, v) = (u, v, g(u, v))$. Thus the tangent plane of S at $(u, v, g(u, v))$ is the plane spanned by $(1, 0, g_u)$ and $(0, 1, g_v)$.

As a specific example, if $U = \{(u, v) \in \mathbb{R}^2 : u^2 + v^2 < 1\}$ and if $g : U \to \mathbb{R}$ is given by $g(u, v) = \sqrt{1 - u^2 - v^2}$, then S is the upper hemisphere $x^2 + y^2 + z^2 = 1$, $z > 0$, of the unit sphere. If $p = (u, v, g(u, v))$ then $p = \boldsymbol{x}(u, v)$ so that T_pS is spanned by $\left(1, 0, -\dfrac{u}{\sqrt{1 - u^2 - v^2}}\right)$ and $\left(0, 1, -\dfrac{v}{\sqrt{1 - u^2 - v^2}}\right)$. In particular, if $p = (0, 0, 1)$ is the north pole, then T_pS is spanned by $(1, 0, 0)$ and $(0, 1, 0)$, and so is horizontal, in agreement with intuition (Figure 3.3).

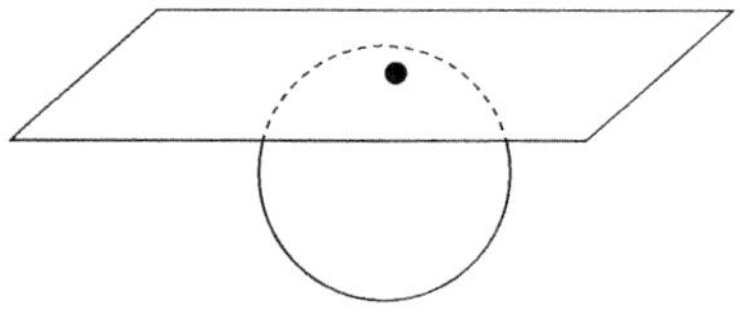

Figure 3.3 Tangent plane to $S^2(1)$ at the north pole

The tangent plane of a surface at a point p, being a vector subspace of $\mathbb{R}^n$, passes through the origin of $\mathbb{R}^n$. However (as in Figures 3.2 and 3.3), it is often helpful to visualise this plane moved parallel to itself so that the tangent vectors are all based at p. This is the plane which best approximates S near p in the sense that any smooth curve on S through p has first order contact with this plane; the curve touches the plane there and its tangent vector lies in the plane.

Example 4 (Cone) We mentioned in §2.1 that the cone $z^2 = x^2 + y^2$ isn't a surface because it doesn't look locally like an open subset of the plane around the vertex $(0, 0, 0)$. This is not hard to prove using a connectivity argument (removing the vertex disconnects the cone, but an open subset of the plane cannot be disconnected by removing a point), but we may also prove it using Proposition 2 by writing down three curves through $(0, 0, 0)$ which lie on the cone but whose tangent vectors do not all lie in a plane. Specifically, if we take $\boldsymbol{\alpha}(t) = (t, 0, t)$, $\boldsymbol{\beta}(t) = (t, 0, -t)$, and $\boldsymbol{\gamma}(t) = (0, t, t)$, then $\boldsymbol{\alpha}'(0)$, $\boldsymbol{\beta}'(0)$ and $\boldsymbol{\gamma}'(0)$ are linearly independent.

As we have seen in Example 3, it is easy to find the tangent plane at a point of a surface which is the graph of a function because such a surface has a natural parametrisation. A similar comment holds for surfaces of revolution, which may be parametrised as described in §2.3, but does not apply to surfaces defined by equations (as discussed in §2.4). However, a rather different method works nicely in this case.

Proposition 5 *Let $f : W \to \mathbb{R}$ be a smooth function defined on an open subset W of $\mathbb{R}^3$, and let k be a real number in the image of f. Assume that grad f is never zero on the subset S of $\mathbb{R}^3$ defined by the equation $f(x, y, z) = k$ (so that, by Theorem 1 of §2.4, S is a surface in $\mathbb{R}^3$). If $p \in S$, then T_pS is the plane of vectors orthogonal to (grad f)(p).*

Proof Let $\boldsymbol{\alpha}(t) = (x(t), y(t), z(t))$ be a smooth curve on S with $\boldsymbol{\alpha}(0) = p$. Then $f(x(t), y(t), z(t)) = k$, so, differentiating, we obtain

$$f_x x' + f_y y' + f_z z' = 0 . \tag{3.2}$$

However, grad $f = (f_x, f_y, f_z)$, and $\boldsymbol{\alpha}'(t) = \big(x'(t), y'(t), z'(t)\big)$, so, using the inner product in $\mathbb{R}^3$, we see that (3.2) is equivalent to

$$\boldsymbol{\alpha}'.\text{grad } f = 0 ,$$

that is to say, $\boldsymbol{\alpha}'$ is orthogonal to grad f (Figure 3.4).

Since, by definition, every element of T_pS is equal to $\boldsymbol{\alpha}'(0)$ for some smooth curve $\boldsymbol{\alpha}$ on S with $\boldsymbol{\alpha}(0) = p$, we now see that every element of T_pS is orthogonal to (grad f)(p). The result now follows since T_pS is a 2-dimensional vector space and hence is the whole of the orthogonal complement of (grad f)(p). □

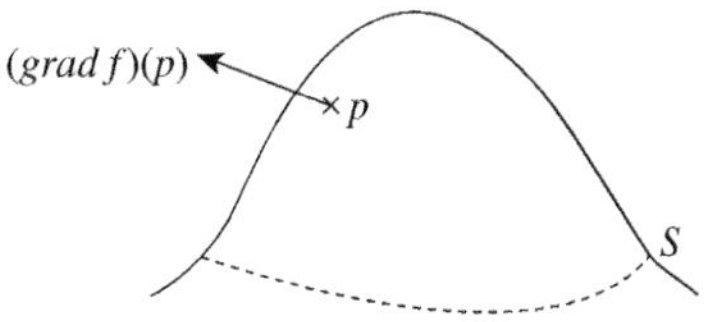

Figure 3.4 grad f is orthogonal to S

Example 6 (Sphere) Let $S^2(r)$ be the sphere of radius $r > 0$ with equation

$$x^2 + y^2 + z^2 = r^2 .$$

If $p = (x, y, z) \in S^2(r)$, then $T_p S^2(r)$ consists of those vectors orthogonal to (the position vector of) p.

Example 7 (Ellipsoid) Let S be the ellipsoid with equation

$$\frac{x^2}{a^2} + \frac{y^2}{b^2} + \frac{z^2}{c^2} = 1 .$$

Then the tangent plane at $(x_0, y_0, z_0) \in S$ consists of those vectors orthogonal to $(2x_0/a^2, 2y_0/b^2, 2z_0/c^2)$ and hence has equation

$$\frac{x_0}{a^2}x + \frac{y_0}{b^2}y + \frac{z_0}{c^2}z = 0 .$$

The parallel translate of this plane to the plane of vectors based at (x_0, y_0, z_0) has equation

$$\frac{x_0}{a^2}x + \frac{y_0}{b^2}y + \frac{z_0}{c^2}z = 1 , \tag{3.3}$$

and this is the plane with which the smooth curves on S through (x_0, y_0, z_0) have first order contact. As a particular example, we see from (3.3) that the tangent plane based at $(0, 0, c)$ is the horizontal plane $z = c$, in agreement with intuition.

3.2 The first fundamental form

Let p be a point on a surface S in $\mathbb{R}^n$. The restriction to the tangent plane $T_p S$ of the inner product on $\mathbb{R}^n$ gives an inner product on $T_p S$ (that is to say, a symmetric positive-definite bilinear form on $T_p S$). Thus, if $\boldsymbol{X}$ and $\boldsymbol{Y}$ are in $T_p S$ then $\boldsymbol{X}.\boldsymbol{X} > 0$ if $\boldsymbol{X}$ is non-zero, $\boldsymbol{X}.\boldsymbol{Y} = \boldsymbol{Y}.\boldsymbol{X}$, and $\boldsymbol{X}.\boldsymbol{Y}$ is linear in both $\boldsymbol{X}$ and $\boldsymbol{Y}$. The corresponding quadratic form given by

$$I(\boldsymbol{X}) = \boldsymbol{X}.\boldsymbol{X} = |\boldsymbol{X}|^2$$

is called the *first fundamental form* or *metric* of S.

The intrinsic properties of S are determined by the first fundamental form, and we now obtain expressions for the inner product and the first fundamental form in terms of a local parametrisation $\boldsymbol{x}(u, v)$ of S. As we saw in Proposition 2 of §3.1, if $\boldsymbol{X}, \boldsymbol{Y}$ are tangent vectors to S at some point then, for some uniquely determined scalars $\lambda_1, \lambda_2, \mu_1, \mu_2$,

$$\boldsymbol{X} = \lambda_1 \boldsymbol{x}_u + \lambda_2 \boldsymbol{x}_v ,$$
$$\boldsymbol{Y} = \mu_1 \boldsymbol{x}_u + \mu_2 \boldsymbol{x}_v .$$

Thus,

$$\begin{aligned}\boldsymbol{X}.\boldsymbol{Y} &= (\lambda_1 \boldsymbol{x}_u + \lambda_2 \boldsymbol{x}_v).(\mu_1 \boldsymbol{x}_u + \mu_2 \boldsymbol{x}_v)\\ &= \lambda_1\mu_1 E + (\lambda_1\mu_2 + \lambda_2\mu_1)F + \lambda_2\mu_2 G ,\end{aligned}$$

where

$$E = \boldsymbol{x}_u.\boldsymbol{x}_u\,, \quad F = \boldsymbol{x}_u.\boldsymbol{x}_v\,, \quad G = \boldsymbol{x}_v.\boldsymbol{x}_v\,.$$

In particular, the first fundamental form I is given by

$$I(\boldsymbol{X}) = |\boldsymbol{X}|^2 = \lambda_1{}^2 E + 2\lambda_1\lambda_2 F + \lambda_2{}^2 G\,. \tag{3.4}$$

The functions E, F and G are determined by $\boldsymbol{x}$ and are called the *coefficients of the first fundamental form* with respect to $\boldsymbol{x}$. Since $\boldsymbol{x}_u$ and $\boldsymbol{x}_v$ are smooth, it follows that E, F and G are smooth functions of u and v.

Example 1 (Graph of a function) Let $S = \Gamma(g)$ be the graph of the smooth function $g : U \to \mathbb{R}$, and let $\boldsymbol{x}(u, v)$ be the (standard) parametrisation of S given by

$$\boldsymbol{x}(u, v) = (u, v, g(u, v))\,.$$

Then

$$\boldsymbol{x}_u = (1, 0, g_u)\,, \quad \boldsymbol{x}_v = (0, 1, g_v)\,,$$

so that

$$E = 1 + g_u{}^2\,, \quad F = g_u g_v\,, \quad G = 1 + g_v{}^2\,.$$

So, for instance, the square of the length of $3\boldsymbol{x}_u + 4\boldsymbol{x}_v$ is $9(1+g_u{}^2)+24g_ug_v+16(1+g_v{}^2)$, while the angle θ betweeen $\boldsymbol{x}_u$ and $\boldsymbol{x}_v$ is given by

$$\cos\theta = \frac{\boldsymbol{x}_u.\boldsymbol{x}_v}{|\boldsymbol{x}_u||\boldsymbol{x}_v|} = \frac{F}{\sqrt{EG}} = \frac{g_u g_v}{\sqrt{(1+g_u{}^2)(1+g_v{}^2)}}\,.$$

Example 2 (Surface of revolution) Let S be the surface of revolution generated by the regular curve $\boldsymbol{\alpha}(v) = (f(v), 0, g(v))$, $f(v) > 0\ \forall v$, and let $\boldsymbol{x}(u, v)$ be the standard local parametrisation given by

$$\boldsymbol{x}(u, v) = (f(v)\cos u, f(v)\sin u, g(v))\,.$$

Then

$$\boldsymbol{x}_u = (-f(v)\sin u, f(v)\cos u, 0)\,, \quad \boldsymbol{x}_v = \left(f'(v)\cos u, f'(v)\sin u, g'(v)\right)\,,$$

so that

$$E = (f(v))^2, \quad F = 0\,, \quad G = \left(f'(v)\right)^2 + \left(g'(v)\right)^2 = |\boldsymbol{\alpha}'(v)|^2\,.$$

Note that in this example the coefficients of the first fundamental form depend on only v, and are particularly simple if the generating curve $\boldsymbol{\alpha}$ is parametrised by arc length.

As already noted, the inner product on each tangent space is positive definite. This leads to the following lemma.

Lemma 3 *Let E, F and G be the coefficients of the first fundamental form of a surface S with respect to some local parametrisation $\boldsymbol{x}$. Then E, G and $EG - F^2$ are all positive.*

Proof It is clear that $E > 0$ and $G > 0$ since each is the square of the length of a non-zero vector. Also, if θ is the angle between $\boldsymbol{x}_u$ and $\boldsymbol{x}_v$, then

$$EG - F^2 = |\boldsymbol{x}_u|^2|\boldsymbol{x}_v|^2(1 - \cos^2\theta) = |\boldsymbol{x}_u|^2|\boldsymbol{x}_v|^2 \sin^2\theta > 0 . \qquad \square$$

3.3 Arc length and angle

As already mentioned, the intrinsic properties of a surface S are those that depend on only the inner product on the tangent planes of the surface; or, equivalently once a local parametrisation has been chosen, depend on only the coefficients E, F and G of the first fundamental form. The intrinsic properties we study in this chapter are the length of curves on S, the angle of intersection of two curves on S, and the area of suitable regions of S.

The crucial point is that, once you know the coefficients of the first fundamental form, you can carry out intrinsic metric geometry on the corresponding coordinate neighbourhood of S without needing to know the actual shape of the surface itself. Indeed, two surfaces having local parametrisations with the same coefficients E, F and G of the first fundamental form have the same (local) intrinsic geometry. So, for instance, you cannot tell whether you are on a plane or (a sufficiently small part of) a cylinder solely by comparing lengths of curves and angles of intersection of curves on these two surfaces. Physically speaking, you can mould a piece of paper round a cylinder without wrinkles or folds, but, for instance, you can not do this round a sphere; the intrinsic geometry of the sphere is different from that of the plane.

In this section we illustrate the use of the coefficients of the first fundamental form in determining the intrinsic geometry of a surface S by describing how to use them to find the length of a curve given in terms of a local parametrisation on S. We then show how to determine the angle of intersection of two curves on S.

Let $\boldsymbol{x} : U \to \mathbb{R}^n$ be a local parametrisation of a surface S in $\mathbb{R}^n$, and let $\boldsymbol{\alpha}(t) = \boldsymbol{x}(u(t), v(t))$ be a smooth curve on S lying on the coordinate neighbourhood $\boldsymbol{x}(U)$. Then

$$\boldsymbol{\alpha}' = u'\boldsymbol{x}_u + v'\boldsymbol{x}_v ,$$

so that

$$\begin{aligned} |\boldsymbol{\alpha}'|^2 &= \boldsymbol{\alpha}'.\boldsymbol{\alpha}' = (u'\boldsymbol{x}_u + v'\boldsymbol{x}_v).(u'\boldsymbol{x}_u + v'\boldsymbol{x}_v) \\ &= Eu'^2 + 2Fu'v' + Gv'^2 . \end{aligned} \tag{3.5}$$

Hence, using the definition given in equation (1.1) of the arc length function s measured from some point $\boldsymbol{\alpha}(t_0)$ on $\boldsymbol{\alpha}$, we have

$$\begin{aligned} s(t_1) &= \int_{t_0}^{t_1} |\boldsymbol{\alpha}'(t)| dt \\ &= \int_{t_0}^{t_1} (Eu'^2 + 2Fu'v' + Gv'^2)^{1/2} dt . \end{aligned} \tag{3.6}$$

The following example illustrates the remark at the beginning of this section in that, if we know E, F and G, we can find arc length along a curve on a surface without knowing the shape of either the surface or the curve.

Example 1 Let $\boldsymbol{x}(u, v)$ be a local parametrisation of a surface S in $\mathbb{R}^n$, with coefficients of the first fundamental form given by

$$E = 1 + 4u^2 , \quad F = -4uv , \quad G = 1 + 4v^2 , \tag{3.7}$$

and let $\boldsymbol{\alpha}(t)$ be the curve on S given by

$$\boldsymbol{\alpha}(t) = \boldsymbol{x}\,(u(t), v(t)) , \quad t \in [-1, 1] , \tag{3.8}$$

where

$$u(t) = t , \quad v(t) = t . \tag{3.9}$$

On the curve $\boldsymbol{\alpha}$, we then have

$$u' = v' = 1 , \quad E = 1 + 4t^2 , \quad F = -4t^2, \quad G = 1 + 4t^2 ,$$

so that, using (3.6), the length of $\boldsymbol{\alpha}$ is given by

$$\int_{-1}^{1} \left(1 + 4t^2 + 2(-4t^2) + 1 + 4t^2\right)^{1/2} dt$$
$$= \sqrt{2} \int_{-1}^{1} dt = 2\sqrt{2} .$$

In fact, the above calculation may be verified by taking a specific surface with the above E, F and G as follows.

Example 2 (Hyperbolic paraboloid) Let S be the hyperbolic paraboloid with equation $z = x^2 - y^2$, and parametrise it as a graph in the usual way by taking

$$\boldsymbol{x}(u, v) = (u, v, u^2 - v^2) .$$

Then

$$\boldsymbol{x}_u = (1, 0, 2u) , \quad \boldsymbol{x}_v = (0, 1, -2v) ,$$

so that E, F and G are given by (3.7). The line segment joining $(-1, -1, 0)$ to $(1, 1, 0)$ lies on the surface, and is parametrised by (3.8) and (3.9). The length of the line segment is (by Pythagoras' Theorem) equal to $2\sqrt{2}$, which agrees with the answer obtained in Example 1.

We now illustrate how a knowledge of E, F and G enables us to find the angle of intersection of two regular curves on a surface S. We use the fact that, if $\boldsymbol{\alpha}(t)$ and $\boldsymbol{\beta}(r)$ are regular curves on S with $\boldsymbol{\alpha}(t_1) = \boldsymbol{\beta}(r_1) = p$, then the angle θ at which they intersect at p is given by

$$\boldsymbol{\alpha}'(t_1).\boldsymbol{\beta}'(r_1) = |\boldsymbol{\alpha}'(t_1)|\,|\boldsymbol{\beta}'(r_1)| \cos\theta , \tag{3.10}$$

where, as usual, we use $'$ to denote differentiation with respect to the appropriate variable.

Example 3 Let $\boldsymbol{x}(u, v)$ be a local parametrisation of a surface S with coefficients of the first fundamental form given, as in Example 1, by

$$E = 1 + 4u^2 , \quad F = -4uv , \quad G = 1 + 4v^2 . \tag{3.11}$$

We shall find the angle θ of intersection of the curves

$$\boldsymbol{\alpha}(t) = \boldsymbol{x}(2t, t^3) , \quad \boldsymbol{\beta}(r) = \boldsymbol{x}(r, 1) .$$

We start by noting that since $\boldsymbol{x}$ is a local parametrisation it is also injective, so that the two curves intersect at the single point $\boldsymbol{\alpha}(1) = \boldsymbol{\beta}(2) = \boldsymbol{x}(2, 1)$. At this point

$$E = 17 , \quad F = -8 , \quad G = 5 .$$

Also

$$\boldsymbol{\alpha}'(1) = 2\boldsymbol{x}_u + 3\boldsymbol{x}_v , \quad \boldsymbol{\beta}'(2) = \boldsymbol{x}_u ,$$

so that

$$\boldsymbol{\alpha}'(1).\boldsymbol{\beta}'(2) = 2E + 3F = 10 ,$$

and

$$|\boldsymbol{\alpha}'(1)|^2 = 4E + 12F + 9G = 17 , \quad |\boldsymbol{\beta}'(2)|^2 = E = 17 .$$

It now follows from (3.10) that

$$\cos\theta = \frac{10}{\sqrt{17.17}} = \frac{10}{17} , \tag{3.12}$$

so the curves $\boldsymbol{\alpha}$ and $\boldsymbol{\beta}$ intersect at an angle of 0.94 radians (to two significant figures) or 54°.

The above calculation holds for any surface with a parametrisation having coefficients of the first fundamental form given by (3.11). In particular, if we take S to be the hyperbolic paraboloid $z = x^2 - y^2$ with parametrisation as in Example 2, then the curves $\boldsymbol{\alpha}$ and $\boldsymbol{\beta}$ are given by $\boldsymbol{\alpha}(t) = (2t, t^3, 4t^2 - t^6)$ and $\boldsymbol{\beta}(r) = (r, 1, r^2 - 1)$ from which the expression for their angle of intersection in $\mathbb{R}^3$ may be calculated directly and seen to agree with (3.12).

We finish the section with a slight digression in order to explain some notation used by many authors for the first fundamental form. **This material can be safely omitted** since it will not be used elsewhere in the book.

It follows from equation (3.6) that, along a smooth curve $\boldsymbol{\alpha}(t) = \boldsymbol{x}\,(u(t), v(t))$,

$$\left(\frac{ds}{dt}\right)^2 = E\left(\frac{du}{dt}\right)^2 + 2F\left(\frac{du}{dt}\right)\left(\frac{dv}{dt}\right) + G\left(\frac{dv}{dt}\right)^2 .$$

This is sometimes written in the form

$$ds^2 = Edu^2 + 2Fdudv + Gdv^2 , \tag{3.13}$$

and ds^2 is then referred to as the first fundamental form. We can make sense of this equation of 'infinitesimals' if we consider du and dv as the linear functions defined on the tangent plane of S by

$$du(\boldsymbol{x}_u) = 1\,, \quad du(\boldsymbol{x}_v) = 0\,,$$
$$dv(\boldsymbol{x}_u) = 0\,, \quad dv(\boldsymbol{x}_v) = 1\,.$$

The right hand side of (3.13) is then precisely the first fundamental form I as given in (3.4).

3.4 Isothermal parametrisations

We begin this section by stressing, once again, that the geometry of a surface (and the curves on it) are the important considerations. A local parametrisation $\boldsymbol{x}(u, v)$ is just a convenient device for investigating these.

However, it is clearly desirable to choose a local parametrisation whose coefficients of the first fundamental form are as simple as possible, so it is an interesting question as to how simple E, F and G can be made for a given surface by choosing a suitable local parametrisation. For instance, can we prove that any surface may be covered using local parametrisations for which $E = G = 1$, $F = 0$? The answer, not surprisingly, is "no"! In fact, since E, F and G determine the local intrinsic properties of a surface the existence of a local parametrisation with $E = G = 1$, $F = 0$ would imply that the surface was (locally at least) metrically indistinguishable from a flat plane.

We often aim, wherever practicable, to choose local parametrisations with $F = 0$ (as described in Example 2 in §3.2 for surfaces of revolution, for instance). Such parametrisations are called *orthogonal* parametrisations because at each point the vectors $\boldsymbol{x}_u$ and $\boldsymbol{x}_v$ are orthogonal. Even better are *isothermal* parametrisations, which have $E = G$ and $F = 0$; that is to say, at each point the coordinate vectors $\boldsymbol{x}_u$ and $\boldsymbol{x}_v$ are orthogonal and have the same length.

Example 1 (Catenoid) We consider the catenoid S which may be parametrised as a surface of revolution by taking

$$\boldsymbol{x}(u, v) = (\cosh v \cos u, \cosh v \sin u, v)\,, \qquad -\pi < u < \pi\,, \quad v \in \mathbb{R}\,.$$

It follows quickly from Example 2 of §3.2 that

$$E = G = \cosh^2 v\,, \quad F = 0\,,$$

so that $\boldsymbol{x}$ is an isothermal parametrisation of S.

Example 2 (Sphere) Let $S^2(1)$ denote the unit sphere in $\mathbb{R}^3$, and, for $(u, v) \in \mathbb{R}^2$, let $\boldsymbol{x}(u, v)$ be the point of intersection with $S^2(1) \setminus \{(0, 0, 1)\}$ of the line in $\mathbb{R}^3$ through $(u, v, 0)$ and $(0, 0, 1)$ (Figure 3.5).

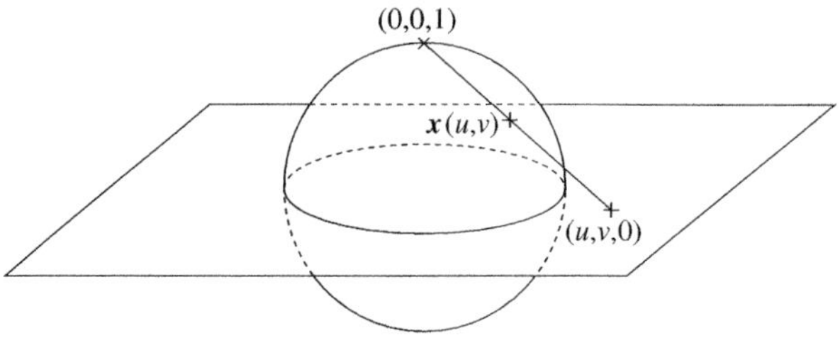

Figure 3.5 An isothermal parametrisation of the sphere $S^2(1)$

In Exercise 2.1 you were invited to prove that $\boldsymbol{x}(u, v)$ is given by

$$\boldsymbol{x}(u, v) = \frac{(2u, 2v, u^2 + v^2 - 1)}{u^2 + v^2 + 1},$$

and that $\boldsymbol{x}$ is a local parametrisation of $S^2(1)$ which covers $S^2(1) \setminus \{(0, 0, 1)\}$.

A routine computation shows that, for this local parametrisation $\boldsymbol{x}$,

$$E = G = \frac{4}{(u^2 + v^2 + 1)^2}, \qquad F = 0, \tag{3.14}$$

so that $\boldsymbol{x}$ is an isothermal parametrisation of $S^2(1) \setminus \{(0, 0, 1)\}$. We may carry out a similar process but considering lines from $(0, 0, -1)$, thus covering $S^2(1)$ with two isothermal parametrisations.

We shall not prove the following proposition (since the proof is rather delicate; ultimately using the theory of existence of solutions of elliptic partial differential equations).

Proposition 3 *Let S be a surface in $\mathbb{R}^n$ and let $p \in S$. Then there exists an isothermal local parametrisation of S whose image contains p.*

In some ways, Examples 1 and 2 are rather misleading, since finding an explicit isothermal local parametrisation is impracticable for most examples (you may remember a similar remark in Chapter 1 concerning arc length parametrisations for regular curves). However, the existence of isothermal local parametrisations is often useful for theoretical work.

The following lemma gives the main geometrical property of isothermal parametrisations.

Lemma 4 *Let $\boldsymbol{x}(u, v)$ be an isothermal local parametrisation (so that $E = G$ and $F = 0$). Then $\boldsymbol{x}$ preserves angles in the sense that if $\boldsymbol{\beta}_1(t)$ and $\boldsymbol{\beta}_2(r)$ are regular curves in $\mathbb{R}^2$ in the domain of $\boldsymbol{x}$ intersecting at an angle ϕ then $\boldsymbol{\alpha}_1 = \boldsymbol{x}\boldsymbol{\beta}_1$ and $\boldsymbol{\alpha}_2 = \boldsymbol{x}\boldsymbol{\beta}_2$ intersect on the surface at the same angle ϕ.*

Proof In the following proof, $\boldsymbol{x}_u$, $\boldsymbol{x}_v$, $\boldsymbol{\alpha}_1'$, $\boldsymbol{\alpha}_2'$ and E are all evaluated at the point of intersection of the curves. Assume that, at their point of intersection, $\boldsymbol{\beta}_1' = (\lambda_1, \mu_1)$ and $\boldsymbol{\beta}_2' = (\lambda_2, \mu_2)$. Then $\boldsymbol{\alpha}_1' = \lambda_1 \boldsymbol{x}_u + \mu_1 \boldsymbol{x}_v$, and $\boldsymbol{\alpha}_2' = \lambda_2 \boldsymbol{x}_u + \mu_2 \boldsymbol{x}_v$. The angle of intersection θ of $\boldsymbol{\alpha}_1$ and $\boldsymbol{\alpha}_2$ is given by (Figure 3.6)

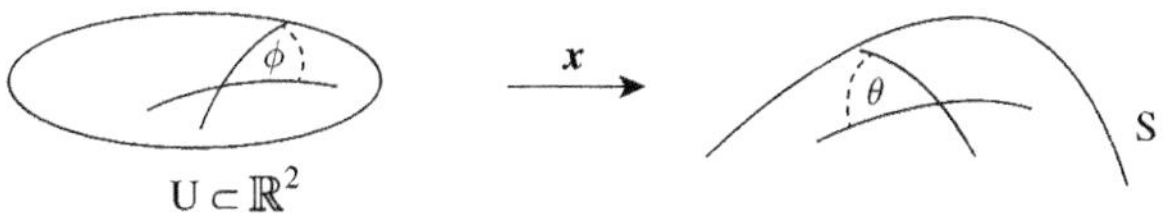

Figure 3.6 An isothermal parametrisation preserves angle

$$\cos\theta = \frac{(\lambda_1 \boldsymbol{x}_u + \mu_1 \boldsymbol{x}_v).(\lambda_2 \boldsymbol{x}_u + \mu_2 \boldsymbol{x}_v)}{|\lambda_1 \boldsymbol{x}_u + \mu_1 \boldsymbol{x}_v||\lambda_2 \boldsymbol{x}_u + \mu_2 \boldsymbol{x}_v|} = \frac{E(\lambda_1\lambda_2 + \mu_1\mu_2)}{\sqrt{E({\lambda_1}^2 + {\mu_1}^2)}\sqrt{E({\lambda_2}^2 + {\mu_2}^2)}}$$
$$= \frac{\lambda_1\lambda_2 + \mu_1\mu_2}{\sqrt{({\lambda_1}^2 + {\mu_1}^2)}\sqrt{({\lambda_2}^2 + {\mu_2}^2)}} = \cos\phi \,. \qquad \square$$

We shall see other examples of angle preserving maps in Chapter 4.

We now give a more substantial example of computing lengths and angles on a surface using only the coefficients E, F and G of the first fundamental form. This example may be omitted by those who are only interested in surfaces in $\mathbb{R}^3$ (although this would be a pity because the example is interesting and historically important!).

Example 5† (Hyperbolic plane) Let S be a surface in $\mathbb{R}^n$ covered by the image of a single parametrisation $\boldsymbol{x} : U \to S$, where $U = \{(u, v) \in \mathbb{R}^2 : v > 0\}$, and assume that the coefficients of the first fundamental form are given by

$$E = \frac{1}{v^2}\,, \quad F = 0\,, \quad G = \frac{1}{v^2}\,, \tag{3.15}$$

(so that $\boldsymbol{x}$ is an isothermal parametrisation). Then, by Lemma 4, the angle of intersection of any two curves on S is equal to the angle of intersection of the corresponding curves in the upper half-plane U.

We now consider the lengths of two particular curves. For any positive real number v_0, we consider the curves on S given by

$$\boldsymbol{\alpha}(t) = \boldsymbol{x}(0, t)\,, \quad t \geq v_0\,,$$

and

$$\boldsymbol{\beta}(r) = \boldsymbol{x}(r, v_0)\,, \quad -\pi \leq r \leq \pi\,.$$

These curves are the images under $\boldsymbol{x}$ of the lines in the upper half-plane illustrated in Figure 3.7.

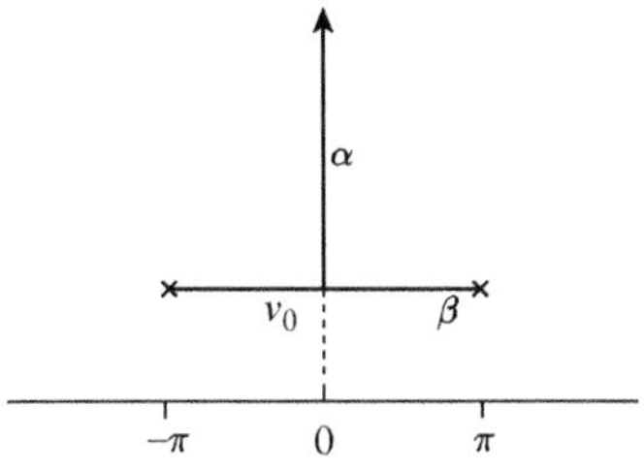

Figure 3.7 Hyperbolic plane

Since $\boldsymbol{x}$ preserves angles, the curves $\boldsymbol{\alpha}$ and $\boldsymbol{\beta}$ intersect orthogonally on S at $\boldsymbol{x}(0, v_0)$, but arc length measured along $\boldsymbol{\alpha}$ and $\boldsymbol{\beta}$ is different from arc length measured along the corresponding curves in the plane.

On the curve $\boldsymbol{\alpha}$, we have

$$u' = 0\,,\ v' = 1\,,\quad E = G = 1/t^2\,,\ F = 0\,,$$

so that, for a real number v_1 with $v_1 > v_0$, the arc length of $\boldsymbol{\alpha}$ between parameter values v_0 and v_1 is

$$\int_{v_0}^{v_1} (1/t)dt = [\log t]_{v_0}^{v_1} = \log(v_1/v_0)\,.$$

Note that, as $v_0 \to 0$ or as $v_1 \to \infty$ the arc length of the restriction of $\boldsymbol{\alpha}$ to $[v_0, v_1]$ tends to ∞.

Similar calculations for $\boldsymbol{\beta}$ show that the length of this curve is given by $2\pi/v_0$, so that as $v_0 \to \infty$ the length tends to 0, while as $v_0 \to 0$, the length of $\boldsymbol{\beta}$ tends to ∞.

Finally, we shall consider curves on S which are images under $\boldsymbol{x}$ of arcs of concentric semicircles in the upper half-plane which intersect the u-axis orthogonally (Figure 3.8). Specifically, for $r_0 > 0$, let $\boldsymbol{\gamma}(t) = \boldsymbol{x}(r_0 \cos t, r_0 \sin t)$. If $0 < \theta_0 < \theta_1 < \pi$, then the length of $\boldsymbol{\gamma}$ from $t = \theta_0$ to $t = \theta_1$ is

$$\int_{\theta_0}^{\theta_1} \frac{1}{\sin t} dt = \left[\log \tan \frac{t}{2}\right]_{\theta_0}^{\theta_1} = \log\left(\frac{\tan \frac{\theta_1}{2}}{\tan \frac{\theta_0}{2}}\right)\,.$$

Note that this length is independent of the radius r_0 of the circular arc in the (u, v)-plane.

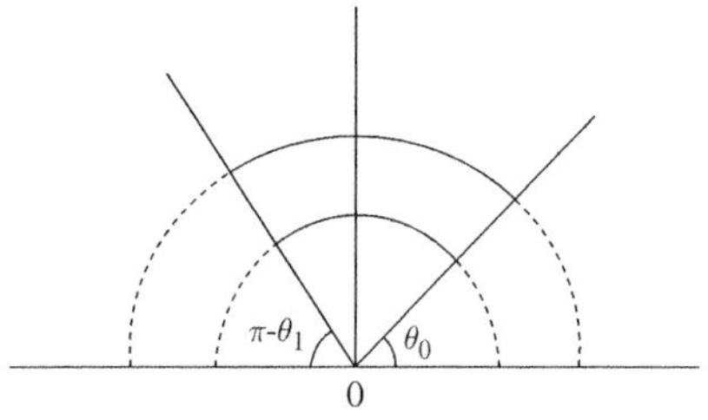

Figure 3.8 The circular arcs have the same hyperbolic length

In all the above calculations, we do not need a formula for the parametrisation $\boldsymbol{x}$. The actual shape of the surface in $\mathbb{R}^n$ is completely irrelevant for the calculations, what matters is the metric. Changing our viewpoint by suppressing $\boldsymbol{x}$, we may regard this whole example as simply the upper half-plane $\{(u, v) \in \mathbb{R}^2 : v > 0\}$ but with a non-standard metric (that is to say, not coming from the standard inner product on the plane). This is called the *hyperbolic plane* and is usually denoted by the letter H. The hyperbolic length of $\boldsymbol{\alpha}$ between parameter values v_0 and v_1 is then $\log(v_1/v_0)$, with similar comments for the other curves $\boldsymbol{\beta}$ and $\boldsymbol{\gamma}$. However, as mentioned earlier, since $E = G$ and $F = 0$, angles are the same in both the hyperbolic and standard metrics.

It turns out that the circular arcs $\boldsymbol{\gamma}(t)$, $\theta_0 \le t \le \theta_1$, are curves on H of shortest hyperbolic length between their endpoints (there is more on this in Example 10 of §7.3), so their length gives the hyperbolic distance apart of these endpoints. As noted earlier, each of these arcs has the same hyperbolic length, so if the defining property of a railway track is that the rails stay a constant distance apart, then for angles θ_0 and θ_1, the lines $y = x \tan \theta_0$ and $y = x \tan \theta_1$ are the two rails of a hyperbolic railway track!

Although not the standard metric, the hyperbolic metric on the upper half-plane may be given a practical interpretation. For instance, if you are in a field bounded by a straight impenetrable hedge and if the field gets more and more muddy the nearer you get to the hedge in such a way that the effort required to make progress is inversely proportional to the distance from the hedge, then the hyperbolic length of a path in the field will measure the effort required to traverse that path. In particular, the path of least effort between two points in the field will be the appropriate circular arc γ as described above.

A deep result of Hilbert says that the whole of the hyperbolic plane cannot be realised as a surface in $\mathbb{R}^3$. However, the part corresponding to $\{(u, v) \in \mathbb{R}^2 : -\pi < u < \pi,\ v > 1\}$ may be put in $\mathbb{R}^3$ as the *pseudosphere* (see Figure 2.10), which, as we saw in Example 2 of §2.3, is the surface of revolution obtained by rotating a suitable tractrix around the z-axis.

This tractrix may be parametrised by

$$\boldsymbol{\gamma}(v) = \left(\frac{1}{v}\,,\ 0\,,\ \operatorname{arccosh} v - \frac{(v^2-1)^{1/2}}{v} \right), \qquad v > 1\,, \tag{3.16}$$

where arccosh v is taken to be the positive number w with $\cosh w = v$. A calculation using the expressions for E, F and G of a surface of revolution found in Example 2 of §3.2 soon shows that E, F and G for the resulting parametrisation of the pseudosphere as a surface of revolution are given by (3.15). If we take the start point of $\boldsymbol{\alpha}$ to be $v_0 = 1$, then $\boldsymbol{\alpha}$ gives the generating curve (3.16) of the pseudosphere, while, for each $v_0 > 1$, $\boldsymbol{\beta}$ gives a parallel (Figure 3.9).

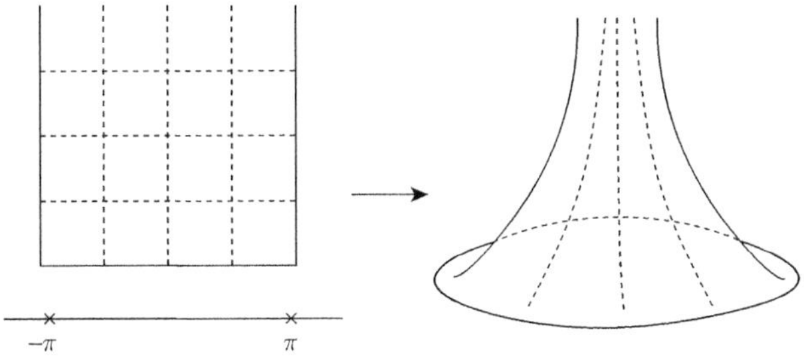

Figure 3.9 Correspondence with pseudosphere

Although Hilbert's Theorem says that the hyperbolic plane cannot be realised as a surface in $\mathbb{R}^3$, it follows from a theorem of Nash (known as the *Nash Embedding Theorem*) that the hyperbolic plane can be realised as a surface in $\mathbb{R}^n$ for some sufficiently large value of n.

3.5 Families of curves

Let $\boldsymbol{x} : U \to \mathbb{R}^n$ be a local parametrisation of a surface S in $\mathbb{R}^n$, and let $\phi(u, v)$ be a smooth real-valued function defined on U with grad ϕ never zero. The level curves $\phi(u, v) =$

constant give a family of curves in U and hence, by applying $\boldsymbol{x}$, a family of curves on $\boldsymbol{x}(U) \subseteq S$.

Example 1 (Coordinate curves) If $\phi(u, v) = u$, then $\text{grad}\,\phi = \left(\frac{\partial\phi}{\partial u}, \frac{\partial\phi}{\partial v}\right) = (1, 0)$ is never zero and the family of curves in U consists of the lines u = constant. The image of this family under $\boldsymbol{x}$ is the corresponding family of *coordinate curves* (Figure 3.10). There are two such families, the other family being given by taking v = constant. The coordinate curve $u = u_0$, where u_0 is a constant, has parametrisation of the form $\boldsymbol{\alpha}(t) = \boldsymbol{x}\,(u_0, v(t))$, so the vectors tangential to this family are scalar multiples of the coordinate vector $\boldsymbol{x}_v$. The members of the family v = constant may be parametrised in a similar way, so their tangent vectors are scalar multiples of $\boldsymbol{x}_u$.

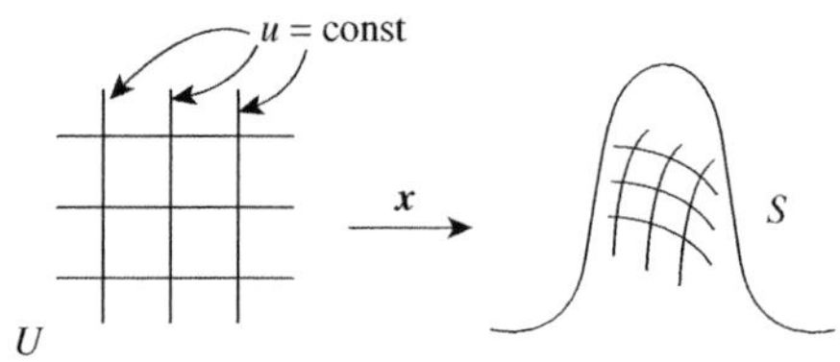

Figure 3.10 Families of coordinate curves

In particular, the angle of intersection θ of the two families of coordinate curves is given by

$$\boldsymbol{x}_u.\boldsymbol{x}_v = |\boldsymbol{x}_u|\,|\boldsymbol{x}_v| \cos\theta\ ,$$

so that

$$\cos\theta = \frac{F}{\sqrt{EG}}\ . \tag{3.17}$$

Example 2 We consider a surface S which admits a local parametrisation with coefficients of the first fundamental form given by

$$E = 1 + 4u^2\ , \quad F = -4uv\ , \quad G = 1 + 4v^2\ ,$$

(see, for instance, Examples 1 and 2 of §3.3).

Let $\mathcal{F}$ be the family of curves on S determined by $u^2 - v^2$ = constant. We shall find the family of curves on S everywhere orthogonal to the curves in $\mathcal{F}$. These are called the *orthogonal trajectories* of $\mathcal{F}$.

We begin by finding the tangent vectors to the family $\mathcal{F}$. If $\boldsymbol{\alpha}(t) = \boldsymbol{x}(u(t), v(t))$ is a member of $\mathcal{F}$, then $u^2(t) - v^2(t)$ = constant so that

$$\boldsymbol{\alpha}'(t) = u'\boldsymbol{x}_u + v'\boldsymbol{x}_v$$

with $uu' - vv' = 0$. Hence $u'/v' = v/u$ so that tangent vectors to the family $\mathcal{F}$ at $\boldsymbol{x}(u, v)$ are the scalar multiples of $v\boldsymbol{x}_u + u\boldsymbol{x}_v$.

Let $\boldsymbol{\beta}(r)$ be a curve on S everywhere orthogonal to the curves of $\mathcal{F}$. We now write the coordinates u and v as functions of r (which will be different from the functions $u(t)$, $v(t)$ used above for $\boldsymbol{\alpha}$) such that $\boldsymbol{\beta}(r) = \boldsymbol{x}\,(u(r), v(r))$. Then $\boldsymbol{\beta}(r)$ is an orthogonal trajectory of $\mathcal{F}$ if and only if

$$\left(\frac{du}{dr}\boldsymbol{x}_u + \frac{dv}{dr}\boldsymbol{x}_v\right) \cdot (v\boldsymbol{x}_u + u\boldsymbol{x}_v) = 0\,.$$

Using the expressions for E, F and G given above, this equation simplifies to give

$$\frac{du}{dr}v + u\frac{dv}{dr} = 0\,,$$

so, integrating up, the orthogonal trajectories of $\mathcal{F}$ are given by $uv = $ constant.

This example may be given geometrical meaning if we note from Example 2 of §3.3 that the above coefficients of the first fundamental form are those of the standard parametrisation of the hyperbolic paraboloid $z = x^2 - y^2$ as a graph. In this case, the family $\mathcal{F}$ of curves $u^2 - v^2 = $ constant consists of the contour lines on the surface, and the family $uv = $ constant consists of the paths of steepest descent (or ascent).

A more general version of the above example is outlined in Exercise 3.22.

3.6 Ruled surfaces

A *ruled surface* is a surface S in $\mathbb{R}^n$ which may be covered by a family of line segments; that is to say, through each point of S there passes a line segment which stays on the surface. Intuitively, a ruled surface in $\mathbb{R}^n$ is the surface in $\mathbb{R}^n$ swept out as a straight line is moved around. A cylinder in $\mathbb{R}^3$ is an obvious example, and other examples appeared in Exercises 2.6, 2.8, 2.9 and 2.10. In this section, we consider two of these examples in more detail.

We first give a mathematical description of a ruled surface. Let I be an open interval and let $\boldsymbol{\alpha} : I \to \mathbb{R}^n$ be a regular curve without self-intersections. Also, let $\boldsymbol{\beta} : I \to \mathbb{R}^n$ be a smooth map which is nowhere zero. If J is also an open interval, we define $\boldsymbol{x} : I \times J \to \mathbb{R}^n$ by

$$\boldsymbol{x}(u, v) = \boldsymbol{\alpha}(u) + v\boldsymbol{\beta}(u)\,, \quad u \in I\,, \ v \in J\,. \tag{3.18}$$

For each fixed u,

$$\boldsymbol{\alpha}(u) + v\boldsymbol{\beta}(u)\,, \quad v \in J\,,$$

is the line segment through $\boldsymbol{\alpha}(u)$ in the direction $\boldsymbol{\beta}(u)$. As u varies, this line segment sweeps out a subset of $\mathbb{R}^n$ which, in many cases, is a surface (Figure 3.11).

Example 1 (Helicoid) Let b be a non-zero real number and let S be the surface in $\mathbb{R}^3$ with equation

$$x \sin\frac{z}{b} = y \cos\frac{z}{b}\,.$$

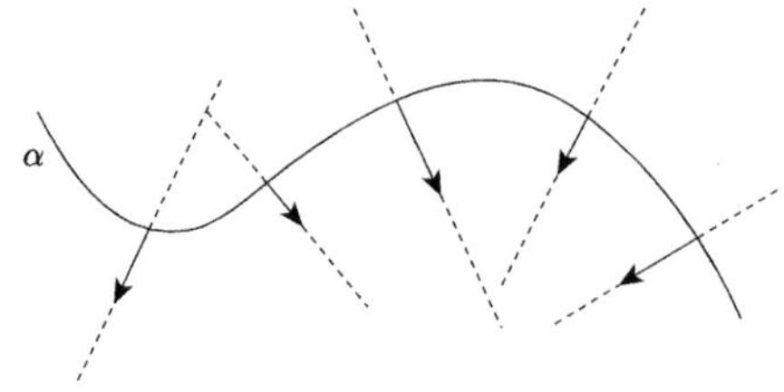

Figure 3.11 A ruled surface

Let

$$\boldsymbol{\alpha}(u) = (0, 0, bu), \quad \boldsymbol{\beta}(u) = (\cos u, \sin u, 0),$$

and let $\boldsymbol{x}(u, v) = \boldsymbol{\alpha}(u) + v\boldsymbol{\beta}(u)$. Then

$$\boldsymbol{x}(u, v) = (v \cos u, v \sin u, bu), \quad (u, v) \in \mathbb{R}^2,$$

and an easy check shows that the image of $\boldsymbol{x}$ is contained in S (Figure 3.12).

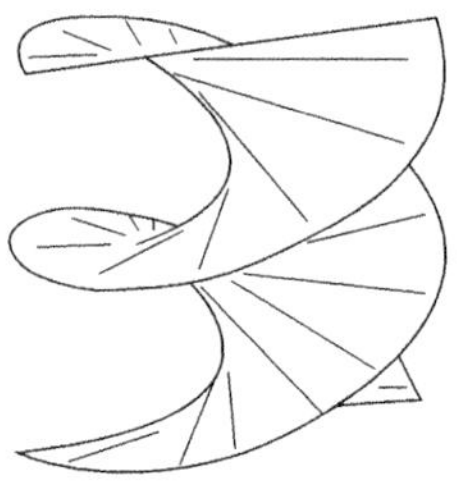

Figure 3.12 Helicoid as a ruled surface

We now show that $\boldsymbol{x}$ is a parametrisation of S. We could do this using the Parametrisation Recognition Theorem (Theorem 4 of §2.5), but we choose to use the original definition of local parametrisation given near the beginning of Chapter 2. To do this we define a smooth map $\boldsymbol{F} : \mathbb{R}^3 \to \mathbb{R}^2$ by

$$\boldsymbol{F}(x, y, z) = \left(\frac{z}{b}, x \cos \frac{z}{b} + y \sin \frac{z}{b}\right),$$

and note that, for all $(u, v) \in \mathbb{R}^2$, $\boldsymbol{F}\boldsymbol{x}(u, v) = (u, v)$. Hence $\boldsymbol{x}$ is a local parametrisation of S, and if $(x, y, z) \in S$ then

$$\boldsymbol{x}\left(\frac{z}{b}, x \cos \frac{z}{b} + y \sin \frac{z}{b}\right) = (x, y, z),$$

so that the image of $\boldsymbol{x}$ is the whole of S.

It follows from the description of S as a ruled surface that S is shaped like a "spiral" staircase of infinite width and height. The surface S is a *helicoid*.

Note that, for all parametrisations of the form given in (3.18), the coordinate curves $u =$ constant are just the lines of the ruling. In the case of the helicoid, the coordinate curves $u =$ constant are the "treads" of the staircase, and the coordinate curves $v =$ constant are helices except for $v = 0$ which is the z-axis.

In fact, we may define a rather nicer parametrisation of the helicoid by taking

$$\tilde{\boldsymbol{x}}(u, v) = (b \sinh v \cos u, b \sinh v \sin u, bu)\,, \quad (u, v) \in \mathbb{R}^2\,, \tag{3.19}$$

which has coefficients of the first fundamental form given by

$$\tilde{E} = \tilde{G} = b^2 \cosh^2 v\,, \quad \tilde{F} = 0\,.$$

Hence $\tilde{\boldsymbol{x}}$ is an isothermal parametrisation, and so gives an angle-preserving correspondence between $\mathbb{R}^2$ and the whole of the helicoid.

Incidentally, if we take $b = 1$ in the above example, the corresponding helicoid has the same coefficients of the first fundamental form as the catenoid (found in Example 1 of §3.4). This shows that, locally, the two surfaces have the same intrinsic geometry. It also shows that the property of being a ruled surface is not intrinsic; you cannot tell by looking solely at the first fundamental form whether or not a surface in $\mathbb{R}^n$ is ruled.

Example 2 (Hyperboloid of one sheet) Let a, b, c be non-zero real numbers, and let S be the surface in $\mathbb{R}^3$ with equation

$$\frac{x^2}{a^2} + \frac{y^2}{b^2} - \frac{z^2}{c^2} = 1\,.$$

We let

$$\boldsymbol{\alpha}(u) = (a \cos u, b \sin u, 0)\,,$$

and seek to find a smooth map $\boldsymbol{\beta}(u)$ such that $\boldsymbol{\alpha}(u) + v\boldsymbol{\beta}(u)$ lies on S for all $(u, v) \in \mathbb{R}^2$ (Figure 3.13). To make the equations simpler, it makes sense in this case to write $\boldsymbol{\beta}(u) = (a\lambda_1(u), b\lambda_2(u), c\lambda_3(u))$ for smooth functions $\lambda_1(u)$, $\lambda_2(u)$ and $\lambda_3(u)$. Then $\boldsymbol{\alpha}(u) + v\boldsymbol{\beta}(u)$ lies on S for all $(u, v) \in \mathbb{R}^2$ if and only if

$$(v\lambda_1 + \cos u)^2 + (v\lambda_2 + \sin u)^2 - (v\lambda_3)^2 = 1\,, \quad \forall (u, v) \in \mathbb{R}^2\,.$$

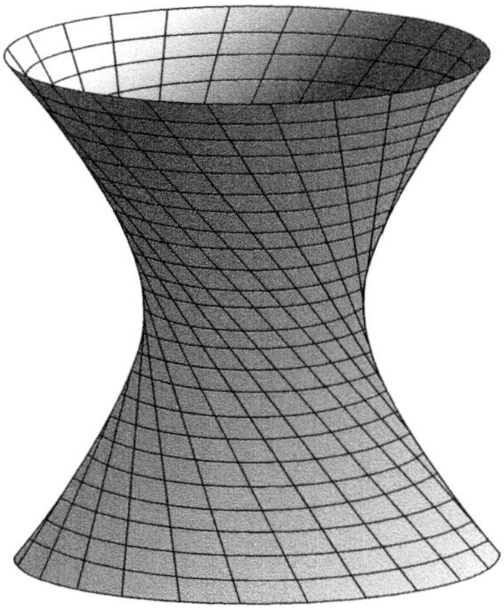

Figure 3.13 Hyperboloid of one sheet as a ruled surface

This holds if and only if the coefficients of v and of v^2 in the above equation are both zero, which gives

$$\lambda_1 \cos u + \lambda_2 \sin u = 0 \quad \text{and} \quad {\lambda_1}^2 + {\lambda_2}^2 - {\lambda_3}^2 = 0 .$$

Since the vectors $\boldsymbol{\beta}(u)$ are determined only up to a scalar multiple, we may take $\lambda_3 = 1$, in which case the above equations give that $\lambda_1 = \pm \sin u$ and $\lambda_2 = \mp \cos u$. It now follows that if we define two smooth maps $\boldsymbol{\beta}^+, \boldsymbol{\beta}^-$ by

$$\boldsymbol{\beta}^{\pm}(u) = (a \sin u, -b \cos u, \pm c) ,$$

then $\boldsymbol{\alpha}(u) + v\boldsymbol{\beta}^{\pm}(u)$ lies on S for all $(u, v) \in \mathbb{R}^2$.

We now define two maps $\boldsymbol{x}^{\pm}$ by putting

$$\boldsymbol{x}^{\pm}(u, v) = \boldsymbol{\alpha}(u) + v\boldsymbol{\beta}^{\pm}(u) = \big(a(\cos u + v \sin u),\ b(\sin u - v \cos u),\ \pm cv\big) ,$$

and show that, if we restrict u to lie in an open interval of length 2π, then $\boldsymbol{x}^{\pm}$ are both local parametrisations of S whose images are the whole of S with just one line omitted (and, of course, we can cover the omitted line by simply altering the domain of definition of $\boldsymbol{x}^{\pm}(u, v)$).

We use the Parametrisation Recognition Theorem to do this. First note that, as shown above, $\boldsymbol{x}^{\pm}(u, v)$ lies on S for all $(u, v) \in \mathbb{R}$. The injectivity of both of $\boldsymbol{x}^{\pm}$ is not hard to check, as is the linear independence of $\boldsymbol{x}_u^{\pm}$ and $\boldsymbol{x}_v^{\pm}$. Finally, it is an interesting exercise to check that any point (x, y, z) on S lies on a line of both the family of lines determined by $\boldsymbol{x}^+$ and the family determined by $\boldsymbol{x}^-$. Thus S is a **doubly ruled** surface.

In Exercise 3.24, you are asked to show that the hyperbolic paraboloid $z = xy$ is also a doubly ruled surface (that this is a ruled surface was first mentioned in Exercise 2.8).

There are some cases where $\boldsymbol{x}(I \times J)$ is not a surface (see, for example, Exercise 3.23). However, as long as the intervals I and J are not too large, then we do obtain a surface when we move a line segment in a direction tranverse to itself (so that $\boldsymbol{\alpha}'$ is never a scalar multiple of $\boldsymbol{\beta}$). In fact,

$$\boldsymbol{x}_u = \boldsymbol{\alpha}' + v\boldsymbol{\beta}' , \quad \boldsymbol{x}_v = \boldsymbol{\beta} ,$$

so Theorem 1 of §2.5 may be used to show that S is a surface near to any point at which $\boldsymbol{\alpha}' + v\boldsymbol{\beta}'$ isn't a scalar multiple of $\boldsymbol{\beta}$. In particular, if $\boldsymbol{\alpha}'$ is not a scalar multiple of $\boldsymbol{\beta}$ then S is a surface in the vicinity of any point of the base curve $\boldsymbol{\alpha}$ (this is the curve on S given by taking $v = 0$).

3.7 Area

We use local parametrisations and integration of functions over suitable subsets of the plane to define the concept of area for (and, more generally, integration of real-valued functions on) surfaces. Like length and angle of intersection of curves on a surface, area and integration are intrinsic properties.

Let $\boldsymbol{x} : U \to \mathbb{R}^n$ be a local parametrisation of a surface S in $\mathbb{R}^n$, and let Q be a subset of U over which we can integrate continuous real-valued functions (so, for instance, Q could be a closed disc, the interior of a polygon together with the polygon that bounds it, or, more generally, the closure of a bounded open set). We let R denote the image $\boldsymbol{x}(Q)$ of Q under $\boldsymbol{x}$ and define the area $A(R)$ of R by

$$A(R) = \iint_Q \sqrt{EG - F^2}\, du\, dv \ .$$

Our initial motivation for this definition is twofold. Firstly, when taking the standard parametrisation of the plane, the above integral reduces to the standard expression for the area of a region in the plane, and, secondly, as we show in Lemma 2 of §3.9, the above expression for area is independent of choice of local parametrisation $\boldsymbol{x}$.

The expression $\sqrt{EG - F^2}\, du\, dv$ is often called the *element of area* and denoted dA. The above equation is then written

$$A(R) = \iint_R dA = \iint_Q \sqrt{EG - F^2}\, du\, dv \ . \tag{3.20}$$

We may use the above procedure to define area for more general subsets of S; all we need is that the subset may be broken up into (a finite number of) the type of pieces we have considered above. We do not need to worry if curves are omitted (as in Example 2, where we find the area of a torus of revolution) or if curves are covered twice, since these will not contribute to the area.

Example 1 (Area of a graph) Consider a suitable region R of the graph of a smooth real-valued function $g(u, v)$ (Figure 3.14). If we parametrise this in the usual way by

$$\boldsymbol{x}(u, v) = (u, v, g(u, v)) \ ,$$

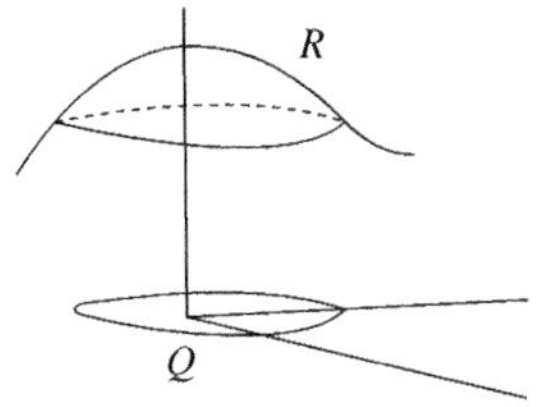

Figure 3.14 Area of a graph

then a short calculation using the expressions for E, F and G found in Example 1 of §3.2 shows that $EG - F^2 = 1 + {g_u}^2 + {g_v}^2$. Hence, if Q is the image of R under orthogonal projection onto the xy-plane, then

$$A(R) = \iint_Q \sqrt{1 + {g_u}^2 + {g_v}^2}\, du\, dv \ ,$$

in accordance with the usual formula for the area of a graph.

We now give a geometrical motivation for the definition of area. Consider a rectangle B in Q (Figure 3.15) with opposite vertices at (u, v) and $(u + \delta u, v + \delta v)$. The parallelogram C in $\mathbb{R}^n$ with adjacent sides $\boldsymbol{x}_u \delta u$ and $\boldsymbol{x}_v \delta v$ is tangential to S at $\boldsymbol{x}(u, v)$ and is, in some sense, the best approximating parallelogram to $\boldsymbol{x}(B)$. The area δA of C is given by

$$\delta A = |\boldsymbol{x}_u \delta u||\boldsymbol{x}_v \delta v| \sin\theta \ ,$$

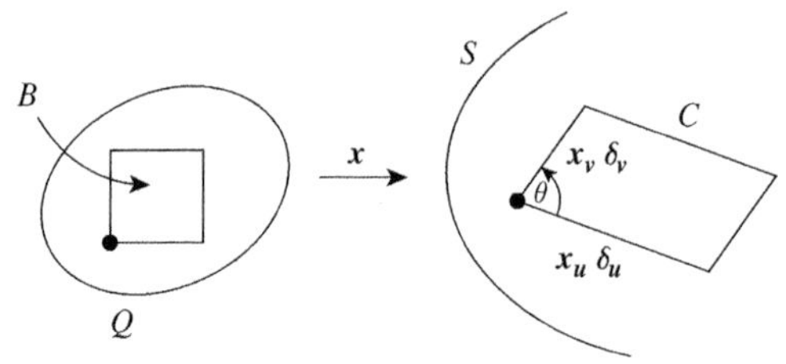

Figure 3.15 Geometrical motivation for area

where θ is the angle between $\boldsymbol{x}_u$ and $\boldsymbol{x}_v$. However, using (3.17), we see that $\sin^2\theta = 1-(F^2/EG) = (EG-F^2)/EG$ from which it follows that

$$\delta A = \sqrt{EG-F^2}\,\delta u\delta v\ .$$

The usual procedure of taking the limit as we consider ever smaller partitions and add the areas of each of the rectangles δA leads to the expression (3.20) for area.

Example 2 (Torus of revolution) We find the area of the torus of revolution $T_{a,b}$ described in Example 4 of §2.3. To do this we use the local parametrisation $\boldsymbol{x} : (0,2\pi)\times(0,2\pi)\to T_{a,b}$ given by

$$\boldsymbol{x}(u,v) = ((a+b\cos v)\cos u,\ (a+b\cos v)\sin u,\ b\sin v)\ .$$

Then

$$E=(a+b\cos v)^2\,,\quad F=0\,,\quad G=b^2\,,$$

so that the element of area is given by

$$dA = b(a+b\cos v)\,du\,dv\ .$$

The image of $\boldsymbol{x}$ is the whole of the torus with just two circles omitted, so

$$\begin{aligned} A(T_{a,b}) &= \int_0^{2\pi}\int_0^{2\pi} b(a+b\cos v)\,du\,dv \\ &= 4\pi^2 ab\ . \end{aligned}$$

More generally than just finding areas, we may integrate a real-valued function f defined on a suitable region R of a surface S. This time we define

$$\iint_R f\,dA = \iint_Q (f\boldsymbol{x})\sqrt{EG-F^2}\,du\,dv\,, \tag{3.21}$$

where, as usual, $f\boldsymbol{x}$ denotes the composite of $\boldsymbol{x}$ and f.

Example 3 (Moment of inertia of rotating sphere) The moment of inertia of an object about a given axis gives a measure of the difficulty of changing the angular motion of the object about that axis. For an object of unit density, it is obtained by integrating the square of the perpendicular distance of each point of the object from the axis. We calculate the moment of inertia for the unit sphere about the z-axis. If we parametrise the unit sphere $S^2(1)$ as a surface of revolution in the usual way via

$$\boldsymbol{x}(u,v) = (\cos v\cos u, \cos v\sin u, \sin v)\,,\quad -\pi<u<\pi\,,\quad -\pi/2<v<\pi/2\,,$$

then we quickly compute that $E = \cos^2 v$, $F = 0$, $G = 1$, so that $\sqrt{EG - F^2} = \cos v$ (since $-\pi/2 < v < \pi/2$). The perpendicular distance of $\boldsymbol{x}(u, v)$ from the z-axis is $\cos v$, so the required moment of inertia is given by

$$\int_{-\pi/2}^{\pi/2} \int_{-\pi}^{\pi} \cos^3 v \, du dv ,$$

which evaluates to give $8\pi/3$ for the required moment of inertia.

If S is a surface in $\mathbb{R}^3$, then we may use the vector cross product to determine the element of area. In fact, using the vector algebra identity

$$(\boldsymbol{a} \times \boldsymbol{b}).(\boldsymbol{c} \times \boldsymbol{d}) = (\boldsymbol{a}.\boldsymbol{c})(\boldsymbol{b}.\boldsymbol{d}) - (\boldsymbol{a}.\boldsymbol{d})(\boldsymbol{b}.\boldsymbol{c}) ,$$

we quickly see that

$$\sqrt{EG - F^2} = |\boldsymbol{x}_u \times \boldsymbol{x}_v| , \tag{3.22}$$

so that, for a surface in $\mathbb{R}^3$,

$$dA = |\boldsymbol{x}_u \times \boldsymbol{x}_v| \, du \, dv . \tag{3.23}$$

This gives a proof of the following proposition.

Proposition 4 *Let $\boldsymbol{x}(u, v)$ be a local parametrisation of a surface S in $\mathbb{R}^3$, and let Q be a region in the domain of $\boldsymbol{x}$ over which we can integrate continuous real-valued functions. If $R = \boldsymbol{x}(Q)$, then*

$$A(R) = \iint_R dA = \iint_Q |\boldsymbol{x}_u \times \boldsymbol{x}_v| \, du \, dv , \tag{3.24}$$

while

$$\iint_R f dA = \iint_Q (f\boldsymbol{x}) |\boldsymbol{x}_u \times \boldsymbol{x}_v| \, du \, dv . \tag{3.25}$$

As remarked earlier, we much prefer to define geometrical notions on a surface without using local parametrisations. However, we have not done this in our treatment of integration on surfaces, so we should show that the definition of integration as given above is independent of the particular local parametrisation chosen. We do this in §3.9, but, since this material is not needed in an essential way for the rest of the book, §3.9 may be omitted if desired.

3.8 Change of variables †

As mentioned earlier, a good choice of local parametrisation can often lead to a simplification of a particular problem. In this section we discuss how to obtain a new parametrisation from a given one by a *change of variables* on the domain of the given local parametrisation. As with the next section, the material in this section is not essential for understanding the rest of the book, and so may be omitted if desired.

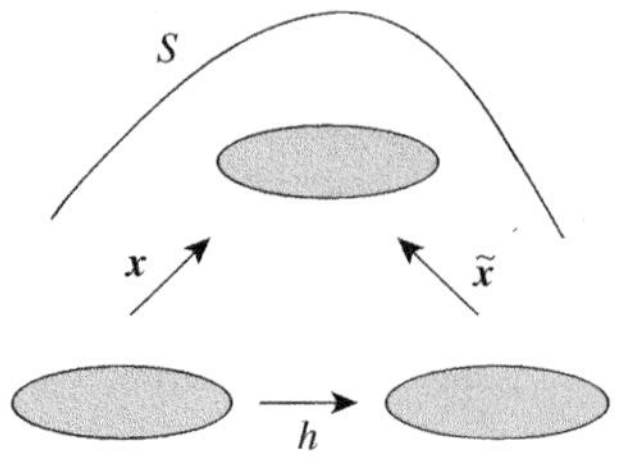

Figure 3.16 For the proof of Proposition 1

We first recall that a smooth map $h : U \to \tilde{U}$ between open subsets $U, \tilde{U}$ of $\mathbb{R}^2$ is called a *diffeomorphism* if there is a smooth map $g : \tilde{U} \to U$ such that $gh = \mathrm{Id} : U \to U$ and $hg = \mathrm{Id} : \tilde{U} \to \tilde{U}$. In this case we call g the *inverse map* of h, and denote it by h^{-1}. It follows from the Inverse Function Theorem that h is a diffeomorphism onto its image if and only if h is injective and the partial derivatives h_u, h_v are linearly independent at each point.

The setup in the following proposition is illustrated in Figure 3.16.

Proposition 1 *Let $\boldsymbol{x}(u, v)$ be a local parametrisation of a surface S in $\mathbb{R}^n$, and let $\tilde{\boldsymbol{x}}(\tilde{u}, \tilde{v})$ be a smooth map into $\mathbb{R}^n$. If*

$$\boldsymbol{x}(u, v) = \tilde{\boldsymbol{x}}\,(\tilde{u}(u, v), \tilde{v}(u, v)) \ , \tag{3.26}$$

where $h(u, v) = (\tilde{u}(u, v), \tilde{v}(u, v))$ is a diffeomorphism from the domain of $\boldsymbol{x}$ onto the domain of $\tilde{\boldsymbol{x}}$, then $\tilde{\boldsymbol{x}}(\tilde{u}, \tilde{v})$ is a local parametrisation of S with the same image as $\boldsymbol{x}$.

Proof It is clear from (3.26) that the image of $\tilde{\boldsymbol{x}}$ is equal to that of $\boldsymbol{x}$. We now produce a map $\tilde{\boldsymbol{F}}$ for $\tilde{\boldsymbol{x}}$ as in condition (S2) of the definition of a local parametrisation of a surface. So, let $\boldsymbol{F}$ be such a map for $\boldsymbol{x}$ and let $\tilde{\boldsymbol{F}} = h\boldsymbol{F}$. Then, noting from (3.26) that $\boldsymbol{x} = \tilde{\boldsymbol{x}}h$, we have

$$\tilde{\boldsymbol{F}}\tilde{\boldsymbol{x}} = h\boldsymbol{F}\tilde{\boldsymbol{x}} = h\boldsymbol{F}\boldsymbol{x}h^{-1} = hh^{-1} = \mathrm{Id}\,. \qquad \square$$

If (3.26) holds, we say that $\tilde{\boldsymbol{x}}$ is obtained from $\boldsymbol{x}$ by the *change of variables* from (u, v) to $(\tilde{u}, \tilde{v})$.

Example 2 (Hyperbolic paraboloid) The map $\boldsymbol{x}(u, v) = (u + v, u - v, u^2 - v^2)$ is a parametrisation of the hyperbolic paraboloid S with equation $z = xy$. If we make the change of variables from (u, v) to $(\tilde{u}, \tilde{v})$, where $\tilde{u} = u + v$, $\tilde{v} = u - v$, then the corresponding local parametrisation $\tilde{\boldsymbol{x}}(\tilde{u}, \tilde{v})$ satisfying (3.26) is given by $\tilde{\boldsymbol{x}}(\tilde{u}, \tilde{v}) = (\tilde{u}, \tilde{v}, \tilde{u}\tilde{v})$.

Returning to the general case, if $\boldsymbol{x}(u, v)$ and $\tilde{\boldsymbol{x}}(\tilde{u}, \tilde{v})$ are related as in (3.26), then the chain rule shows that

$$\boldsymbol{x}_u = \tilde{\boldsymbol{x}}_{\tilde{u}}\tilde{u}_u + \tilde{\boldsymbol{x}}_{\tilde{v}}\tilde{v}_u \ , \quad \boldsymbol{x}_v = \tilde{\boldsymbol{x}}_{\tilde{u}}\tilde{u}_v + \tilde{\boldsymbol{x}}_{\tilde{v}}\tilde{v}_v \ , \tag{3.27}$$

as may be easily checked for Example 2.

Example 3 (Tchebycheff parametrisation) Assume that $\boldsymbol{x}(u, v)$ is a local parametrisation of a surface S with coefficients of the first fundamental form given by

$$E = \operatorname{sech}^2 v\,, \quad F = 0\,, \quad G = \tanh^2 v\,. \tag{3.28}$$

Consider the change of variables given by taking

$$\tilde{u} = \frac{1}{2}(u + v)\,, \quad \tilde{v} = \frac{1}{2}(u - v)\,,$$

and let $\tilde{\boldsymbol{x}}(\tilde{u}, \tilde{v})$ be the corresponding local parametrisation satisfying (3.26).

It follows from the chain rule that

$$2\boldsymbol{x}_u = \tilde{\boldsymbol{x}}_{\tilde{u}} + \tilde{\boldsymbol{x}}_{\tilde{v}} \quad \text{and} \quad 2\boldsymbol{x}_v = \tilde{\boldsymbol{x}}_{\tilde{u}} - \tilde{\boldsymbol{x}}_{\tilde{v}}\,,$$

so that

$$\tilde{\boldsymbol{x}}_{\tilde{u}} = \boldsymbol{x}_u + \boldsymbol{x}_v \quad \text{and} \quad \tilde{\boldsymbol{x}}_{\tilde{v}} = \boldsymbol{x}_u - \boldsymbol{x}_v\,.$$

Hence $\tilde{E} = E + 2F + G = 1$, and $\tilde{G} = E - 2F + G = 1$.

A local parametrisation $\boldsymbol{x}(u, v)$ with coefficients of the first fundamental form satisfying $E = G = 1$ is called a *Tchebycheff parametrisation*. In this case, if u_0 and v_0 are constants then the coordinate curves $u \mapsto \boldsymbol{x}(u, v_0)$ and $v \mapsto \boldsymbol{x}(u_0, v)$ are parametrised by arc length. We also have that $F = \cos\theta$, where θ is the angle of intersection of the coordinate curves. Intuitively, a Tchebycheff parametrisation may be thought of as moulding a piece of fabric over the surface without stretching the fibres but changing the angle θ at which the two sets of fibres (the weft and the warp) meet.

Example 4 (Pseudosphere) We may obtain the pseudosphere by rotating the curve $v \mapsto (\operatorname{sech} v, 0, v - \tanh v)$, $v > 0$, about the z-axis as in Example 2 of §2.3. A short calculation using, for instance, Example 2 of §3.2 shows that the coefficients of the first fundamental form of the corresponding parametrisation of the pseudosphere as a surface of revolution are given by (3.28). Hence, Example 3 shows that the pseudosphere may be covered using Tchebycheff parametrisations.

3.9 Coordinate independence †

We finish the chapter by showing that the definition of integration we gave in §3.7 is independent of the particular local parametrisation chosen. As already mentioned, this material is not needed in an essential way for the rest of the book and so may be omitted if desired.

The set-up for the following lemma is illustrated in Figure 3.17. The lemma is, in some sense, a converse to Proposition 1 of §3.8. Its proof has similarities with that of Lemma 1 of §3.1.

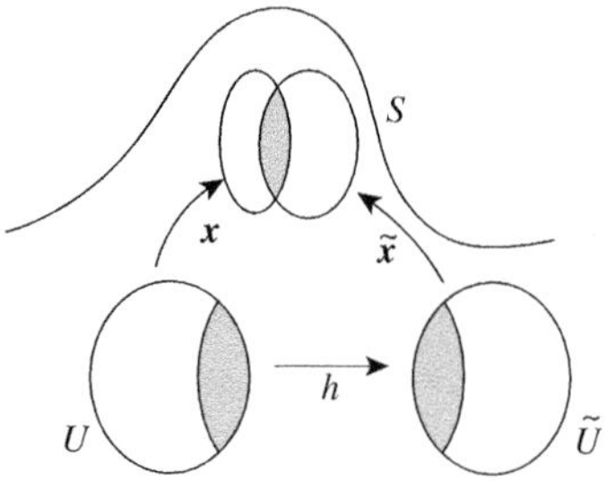

Figure 3.17 For the proof of Lemma 1

Lemma 1 *Let $\boldsymbol{x} : U \to \mathbb{R}^n$, $\tilde{\boldsymbol{x}} : \tilde{U} \to \mathbb{R}^n$ be two local parametrisations of the same surface S. If $p \in \boldsymbol{x}(U) \cap \tilde{\boldsymbol{x}}(\tilde{U})$ and if $p = \boldsymbol{x}(u_0, v_0)$, then there is a diffeomorphism $h(u,v) = (\tilde{u}(u,v), \tilde{v}(u,v))$, defined on an open neighbourhood of (u_0, v_0) in U such that*

$$\boldsymbol{x}(u,v) = \tilde{\boldsymbol{x}}\left(\tilde{u}(u,v), \tilde{v}(u,v)\right) . \tag{3.29}$$

Proof Let $\boldsymbol{F} : W \to \mathbb{R}^2$ be a map for $\boldsymbol{x}$ as in condition (S2) for a surface, and let $\tilde{\boldsymbol{F}} : \tilde{W} \to \mathbb{R}^2$ be such a map for $\tilde{\boldsymbol{x}}$. If we set

$$V = \{(u,v) \in U : \boldsymbol{x}(u,v) \in W \cap \tilde{W}\} ,$$

then continuity of $\boldsymbol{x}$ implies that V is an open subset of $\mathbb{R}^2$, and injectivity of $\tilde{\boldsymbol{x}}$ allows us to define a map $h : V \to \mathbb{R}^2$ by setting

$$\boldsymbol{x}(u,v) = \tilde{\boldsymbol{x}}\left(h(u,v)\right) . \tag{3.30}$$

If we apply $\tilde{\boldsymbol{F}}$ to both sides of (3.30), we obtain h as the composite $\tilde{\boldsymbol{F}}\boldsymbol{x}$ of two smooth maps. Hence h is smooth, and its image is the open set

$$\tilde{V} = \{(\tilde{u}, \tilde{v}) \in \tilde{U} : \tilde{\boldsymbol{x}}(\tilde{u}, \tilde{v}) \in W \cap \tilde{W}\} .$$

We now show that $h : V \to \tilde{V}$ is a diffeomorphism. To do this we use injectivity of $\boldsymbol{x}$ to define $\tilde{h} : \tilde{V} \to V$ by setting

$$\tilde{\boldsymbol{x}}(\tilde{u}, \tilde{v}) = \boldsymbol{x}\left(\tilde{h}(\tilde{u}, \tilde{v})\right) .$$

Then $\tilde{h}$ is smooth, and, on V,

$$\boldsymbol{x} = \tilde{\boldsymbol{x}}h = \boldsymbol{x}\tilde{h}h .$$

Injectivity of $\boldsymbol{x}$ now shows that $\tilde{h}h$ is the identity map on V, and similar reasoning shows that $h\tilde{h}$ is the identity map on $\tilde{V}$. Hence h is a diffeomorphism, and the lemma is proved. □

The diffeomorphism h in Lemma 1 is called the *transition function* from $\boldsymbol{x}$ to $\tilde{\boldsymbol{x}}$. As in the previous section, we have

$$\boldsymbol{x}_u = \tilde{\boldsymbol{x}}_{\tilde{u}}\tilde{u}_u + \tilde{\boldsymbol{x}}_{\tilde{v}}\tilde{v}_u , \quad \boldsymbol{x}_v = \tilde{\boldsymbol{x}}_{\tilde{u}}\tilde{u}_v + \tilde{\boldsymbol{x}}_{\tilde{v}}\tilde{v}_v , \tag{3.31}$$

so, if E, F, G and $\tilde{E}$, $\tilde{F}$, $\tilde{G}$ are the coefficients of the first fundamental form of S with respect to the two local parametrisations $\boldsymbol{x}$ and $\tilde{\boldsymbol{x}}$ respectively, then

$$\begin{aligned} E &= \tilde{E}\tilde{u}_u{}^2 + 2\tilde{F}\tilde{u}_u\tilde{v}_u + \tilde{G}\tilde{v}_u{}^2 \,, \\ F &= \tilde{E}\tilde{u}_u\tilde{u}_v + \tilde{F}(\tilde{u}_u\tilde{v}_v + \tilde{u}_v\tilde{v}_u) + \tilde{G}\tilde{v}_u\tilde{v}_v \,, \\ G &= \tilde{E}\tilde{u}_v{}^2 + 2\tilde{F}\tilde{u}_v\tilde{v}_v + \tilde{G}\tilde{v}_v{}^2 \,, \end{aligned}$$

which we may write in matrix notation as follows

$$\begin{pmatrix} E & F \\ F & G \end{pmatrix} = \begin{pmatrix} \tilde{u}_u & \tilde{v}_u \\ \tilde{u}_v & \tilde{v}_v \end{pmatrix} \begin{pmatrix} \tilde{E} & \tilde{F} \\ \tilde{F} & \tilde{G} \end{pmatrix} \begin{pmatrix} \tilde{u}_u & \tilde{u}_v \\ \tilde{v}_u & \tilde{v}_v \end{pmatrix} . \tag{3.32}$$

Lemma 2 *The expression* (3.20) *for area and* (3.21) *for integration of a function are independent of choice of local parametrisation* $\boldsymbol{x}$.

Proof Taking determinants in (3.32), we find that

$$EG - F^2 = (\tilde{E}\tilde{G} - \tilde{F}^2)\left|\frac{\partial(\tilde{u},\tilde{v})}{\partial(u,v)}\right|^2 , \tag{3.33}$$

where $\left|\frac{\partial(\tilde{u},\tilde{v})}{\partial(u,v)}\right|$ is the modulus of the determinant of the Jacobian matrix $\begin{pmatrix} \tilde{u}_u & \tilde{u}_v \\ \tilde{v}_u & \tilde{v}_v \end{pmatrix}$.

We now recall the formula for the change of variables in integration, namely, if $\tilde{Q} = h(Q)$, then for a function $\tilde{f}(\tilde{u},\tilde{v})$,

$$\iint_{\tilde{Q}} \tilde{f}(\tilde{u},\tilde{v})\, d\tilde{u}\, d\tilde{v} = \iint_Q \tilde{f}\,(\tilde{u}(u,v),\tilde{v}(u,v))\left|\left(\frac{\partial(\tilde{u},\tilde{v})}{\partial(u,v)}\right)\right| du\, dv \,,$$

from which we see, using (3.33) for the second equality, that

$$\begin{aligned} \iint_{\tilde{Q}} \sqrt{\tilde{E}\tilde{G} - \tilde{F}^2}\, d\tilde{u}\, d\tilde{v} &= \iint_Q \sqrt{\tilde{E}\tilde{G} - \tilde{F}^2}\left|\left(\frac{\partial(\tilde{u},\tilde{v})}{\partial(u,v)}\right)\right| du\, dv & (3.34) \\ &= \iint_Q \sqrt{EG - F^2}\, du\, dv \,. & (3.35) \end{aligned}$$

This shows that the definition of area we gave in §3.7 is independent of the choice of local parametrisation. A similar method may be used to show that our definition of integration of functions is also independent of choice of local parametrisation. □

Remark 3 If we had chosen to define the tangent plane at a point p to be the 2-plane spanned by $\boldsymbol{x}_u$ and $\boldsymbol{x}_v$ at that point, then (3.31) would show that this definition was independent of the choice of local parametrisation.

Exercises

3.1 Assume that $\boldsymbol{x}(u,v) = (u, v, u^2 + v^3)$ is a parametrisation of a surface S in $\mathbb{R}^3$. Is there a point on S at which the tangent plane to S is perpendicular to $(-1, 1, 0)$?

3.2 Let $\boldsymbol{x}(u, v)$ and S be as in Exercise 3.1. Show that $(2, -1, 3) \in S$ and that $(1, -1, 1) \in T_{(2,-1,3)}S$. Find a vector (a, b, c) in $T_{(2,-1,3)}S$ which is orthogonal to $(1, -1, 1)$.

3.3 Assuming that the equation $2x^2 - xy + 4y^2 = 1$ defines a surface S in $\mathbb{R}^3$, find a unit normal vector and a basis for the tangent plane at the point $p = (0, 1/2, 2)$.

3.4 Find those points on the ellipsoid

$$\frac{x^2}{a^2} + \frac{y^2}{b^2} + \frac{z^2}{c^2} = 1$$

at which the tangent plane is orthogonal to $(1, 1, 1)$.

3.5 Let S be the surface with equation $x^2 + y^2 - z^2 = 1$. Is there a point of S at which the tangent plane is orthogonal to $(1, 0, -1)$? Find all points of S at which the tangent plane is orthogonal to $(1, 1, 1)$.

3.6 Let a, b, c, be non-zero real numbers. Show that each of the equations

$$x^2 + y^2 + z^2 = ax\ ,$$
$$x^2 + y^2 + z^2 = by\ ,$$
$$x^2 + y^2 + z^2 = cz\ ,$$

defines a surface and that each pair of surfaces intersects orthogonally at all points of intersection. (Note, incidentally, that each of these surfaces is a sphere.)

3.7 Find the equation of the tangent plane based at the point $(a/2, b/2, c/\sqrt{2})$ of the ellipsoid

$$\frac{x^2}{a^2} + \frac{y^2}{b^2} + \frac{z^2}{c^2} = 1\,.$$

3.8 Let S be a surface parametrised by

$$\boldsymbol{x}(u, v) = (u \cos v, u \sin v, \log \cos v + u)\,, \quad -\frac{\pi}{2} < v < \frac{\pi}{2}\,, \quad u \in \mathbb{R}\,.$$

Find the coefficients E, F and G of the first fundamental form.

3.9 *(This exercise uses the optional material in the second half of §2.4.)* Let $f : \mathbb{R}^4 \to \mathbb{R}^2$ be given by

$$f(x_1, x_2, x_3, x_4) = ({x_1}^2 + {x_2}^2, {x_3}^2 + {x_4}^2)\,.$$

For each pair of positive real numbers r_1, r_2 show that $({r_1}^2, {r_2}^2) \in \mathbb{R}^2$ is a regular value of f. Let S be the surface in $\mathbb{R}^4$ determined by the equations ${x_1}^2 + {x_2}^2 = {r_1}^2$, ${x_3}^2 + {x_4}^2 = {r_2}^2$, and let

$$\boldsymbol{x}(u, v) = (r_1 \cos u, r_1 \sin u, r_2 \cos v, r_2 \sin v)\,, \quad 0 < u, v < 2\pi\,.$$

Use the Parametrisation Recognition Theorem (Theorem 4 in §2.5) to show that $\boldsymbol{x}(u, v)$ is a local parametrisation of S, and compute the coefficients of the first fundamental form.

3.10 Assume that

$$\boldsymbol{x}(u,v) = (\sinh v \sin u, -\sinh v \cos u, u)\,, \quad -\pi < u < \pi\,, \quad v \in \mathbb{R}\,,$$

is a local parametrisation of (part of) the surface S in $\mathbb{R}^3$ with equation $x \cos z + y \sin z = 0$ (which is easily checked using the Parametrisation Recognition Theorem), and let $\tilde{\boldsymbol{x}}(u,v)$ be the local parametrisation of a catenoid given by

$$\tilde{\boldsymbol{x}}(u,v) = (\cosh v \cos u, \cosh v \sin u, v)\,, \quad -\pi < u < \pi\,, \quad v \in \mathbb{R}\,.$$

Show that

$$\boldsymbol{x}_u = \tilde{\boldsymbol{x}}_v\,, \quad \boldsymbol{x}_v = -\tilde{\boldsymbol{x}}_u\,,$$

and, for each $\theta \in \mathbb{R}$, find the coefficients of the first fundamental form of the surface S_θ parametrised by

$$\boldsymbol{x}_\theta(u,v) = \cos\theta\, \boldsymbol{x}(u,v) + \sin\theta\, \tilde{\boldsymbol{x}}(u,v)\,, \quad -\pi < u < \pi\,, \quad v \in \mathbb{R}\,.$$

In particular, show that the coefficients of the first fundamental form of S_θ are independent of θ, and show also that the tangent planes to each of the surfaces S_θ at the point determined by (u, v) are parallel.

(The surface S is an example of a *helicoid* or *spiral staircase* surface, as discussed in Example 1 of §3.6. It turns out that both the helicoid and the catenoid are *minimal surfaces*, and, as we discuss in Chapter 9, this exercise illustrates a general property of such surfaces.)

3.11 Let $\boldsymbol{x}(u,v)$ and S be as in Exercise 3.8, and, for $v \in (-\frac{\pi}{2}, \frac{\pi}{2})$, let $\boldsymbol{\alpha}_v(t) = \boldsymbol{x}(t,v)$. Show that the length of $\boldsymbol{\alpha}_v$ from $t = u_0$ to $t = u_1$ is independent of v.

3.12 Let $\boldsymbol{x}(u,v)$ be a local parametrisation of a surface S with coefficients of the first fundamental form given by

$$E = 2 + \sinh^2 u\,, \quad F = \sinh u \sinh v\,, \quad G = 2 + \sinh^2 v\,.$$

(i) If $\boldsymbol{\alpha}(t) = \boldsymbol{x}(t,t)$, find the length of $\boldsymbol{\alpha}$ between $t = 0$ and $t = 1$.
(ii) If $\boldsymbol{\beta}(r) = \boldsymbol{x}(r,-r)$, show that $\boldsymbol{\alpha}$ and $\boldsymbol{\beta}$ intersect orthogonally.

3.13 Let $\boldsymbol{x}(u,v)$ be a local parametrisation of a surface S with coefficients of the first fundamental form given by

$$E = 1 + 4u^2\,, \quad F = \frac{4}{3}uv\,, \quad G = 1 + \frac{4}{9}v^2\,.$$

If θ is the angle of intersection of the curves

$$\boldsymbol{\alpha}(t) = \boldsymbol{x}(\cos t, \sin t)\,, \quad 0 < t < 2\pi\,,$$

and

$$\boldsymbol{\beta}(r) = \boldsymbol{x}(r, \sqrt{3}r)\,, \quad 0 < r\,,$$

show that $\cos\theta = -\dfrac{1}{2\sqrt{2}}$.

3.14 Let $\boldsymbol{x}(u, v)$ be the local parametrisation of the ellipsoid

$$\frac{x^2}{a^2} + \frac{y^2}{b^2} + \frac{z^2}{c^2} = 1$$

given by

$$\boldsymbol{x}(u, v) = (a \cos v \cos u, b \cos v \sin u, c \sin v), \quad -\frac{3\pi}{4} < u < \frac{5\pi}{4}, \ -\pi/2 < v < \pi/2.$$

If $L(u)$ is the length of the curve $\boldsymbol{\alpha}_u(t) = \boldsymbol{x}(u, t)$, $-\pi/2 < t < \pi/2$, show that L has stationary values at $u = -\pi/2, 0, \pi/2$ and π, and interpret this result geometrically. **It may help to use the result (often called 'differentiating under the integral sign') which says that if $f(u, v)$ is smooth then**

$$\frac{d}{du}\left(\int_a^b f(u, v)\, dv\right) = \int_a^b \frac{\partial f}{\partial u}\, dv\,.$$

3.15 Let $\boldsymbol{x} : \mathbb{R}^2 \to \mathbb{R}^3$ be the local parametrisation of the unit sphere $S^2(1)$ obtained in Example 2 of §3.4. Verify that the coefficients of the first fundamental form of $S^2(1)$ with respect to this local parametrisation are as given in that example.

3.16 Let $\boldsymbol{x} : \mathbb{R}^2 \to \mathbb{R}^3$ be the local parametrisation of the unit sphere $S^2(1)$ obtained in Example 2 of §3.4. By performing calculations as described in §3.3, and using the expressions for E, F and G given in (3.14), find the length of the image under $\boldsymbol{x}$ of the coordinate axis $v = 0$. (In fact, this image is a unit circle on $S^2(1)$ (with one point omitted), so you should get 2π for your answer!)

3.17 Let $f(z)$ be a holomorphic function of the complex variable z. If $\mathbb{C}^2$ is identified with $\mathbb{R}^4$ in the usual way then the graph of f is a surface in $\mathbb{R}^4$ which may be parametrised by $\boldsymbol{x}(z) = (z, f(z))$. Use the Cauchy–Riemann equations to show that $\boldsymbol{x}$ is an isothermal parametrisation.

3.18 Let

$$\boldsymbol{x}(u, v) = (v \cos u, v \sin u, u + v)\,, \quad u, v \in \mathbb{R}\,,$$

be a parametrisation of a surface S in $\mathbb{R}^3$, and let $\mathcal{F}$ be the family of curves obtained by intersecting S with the planes $z =$ constant.

Show that the angle of intersection θ of the coordinate curve $v =$ constant with the curve in the family $\mathcal{F}$ at $\left(\frac{\pi}{2\sqrt{3}}, \frac{\pi}{6}, \frac{\pi}{2}\right)$ satisfies $\cos\theta = \pm\frac{\pi^2}{\pi^2 + 9}$.

Find the orthogonal trajectories of $\mathcal{F}$ in S, and decide whether the orthogonal trajectory through $(1, 0, 1)$ passes through the point $(0\,, -\pi/2\,, 0)$.

3.19 Let S be the surface in $\mathbb{R}^3$ defined by the equation $x^2 + (y - 1)^2 = 1$, and let $\boldsymbol{x}(u, v)$ be the local parametrisation of S given by

$$\boldsymbol{x}(u, v) = (\sin u, 1 - \cos u, v)\,, \quad 0 < u < 2\pi\,, \quad v \in \mathbb{R}\,.$$

Let $\mathcal{F}$ be the family of curves on S, each member of which is obtained by intersecting S with the paraboloid $xz = \lambda y$, where λ is a constant.

Find a function $\psi(u, v)$ such that the family of curves given by $\psi(u, v) =$ constant gives the orthogonal trajectories of $\mathcal{F}$. Show that each curve in this family of orthogonal trajectories lies on a sphere with centre the origin in $\mathbb{R}^3$.

3.20 Let $\boldsymbol{x}(u, v)$ be a local parametrisation of a surface S. Show that, in the usual notation, the vector $\alpha\boldsymbol{x}_u + \beta\boldsymbol{x}_v$ bisects the angle between the coordinate curves if and only if

$$\sqrt{G}(\alpha E + \beta F) = \pm\sqrt{E}(\alpha F + \beta G)\,.$$

If

$$\boldsymbol{x}(u, v) = (u, v, u^2 - v^2)\,,$$

find a vector tangential to S which bisects the angle between the coordinate curves at the point $(1, 1, 0)$.

3.21 Let

$$\boldsymbol{x}(u, v) = (v\cos u, v\sin u, u)$$

be a local parametrisation of a helicoid. Find two families of curves on the helicoid which, at each point, bisect the angles between the coordinate curves.

3.22 This exercise provides a generalisation of Example 2 of §3.5. Let $\boldsymbol{x}(u, v)$ be a local parametrisation of a surface S and let $\mathcal{F}$ be the family of curves on S determined by $\phi(u, v) =$ constant. Show that the tangent vectors to the members of $\mathcal{F}$ are scalar multiples of $\phi_v\boldsymbol{x}_u - \phi_u\boldsymbol{x}_v$. Hence show that if $\boldsymbol{x}(u, v) = (u, v, \phi(u, v))$ is the standard parametrisation of the graph S of the smooth function $\phi(u, v)$ then the paths $\boldsymbol{x}\,(u(r), v(r))$ of steepest descent on S satisfy the equation

$$\frac{du}{dr}\phi_v - \frac{dv}{dr}\phi_u = 0\,.$$

As a particular example, find a smooth function $\psi(u, v)$ so that the paths of steepest descent on the graph of $\phi(u, v) = u^3 + v^3$ are given by $\psi(u, v) =$ constant.

3.23 Let $\boldsymbol{\alpha}(u) = (\cos u, \sin u, -1)$ and $\boldsymbol{\beta}(u) = (-\cos u, -\sin u, 1)$. Sketch the shape swept out by the lines through $\boldsymbol{\alpha}(u)$ in direction $\boldsymbol{\beta}(u)$ as described in §3.6 on ruled surfaces. This shape is the image of the map

$$\boldsymbol{x}(u, v) = \boldsymbol{\alpha}(u) + v\boldsymbol{\beta}(u)\,, \quad u, v \in \mathbb{R}\,.$$

Prove that $\boldsymbol{x}_u$ and $\boldsymbol{x}_v$ are linearly dependent if and only if $v = 1$. Mark the corresponding points on your sketch of the image of $\boldsymbol{x}$.

3.24 Let S be the hyperbolic paraboloid with equation $z = xy$. If $p = (p_1, p_2, p_3) \in S$, find conditions on the vector $\boldsymbol{v} = (v_1, v_2, v_3)$ so that the line in $\mathbb{R}^3$ through p in direction $\boldsymbol{v}$ should lie on S. Hence show that S is a doubly ruled surface. Show that the two rulings through a point p of S are mutually orthogonal if and only if p lies on the intersection of S with the coordinate axes of $\mathbb{R}^3$.

3.25 Let I be an open interval in $\mathbb{R}$ and let $\boldsymbol{\alpha} : I \to \mathbb{R}^3$ be a regular curve parametrised by arc length. Let $\boldsymbol{\beta} : I \to \mathbb{R}^3$ be a smooth map with $|\boldsymbol{\beta}(u)| = 1$ and $\boldsymbol{\beta}'(u)$ never zero.

Consider the ruled surface S which is the image of the map $\boldsymbol{x} : I \times \mathbb{R} \to \mathbb{R}^3$ given by

$$\boldsymbol{x}(u, v) = \boldsymbol{\alpha}(u) + v\boldsymbol{\beta}(u) .$$

(We recall from §3.6 that S is a surface near to any point at which $\boldsymbol{x}_u$ and $\boldsymbol{x}_v$ are linearly independent.)

(i) Show that $\boldsymbol{x}_u$ and $\boldsymbol{x}_v$ are linearly dependent at (u_0, v_0) if and only if

$$\boldsymbol{\alpha}'(u_0) \times \boldsymbol{\beta}(u_0) + v_0\boldsymbol{\beta}'(u_0) \times \boldsymbol{\beta}(u_0) = 0 .$$

(ii) Using the hypotheses on $\boldsymbol{\beta}$, show that there exists a curve $\boldsymbol{\gamma} : I \to \mathbb{R}^3$, with $\boldsymbol{\gamma}(u) \in S$ for all $u \in I$, such that

$$\boldsymbol{\gamma}'(u) \cdot \boldsymbol{\beta}'(u) = 0 , \quad \forall u \in I .$$

The curve $\boldsymbol{\gamma}$ is called the *striction curve* of the ruled surface S. (*Hint*: you may assume $\boldsymbol{\gamma}(u) = \boldsymbol{\alpha}(u) + \lambda(u)\boldsymbol{\beta}(u)$ for some smooth function $\lambda(u)$.)

(iii) Consider the map into S given by

$$\boldsymbol{y}(u, v) = \boldsymbol{\gamma}(u) + v\boldsymbol{\beta}(u) ,$$

where $\boldsymbol{\gamma}$ is the striction curve found in (ii). Show that $\boldsymbol{\beta}'(u)$ is parallel to $\boldsymbol{\gamma}'(u) \times \boldsymbol{\beta}(u)$ for all $u \in I$, and use this fact to show that the points where $\boldsymbol{y}_u$ and $\boldsymbol{y}_v$ are linearly dependent are all located on the striction curve $\boldsymbol{\gamma}$.

3.26 Determine all surfaces of revolution which are also ruled surfaces.

3.27 Compute the area of the southern hemisphere of the unit sphere $S^2(1)$ by parametrising $S^2(1)$ as a surface of revolution.

3.28 Let $U = \{(u, v) \in \mathbb{R}^2 : 0 < u < 1,\ 0 < v < 1\}$ be the open unit square. Let $\boldsymbol{x} : U \to \mathbb{R}^n$ be a local parametrisation of a surface S with coefficients of the first fundamental form given by

$$E = \frac{1}{u+v} + \frac{1}{(1-u)(1-v)} , \quad F = \frac{1}{u+v} , \quad G = \frac{1}{u+v} - \frac{1}{(1+u)(1+v)} .$$

Find the area of the image of U under $\boldsymbol{x}$.

3.29 Let $\tilde{S}$ be the cylinder in $\mathbb{R}^3$ with equation $x^2 + y^2 = 1$, and let f be the map obtained by restricting to $S^2(1)$ (minus the poles) horizontal projection onto $\tilde{S}$ radially away from the z-axis (Figure 3.18).

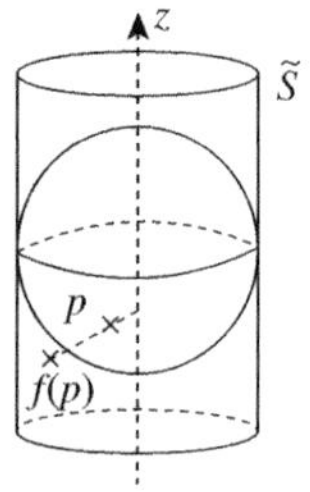

Figure 3.18 For Exercise 3.29

Use the standard parametrisation of $S^2(1)$ as a surface of revolution to show that f is area preserving in the sense that f maps any region in $S^2(1)$ (minus the poles) to a region in $\tilde{S}$ of the same area. This is a theorem due to Archimedes, who liked it so much that he had it engraved on his tombstone.

3.30 *(This exercise uses the optional material in the second half of §2.4.)* Let r_1, r_2 be positive real numbers and let S be the surface in $\mathbb{R}^4$ determined by the equations ${x_1}^2 + {x_2}^2 = {r_1}^2$ and ${x_3}^2 + {x_4}^2 = {r_2}^2$. Then S is the *flat torus* described in Example 4 of §2.4. Assuming that

$$\boldsymbol{x}(u, v) = (r_1 \cos u, r_1 \sin u, r_2 \cos v, r_2 \sin v) , \quad 0 < u < 2\pi , \quad 0 < v < 2\pi ,$$

is a local parametrisation of the whole of S with two circles removed, show that the area of the flat torus is $4\pi^2 r_1 r_2$.

3.31 *(This exercise uses material in the optional §3.8.)* The coordinate curves of a local parametrisation $\boldsymbol{x}(u, v)$ of a surface S form a *Tchebycheff net* if the lengths of the opposite sides of any quadrilateral formed by these curves are equal.

(i) Show that a necessary and sufficient condition for the coordinate curves to form a Tchebycheff net is

$$\frac{\partial E}{\partial v} = \frac{\partial G}{\partial u} = 0 .$$

To do this you may need to 'differentiate under the integral sign' as described in Exercise 3.14.

(ii) Suppose that the coordinate curves of a local parametrisation $\boldsymbol{x}(u, v)$ form a Tchebycheff net and consider the change of variables from (u, v) to $(\tilde{u}, \tilde{v})$ given by

$$\tilde{u}(u, v) = \int_{u_0}^{u} \sqrt{E(t, v)}dt , \quad \tilde{v}(u, v) = \int_{v_0}^{v} \sqrt{G(u, t)}dt ,$$

where (u_0, v_0) is a fixed base point. If $\tilde{\boldsymbol{x}}(\tilde{u}, \tilde{v})$ is the corresponding local parametrisation satisfying $\boldsymbol{x}(u, v) = \tilde{\boldsymbol{x}}\,(\tilde{u}(u, v), \tilde{v}(u, v))$, show that the two families of coordinate curves of $\tilde{\boldsymbol{x}}(\tilde{u}, \tilde{v})$ are the same as the two families of coordinate curves of $\boldsymbol{x}(u, v)$. Show also that if $\tilde{E}$, $\tilde{F}$, $\tilde{G}$ are the coefficients of the first fundamental form with respect to $\tilde{\boldsymbol{x}}(\tilde{u}, \tilde{v})$ then

$$\tilde{E} = 1 , \quad \tilde{F} = \cos\theta , \quad \tilde{G} = 1 ,$$

where θ is the angle between the coordinate curves.

4 Smooth maps

In this chapter we consider smooth maps defined on surfaces. There are two major reasons for doing so. Firstly, the fundamental importance and interest of isometries (which are smooth bijective maps between surfaces which preserve arc length of curves and area of regions) and of conformal maps (which preserve angles at which curves intersect); and secondly, the way in which a surface S curves in $\mathbb{R}^3$ may be studied using the Gauss map $N : S \to S^2(1) \subset \mathbb{R}^3$, which is obtained by taking the unit normal to S.

In Chapter 5 we show how the rate of change of the Gauss map may be used to describe the curvature of a surface in $\mathbb{R}^3$, and the importance of isometries will become clear when we discuss the Theorema Egregium in Chapter 6 and undertake the study of geodesics in Chapter 7.

Every isometry is a conformal map but the converse is false. Isometries are the analogues for surfaces of rigid motions of the plane, while conformal maps are the analogues of complex differentiable functions on the plane, since (away from points where the derivative vanishes) these maps are angle-preserving.

The idea in much of what we do in this chapter (and beyond) is to use local parametrisations to transfer the (local) study of maps defined on surfaces to the more familiar situation of smooth $\mathbb{R}^n$-valued maps defined on open subsets of Euclidean space. In this spirit, we begin the chapter by using local parametrisations to define smoothness for $\mathbb{R}^n$-valued maps defined on surfaces.

However, as we have prevously remarked, it is the surface S that is important; the role of the local parametrisations is to help describe and study S. So, it is important to check that any notions we define on S using a local parametrisation should be independent of the choice of local parametrisation. For instance, having used a local parametrisation to define the notion of a smooth map on a surface, we then obtain a geometrical characterisation which shows that our definition is independent of choice of local parametrisation.

We then define the derivative of a smooth map at a point, which, as will be expected, gives a linear approximation to the map near that point. Isometries and conformal maps are discussed next, with the role of local parametrisations again highlighted. Finally, in two appendices, we give some more substantial examples which may be omitted if time is short, since they are not needed in an essential way for the rest of the book.

Throughout the chapter we endeavour to find criteria, in terms of partial derivatives of maps defined on open subsets of Euclidean space, for the various sorts of maps on surfaces we consider.

4.1 Smooth maps between surfaces

Our first task is to define the notion of smoothness for a map $f : S \to \mathbb{R}^m$, where S is a surface in $\mathbb{R}^n$. For ease of application, we shall do this using a local parametrisation $\boldsymbol{x} : U \to S$, and then give a characterisation of smoothness which will show that our definition is independent of the choice of local parametrisation.

So, we say that a map $f : S \to \mathbb{R}^m$ is *smooth* at $\boldsymbol{x}(u, v) \in S$ if the composite $f\boldsymbol{x}$ is smooth at (u, v), and we say that f is *smooth* if it is smooth at each point of S (Figure 4.1). The important point here is that the concept of smoothness is fine for $f\boldsymbol{x}$ (in terms of partial derivatives as given in §2.1), since $f\boldsymbol{x}$ is an $\mathbb{R}^m$-valued function defined on an open subset (namely U) of a Euclidean space (namely $\mathbb{R}^2$). We are then using this to **define** the concept of smoothness for maps defined on a surface.

The first smooth map we consider is the *Gauss map* of a surface S in $\mathbb{R}^3$, which we now describe. The unit normal at a point $p \in S \subset \mathbb{R}^3$ is unique up to sign (Figure 4.2), and a smooth choice of unit normal $\boldsymbol{N}$ gives the Gauss map of the surface (again, unique up to sign). As already indicated, this is one of the most important smooth maps we consider, since its rate of change may be used to describe how S is curving in $\mathbb{R}^3$.

A smooth choice of $\boldsymbol{N}$ can always be made on a coordinate neighbourhood in S. Indeed, if $\boldsymbol{x} : U \to S$ is a local parametrisation of S, we may take

$$\boldsymbol{N}\boldsymbol{x} = \frac{\boldsymbol{x}_u \times \boldsymbol{x}_v}{|\boldsymbol{x}_u \times \boldsymbol{x}_v|}\,, \tag{4.1}$$

and, since $\boldsymbol{x}_u \times \boldsymbol{x}_v \neq 0$, the right hand side of (4.1) is smooth on U. It now follows from our definition that $\boldsymbol{N}$ is smooth at each point of $\boldsymbol{x}(U)$.

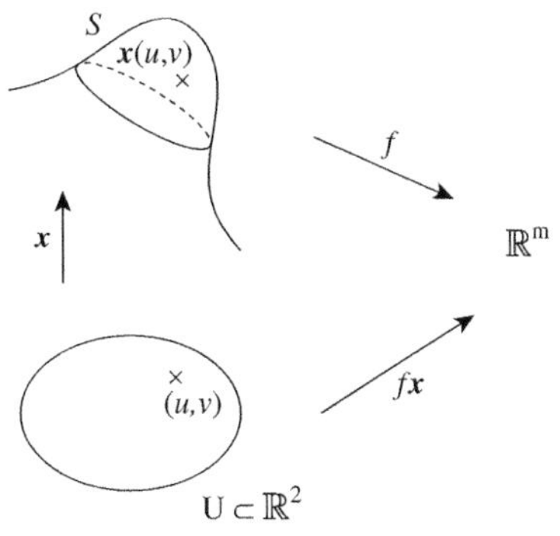

Figure 4.1 Smooth maps

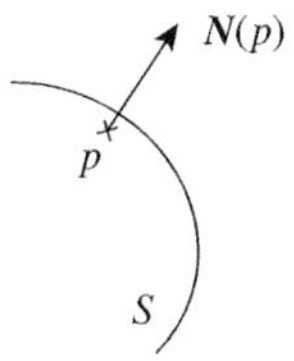

Figure 4.2 Gauss map

Example 1 (Gauss map of catenoid) As we saw in Example 1 of §2.3, the catenoid $x^2 + y^2 = \cosh^2 z$ may be parametrised as a surface of revolution by

$$\boldsymbol{x}(u, v) = (\cosh v \cos u, \cosh v \sin u, v)\,, \quad -\pi < u < \pi,\ v \in \mathbb{R}\,,$$

and a straightforward calculation shows that

$$\boldsymbol{x}_u \times \boldsymbol{x}_v = (\cosh v \cos u, \cosh v \sin u, -\cosh v \sinh v)\,.$$

Thus

$$\boldsymbol{N}\boldsymbol{x} = \frac{\boldsymbol{x}_u \times \boldsymbol{x}_v}{|\boldsymbol{x}_u \times \boldsymbol{x}_v|} = \frac{(\cos u, \sin u, -\sinh v)}{\cosh v}\,.$$

This example may be easily generalised.

Example 2 (Surface of revolution) If we parametrise a surface of revolution in the usual way, namely

$$\boldsymbol{x}(u, v) = (f(v)\cos u, f(v)\sin u, g(v))\,, \quad -\pi < u < \pi\,, \quad f(v) > 0\ \forall v\,,$$

then a short calculation on the lines given in Example 1 shows that

$$\boldsymbol{N}\boldsymbol{x} = \frac{(g' \cos u, g' \sin u, -f')}{(f'^2 + g'^2)^{1/2}}\,.$$

In Exercise 4.2 you are invited to prove that the Gauss map of a surface S of revolution maps parallels of S onto parallels of the unit sphere $S^2(1)$ and meridians to meridians.

It is often possible to define the Gauss map smoothly over the whole of a surface S in $\mathbb{R}^3$. For instance, for the sphere we can pick the outward unit normal, and, more generally, if a surface S in $\mathbb{R}^3$ is defined by an equation of the form $f(x, y, z) =$ constant where $\operatorname{grad} f$ is never zero on S, then $\boldsymbol{N} = \operatorname{grad} f/|\operatorname{grad} f|$ gives a smooth unit normal defined on the whole of S. However, if we form a *Möbius band* by taking a rectangular strip of paper, twisting it once and glueing the ends together (see Exercise 2.10 for a picture and an explicit example), then we cannot define a unit normal smoothly over the whole band. Such a surface is said to be *non-orientable*, while the sphere is an example of an *orientable* surface. A particular choice $\boldsymbol{N}$ of one of the two Gauss maps defined on the whole of an orientable surface S is called an *orientation* of S. We shall return to this topic (for surfaces in $\mathbb{R}^n$) towards the end of §7.1.

We now consider another useful type of smooth map defined on surfaces. Before giving the example, we note that if $\boldsymbol{x} : U \to S \subset \mathbb{R}^n$ is a local parametrisation of a surface S, then the coordinate neighbourhood $\boldsymbol{x}(U)$ is itself a surface (which may be covered by just one local parametrisation, namely $\boldsymbol{x}$).

Example 3 (Inverse of a local parametrisation) Since a local parametrisation $\boldsymbol{x} : U \to S \subset \mathbb{R}^n$ is a bijection onto its image $\boldsymbol{x}(U)$, there is an inverse map $\boldsymbol{x}^{-1} : \boldsymbol{x}(U) \to U$ which assigns to each point $p \in \boldsymbol{x}(U)$ the unique point $q \in U$ such that $\boldsymbol{x}(q) = p$. Then $\boldsymbol{x}\boldsymbol{x}^{-1}(p) = p$ for all $p \in \boldsymbol{x}(U)$ and $\boldsymbol{x}^{-1}\boldsymbol{x}(q) = q$ for all $q \in U$ (Figure 4.3). This latter condition, together with our definition of smoothness, implies that $\boldsymbol{x}^{-1}$ is a smooth map from the surface $\boldsymbol{x}(U)$ to U.

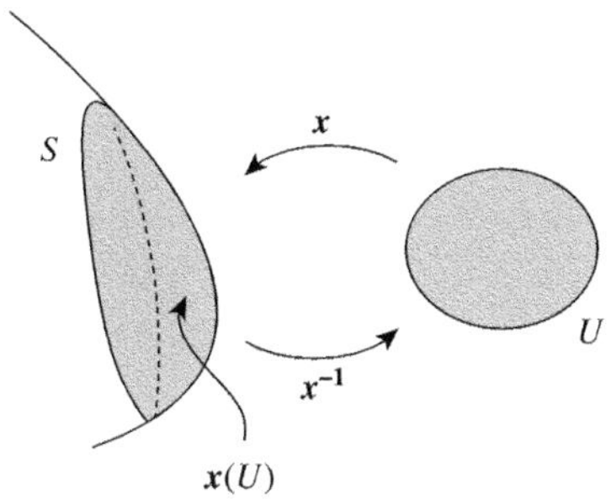

Figure 4.3 x^{-1} is smooth

Before we check that our definition of smoothness for maps defined on a surface S in $\mathbb{R}^n$ is independent of choice of local parametrisation, it will be convenient to recall (from §2.1) that an *open subset* of a surface S is defined to be the intersection of S with an open subset of $\mathbb{R}^n$. For example, the upper hemisphere of the unit sphere $S^2(1)$ is an open subset of $S^2(1)$, being the intersection of $S^2(1)$ with the open set $z > 0$ of $\mathbb{R}^3$. All coordinate neighbourhoods are open subsets of S, and any surface is an open subset of itself. We noted in Lemma 6 of §2.1 that a non-empty open subset of a surface is itself a surface. Finally, an *open neighbourhood* of a point p in S is simply an open subset of S containing p.

We now check that our definition of smoothness for maps defined on a surface S in $\mathbb{R}^n$ is independent of the choice of local parametrisation by showing that, locally at least, $f : S \to \mathbb{R}^m$ is smooth if and only if f is the restriction to S of a smooth map on $\mathbb{R}^n$. Rather more formally we show the following.

Proposition 4 *A map $f : S \to \mathbb{R}^m$ is smooth on an open neighbourhood of a point $p \in S$ if and only if there exists an open set W in $\mathbb{R}^n$ which contains p, and a smooth map $g : W \to \mathbb{R}^m$ such that $f(q) = g(q)$ for all $q \in S \cap W$.*

Proof Once again, it is important to note that we are investigating a new concept, namely smoothness of maps defined on a surface, and we are doing this by using the familiar idea of smoothness for maps defined on open subsets of Euclidean space.

Assume first, then, that a map g exists as in the statement of the proposition. Let $\boldsymbol{x}$ be a local parametrisation of S whose image contains p and assume, without loss of generality, that the image of $\boldsymbol{x}$ lies in $S \cap W$. Then $f\boldsymbol{x} = g\boldsymbol{x}$, so that $f\boldsymbol{x}$ is equal to the composite of two smooth maps defined on open subsets of Euclidean spaces, and so is smooth. Hence, by our definition of smoothness for maps on S, the map f is smooth on the image of $\boldsymbol{x}$.

Conversely, assume that f is smooth on an open neighbourhood of a point $p \in S$. Then there is a local parametrisation $\boldsymbol{x}$ of S whose image contains p and is such that $f\boldsymbol{x}$ is smooth. Let $\boldsymbol{F} : W \to \mathbb{R}^2$ be a smooth map for $\boldsymbol{x}$ as in condition (S2) of §2.1, and let $g = f\boldsymbol{x}\boldsymbol{F}$. Then $g : W \to \mathbb{R}^n$ is the composite of smooth maps $f\boldsymbol{x}$ and $\boldsymbol{F}$ defined on open subsets of Euclidean spaces and hence is smooth. Also, if $q = \boldsymbol{x}(u, v)$ then

$$g(q) = g\boldsymbol{x}(u, v) = f\boldsymbol{x}\boldsymbol{F}\boldsymbol{x}(u, v) = f\boldsymbol{x}(u, v) = f(q) ,$$

so that $f(q) = g(q)$ for all $q \in S \cap W$. □

This characterisation of a smooth map on S is often useful (and is independent of any choice of local parametrisation).

If a smooth map $f : S \to \mathbb{R}^m$ has its image on a surface $\tilde{S}$ in $\mathbb{R}^m$, then we say that f is a smooth map *from S to $\tilde{S}$*. For instance, if S is a surface in $\mathbb{R}^3$, then the Gauss map discussed earlier in this section is a smooth map from S to the unit sphere $S^2(1)$. As a particular example, if the outward unit normal is chosen, the Gauss map of the unit sphere is simply the identity map.

The next result follows quickly from the above geometrical characterisation of smoothness (and the corresponding result for smooth maps between Euclidean spaces).

Lemma 5 *Let $f : S_1 \to S_2$, $g : S_2 \to S_3$ be smooth maps between surfaces. Then the composite $gf : S_1 \to S_3$ is also smooth.*

4.2 The derivative of a smooth map

We begin by recalling the motivation behind the theory of differentiation of a smooth map $f : W \to \mathbb{R}^m$ defined on an open set W of $\mathbb{R}^n$. The idea is that, near any given point $p \in W$, the derivative $df_p : \mathbb{R}^n \to \mathbb{R}^m$ provides a linear approximation to f near p; the hope being that a study of the linear map df_p (linear maps are usually relatively easy to analyse) will yield information concerning the behaviour of the non-linear map f near p (which is usually more difficult to study directly). The Inverse Function Theorem is a classic example of this.

We wish to carry out the same procedure for a smooth map $f : S \to \mathbb{R}^m$, where S is a surface in $\mathbb{R}^n$. Again, perhaps the easiest (but maybe not the most satisfying) way of doing this is to use a local parametrisation $\boldsymbol{x}(u, v)$ of S to define the derivative, and then prove that the definition is independent of the local parametrisation used. So, if $\boldsymbol{x}(q) = p \in S$ then the *derivative* $df_p : T_pS \to \mathbb{R}^m$ of f at p is defined to be the linear map which has the following effect on the basis vectors $\boldsymbol{x}_u, \boldsymbol{x}_v$ of T_pS:

$$df_p(\boldsymbol{x}_u) = \left.\frac{\partial(f\boldsymbol{x})}{\partial u}\right|_q , \qquad df_p(\boldsymbol{x}_v) = \left.\frac{\partial(f\boldsymbol{x})}{\partial v}\right|_q .$$

We often omit to mention the point at which we are differentiating; we simply write

$$df(\boldsymbol{x}_u) = (f\boldsymbol{x})_u , \qquad df(\boldsymbol{x}_v) = (f\boldsymbol{x})_v , \tag{4.2}$$

thus enabling us to find the derivative of a smooth map f defined on a surface by using the partial derivatives of a smooth map, namely $f\boldsymbol{x}$, defined on open subset of the plane.

When a parametrisation $\boldsymbol{x}$ has been chosen it is usual to omit mention of $\boldsymbol{x}$, and write $f(u, v)$ rather than $f\boldsymbol{x}(u, v)$, thus regarding f as a function of u and v. In a similar spirit, we write f_u and f_v rather than $(f\boldsymbol{x})_u$ and $(f\boldsymbol{x})_v$.

For instance, using this convention, we would write (4.1) as

$$\boldsymbol{N} = \frac{\boldsymbol{x}_u \times \boldsymbol{x}_v}{|\boldsymbol{x}_u \times \boldsymbol{x}_v|} . \tag{4.3}$$

Example 1 (Gauss map of catenoid) This continues Example 1 of §4.1, where we found that

$$N = \frac{(\cos u, \sin u, -\sinh v)}{\cosh v} .$$

Thus

$$N_u = \frac{\partial}{\partial u}\left(\frac{(\cos u, \sin u, -\sinh v)}{\cosh v}\right) = \frac{(-\sin u, \cos u, 0)}{\cosh v} ,$$

while a short calculation shows that

$$N_v = \frac{(-\cos u \sinh v, -\sin u \sinh v, -1)}{\cosh^2 v} .$$

As we shall see in §4.4, the Gauss map of the catenoid is very special in that it is an angle-preserving map.

Returning to the general situation, the conventions described above lead us to write (4.2) as

$$df(x_u) = f_u , \qquad df(x_v) = f_v , \tag{4.4}$$

which gives us the following explicit formula for the derivative of the map f applied to a general tangent vector $ax_u + bx_v$:

$$df(ax_u + bx_v) = af_u + bf_v . \tag{4.5}$$

Unless stated otherwise, we shall use the above conventions from now on.

We now show that, as we would hope, the tangent vector of a smooth curve α on S is mapped by df to the tangent vector of the image curve $f\alpha$ (Figure 4.4).

Proposition 2 *Let $f : S \to \mathbb{R}^m$ be a smooth map defined on a surface S in $\mathbb{R}^n$, and let α be a smooth curve on S. Then*

$$df(\alpha') = (f\alpha)' . \tag{4.6}$$

Proof Let $x(u, v)$ be a local parametrisation of S and let $\alpha(t) = x(u(t), v(t))$. Then

$$df(\alpha') = df(u'x_u + v'x_v) .$$

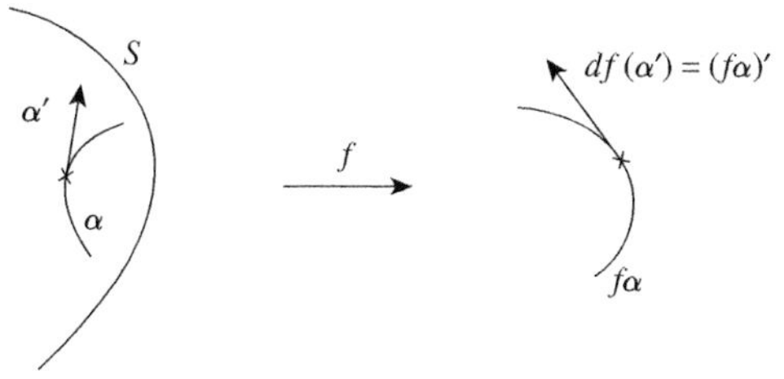

Figure 4.4 df maps tangent vectors to tangent vectors

Also, $f\boldsymbol{\alpha}(t) = f\boldsymbol{x}\,(u(t), v(t))$, so, using the chain rule and the notation mentioned in the paragraph preceding Example 1,

$$(f\boldsymbol{\alpha})' = u' f_u + v' f_v \,.$$

Equation (4.6) now follows immediately from (4.5). □

In a similar spirit to earlier notational conventions, when a regular curve $\boldsymbol{\alpha}(t)$ has been chosen, it is usual to write f' (or df/dt) rather than $(f\boldsymbol{\alpha})'$ for the rate of change of f along $\boldsymbol{\alpha}$. Thus (4.6) will become

$$df(\boldsymbol{\alpha}') = f' \,. \tag{4.7}$$

The geometrical characterisation of derivative given in Proposition 2 shows that our definition of derivative as the linear map defined using (4.2) is independent of choice of local parametrisation. It may also be used to prove the following two propositions.

Proposition 3 *Let S be a surface in $\mathbb{R}^n$ and let f be the restriction to S of a smooth map $g : W \to \mathbb{R}^m$, where W is an open subset of $\mathbb{R}^n$. Then the derivative df_p of f at a point $p \in W \cap S$ is the restriction to T_pS of the derivative of g at p.*

The above proposition sometimes gives a quick way of finding the derivative of a smooth map f defined on S, and may also be useful if we have no convenient local parametrisations of S.

Proposition 4 *Let $f : S \to \mathbb{R}^m$ be smooth. If the image of f is contained in a surface $\tilde{S}$ in $\mathbb{R}^m$, then, for each $p \in S$, the image of df_p lies in $T_{f(p)}\tilde{S}$. Hence the derivative of f at p is a linear map $df_p : T_pS \to T_{f(p)}\tilde{S}$.*

Proof Let $p \in S$ and $\boldsymbol{X} \in T_pS$. Then $\boldsymbol{X} = \boldsymbol{\alpha}'(0)$ for some smooth curve $\boldsymbol{\alpha}(t)$ in S with $\boldsymbol{\alpha}(0) = p$. It follows from (4.6) that $df_p(\boldsymbol{X})$ is tangential to the smooth curve $f\boldsymbol{\alpha}$ in $\tilde{S}$ through $f(p)$. The definition of tangent vectors given at the start of §3.1 now shows that $df_p(\boldsymbol{X}) \in T_{f(p)}\tilde{S}$. □

Example 5 (Sphere and ellipsoid) Let $S^2(1)$ denote the unit sphere in $\mathbb{R}^3$, and, for positive real numbers a, b, c, let $\tilde{S}$ be the ellipsoid with equation $\dfrac{x^2}{a^2} + \dfrac{y^2}{b^2} + \dfrac{z^2}{c^2} = 1$. Let $f : S^2(1) \to \mathbb{R}^3$ be given by $f(x, y, z) = (ax, by, cz)$. Then f is smooth since it is the restriction to $S^2(1)$ of the smooth map $g : \mathbb{R}^3 \to \mathbb{R}^3$ given by the same formula, and it is clear that f gives a bijective correspondence between the points of $S^2(1)$ and $\tilde{S}$. Since g is a linear map, the derivative of g at any point is simply g itself (the best linear approximation to a linear map is the linear map), so the derivative df_p of f at $p \in S^2(1)$ is just the restriction to $T_pS^2(1)$ of g. If $p = (x, y, z) \in S^2(1)$ then, as noted in Example 6 of §3.1, $T_pS^2(1)$ is the plane of vectors orthogonal to p, while, from Example 7 of §3.1, $T_{f(p)}\tilde{S}$ is the plane of vectors orthogonal to $(x/a, y/b, z/c)$. It is a nice exercise (see Exercise 4.8) to check directly that df_p maps $T_pS^2(1)$ to $T_{f(p)}\tilde{S}$.

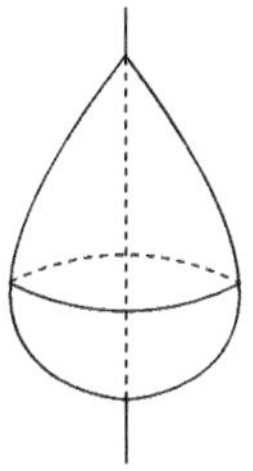

Figure 4.5 Poles of a surface of revolution

Example 6 (Surface of revolution) Recall from Example 3 of §2.3 that a *pole* of a surface of revolution S in $\mathbb{R}^3$ is a point p at which S intersects its axis of rotation, which, as usual, we assume to be the z-axis. Since the pole lies on the z-axis, it is fixed under rotations about the z-axis, and, since these rotations are linear maps and hence are equal to their derivatives, it follows that, if p is not a singular point of the surface, then T_pS is also (setwise) fixed by (the derivatives of) these rotations. Hence the tangent space T_pS at p is orthogonal to the axis of rotation (Figure 4.5). Examples of surfaces of revolution whose poles are not singular points are provided by the spheres, or, more generally, ellipsoids of revolution with equation

$$\frac{1}{a^2}(x^2 + y^2) + \frac{z^2}{c^2} = 1 ,$$

where a and c are positive real numbers.

We saw in Lemma 5 of §4.1 that the composite of smooth maps between surfaces is smooth. We conclude this section by noting that the geometrical characterisation of derivative given in Proposition 2 shows that the usual chain rule holds for smooth maps, that is to say the derivative of the composite is the composite of the derivatives. For future use, we state this formally as a theorem.

Theorem 7 (Chain rule) *Let $f : S_1 \to S_2$, $g : S_2 \to S_3$ be smooth maps between surfaces. Then the composite $gf : S_1 \to S_3$ is smooth, and if $p \in S_1$ then $d(gf)_p = dg_{f(p)}df_p$.*

We end with a summary of §4.1 and §4.2. We used a local parametrisation to transfer the notion of a smooth map and its derivative from the familiar one for maps between Euclidean spaces to the new one for maps defined on surfaces. We then obtained (in Proposition 4 of §4.1 and Proposition 2 of this section) characterisations of these new concepts. In particular, we showed that, locally at least, a smooth map f on a surface is the restriction to the surface of a smooth map g defined on the containing Euclidean space. The derivative of f is then the restriction to the tangent space of the surface of the derivative of g. The characterisations show that our original definition of smooth map and its derivative on a surface is independent of choice of local parametrisation, and also enabled us to deduce several important properties of smooth maps on surfaces and the derivatives of these maps.

4.3 Local isometries

In the next three sections we study two geometrically interesting and important types of map between surfaces. However, for convenience, we begin with a few basic definitions. We first recall from set theory that a bijective map $f : A \to B$ between sets A and B has an inverse map $f^{-1} : B \to A$ which assigns to each point $q \in B$ the unique point $p \in A$ such that $f(p) = q$. Then

$$f\,f^{-1}(q) = q\,,\ \forall q \in B\,, \quad \text{and} \quad f^{-1}f(p) = p\,,\ \forall p \in A\,.$$

In particular, a smooth bijective map $f : S \to \tilde{S}$ between surfaces S and $\tilde{S}$ has an inverse map $f^{-1} : \tilde{S} \to S$, and if f^{-1} is also smooth then f is called a *diffeomorphism*. The Inverse Function Theorem may be used to show that a smooth bijective map f is a diffeomorphism if and only if the derivative df of f is a linear isomorphism at each point of S, or, equivalently once a local parametrisation $\boldsymbol{x}(u, v)$ has been chosen, if and only if f_u and f_v are linearly independent at each point.

The chain rule shows that if $f : S \to \tilde{S}$ is a diffeomorphism and if $p \in S$ then

$$d(f^{-1})|_{f(p)} = (df|_p)^{-1}\,. \tag{4.8}$$

If there exists a diffeomorphism $f : S \to \tilde{S}$, then the surfaces S and $\tilde{S}$ are said to be *diffeomorphic*. As far as properties concerning differentiability are concerned, the surfaces are essentially indistinguishable.

We now consider smooth maps between surfaces which preserve the length of curves on the surfaces. A smooth map $f : S \to \tilde{S}$ is a *local isometry* if whenever $\boldsymbol{\alpha}$ is a smooth curve of finite length on S then $f\boldsymbol{\alpha}$ is a curve on $\tilde{S}$ of the same length. We now find conditions on the derivative of f which enable us to decide whether a given map f is a local isometry.

It follows from (4.6) that f is a local isometry if and only if df preserves the length of tangent vectors, in that

$$|df(\boldsymbol{X})| = |\boldsymbol{X}|\,, \quad \text{for all vectors } \boldsymbol{X} \text{ tangential to } S\,, \tag{4.9}$$

and if we apply (the square of) (4.9) to tangent vectors $\boldsymbol{X}_1$, $\boldsymbol{X}_2$ and $\boldsymbol{X}_1 + \boldsymbol{X}_2$, we obtain the following proposition. Here, as usual, "." denotes the inner product.

Proposition 1 *A smooth map $f : S \to \tilde{S}$ is a local isometry if and only if, for all vectors X_1, X_2 tangential to S at any point $p \in S$,*

$$df(\boldsymbol{X}_1).df(\boldsymbol{X}_2) = \boldsymbol{X}_1.\boldsymbol{X}_2\,. \tag{4.10}$$

For each $p \in S$, $df_p : T_pS \to T_{f(p)}\tilde{S}$ is linear, so, in order to check condition (4.10), it suffices to take a basis of T_pS and check that (4.10) holds whenever $\boldsymbol{X}_1$, $\boldsymbol{X}_2$ are vectors in that basis. Such a basis is provided by $\{\boldsymbol{x}_u, \boldsymbol{x}_v\}$, where $\boldsymbol{x}(u, v)$ is a local parametrisation of S, so the following proposition is immediate from (4.4). This proposition provides a very useful criterion in terms of partial derivatives for deciding whether or not a given map is a local isometry.

Proposition 2 *Let $\boldsymbol{x} : U \to S$ be a local parametrisation of S. A map $f : S \to \tilde{S}$ is a local isometry on $\boldsymbol{x}(U)$ if and only if*

$$f_u \cdot f_u = E\,, \quad f_u \cdot f_v = F\,, \quad f_v \cdot f_v = G\,, \tag{4.11}$$

where E, F and G are the coefficients of the first fundamental form of S.

Example 3 (Plane and cylinder) Let S be the xz-plane in $\mathbb{R}^3$ and let $\tilde{S}$ be the cylinder in $\mathbb{R}^3$ with equation $x^2 + y^2 = 1$. Let $f : S \to \tilde{S}$ be the map defined by

$$f(x, 0, z) = (\cos x, \sin x, z)\,.$$

Then f wraps the xz-plane round the cylinder an infinite number of times (Figure 4.6). In terms of the parametrisation of S given by $\boldsymbol{x}(u, v) = (u, 0, v)$, we find that

$$f(u, v) = (\cos u, \sin u, v)\,,$$

and condition (4.11) is easy to check.

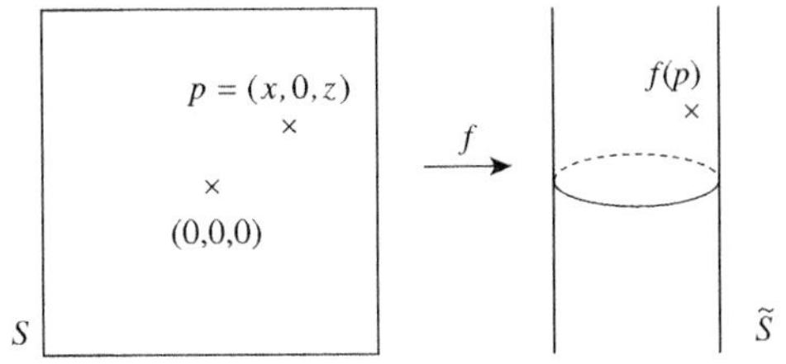

Figure 4.6 Wrapping a plane round a cylinder

Example 4† (Flat torus) This example concerns a surface in $\mathbb{R}^4$, and may be omitted if it is wished to concentrate on the geometry of surfaces in $\mathbb{R}^3$. Let $\tilde{S}$ be the surface in $\mathbb{R}^4$ discussed in Example 4 of §2.4. This is the product of two plane circles, and is defined by the equations $x_1{}^2 + x_2{}^2 = r_1{}^2$, $x_3{}^2 + x_4{}^2 = r_2{}^2$, where r_1 and r_2 are positive real numbers. As mentioned in Example 4 of §2.4, this surface is differentiably equivalent to a torus of revolution in $\mathbb{R}^3$. Let S be the xy-plane in $\mathbb{R}^3$, and let $f : S \to \tilde{S}$ be given by

$$f(x, y, 0) = \left(r_1 \cos \frac{x}{r_1}, r_1 \sin \frac{x}{r_1}, r_2 \cos \frac{y}{r_2}, r_2 \sin \frac{y}{r_2} \right).$$

Using the local parametrisation of S given by $\boldsymbol{x}(u, v) = (u, v, 0)$, we quickly see that $f_u \cdot f_u = f_v \cdot f_v = 1$, while $f_u \cdot f_v = 0$, from which it follows that f is a local isometry from the whole of the plane onto the flat torus. This has the effect of wrapping the plane round the flat torus (a "doubly infinite" number of times) in such a way that lengths of curves are preserved.

A bijective local isometry $f : S \to \tilde{S}$ is called an *isometry*; such a map provides a diffeomorphism between S and $\tilde{S}$ such that corresponding curves have the same lengths, and, as we shall see, corresponding regions have the same area. The inverse of an isometry is also an isometry and, in this situation, the surfaces S and $\tilde{S}$ are said to be *isometric*. As far as intrinsic metric properties are concerned, isometric surfaces are essentially indistinguishable.

Example 5 Let S denote the xz-plane in $\mathbb{R}^3$ and let f be the map on S defined by

$$f(x,0,z) = (\alpha(x), \beta(x), z) \ , \quad z \in \mathbb{R} \, ,$$

where $x \mapsto (\alpha(x), \beta(x))$ is a regular curve in $\mathbb{R}^2$ parametrised by arc length (Figure 4.7).

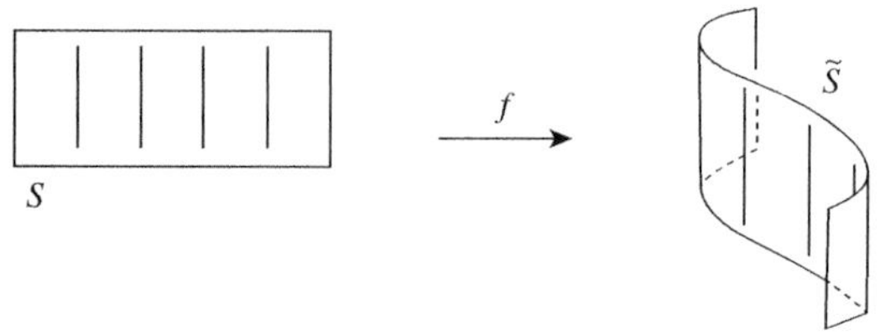

Figure 4.7 Isometric surfaces

If the curve has no self-intersections then the image of f is a surface $\tilde{S}$ in $\mathbb{R}^3$ and $f : S \to \tilde{S}$ is an isometry. (Actually, as a small technicality, for $\tilde{S}$ to be a surface, we need the curve $x \mapsto (\alpha(x), \beta(x))$ to be what is called a *proper map*.)

We now seek to justify the terminology "local isometry". In a natural sense, an isometry $f : S \to \tilde{S}$ preserves distances apart of pairs of points on the two surfaces. However, a local isometry from S to $\tilde{S}$ gives an isometry only between sufficiently small open subsets of S and $\tilde{S}$, and, as can be seen from Example 3, only preserves the distance apart of points locally. Similarly, if you have read Example 4 you will see that the local metric geometry of the flat torus in that example is the same as that of the plane (which is the reason for the name "flat" torus), but globally the metric geometry is very different. As an example of this (see Exercise 4.12), any two points of $\tilde{S}$ may be joined by a curve on $\tilde{S}$ of length at most $\pi\sqrt{{r_1}^2 + {r_2}^2}$.

We remark that the *rigidity* of surfaces in $\mathbb{R}^3$ is an interesting question with a long history and an extensive associated literature. A surface S in $\mathbb{R}^3$ is *rigid* if, whenever there is an isometry f between S and a surface $\tilde{S}$ in $\mathbb{R}^3$, then this isometry is the restriction to S of a rigid motion of $\mathbb{R}^3$ (possibly followed by a reflection). It is clear from Example 5 that a plane is not rigid; however, ellipsoids are rigid, which helps explain why an egg is strong but a piece of paper is very floppy!

We finish this section by showing that isometries are area-preserving maps.

Proposition 6 *Let $f : S \to \tilde{S}$ be an isometry between surfaces S and $\tilde{S}$. Then, for a region R of S, the area of the image $f(R)$ is equal to the area of R.*

Proof Let $\boldsymbol{x} : U \to S$ be a local parametrisation of S, and let $q \in$ U. It follows from Proposition 2, and from Theorem 3 of §2.5, that there is an open neighbourhood U_0 of q in U such that the restriction $\tilde{\boldsymbol{x}}$ of $f\boldsymbol{x}$ to U_0 is a local parametrisation of $\tilde{S}$ whose coefficients of the first fundamental form are the same as those of $\boldsymbol{x}$. The proposition now follows from the discussion of area in §3.7. □

The converse of Proposition 6 is not true; Exercise 3.29 gives an example of an area-preserving map which is not an isometry.

4.4 Conformal maps

We begin by defining the main objects of study in this section. Let $f : S \to \tilde{S}$ be a smooth map between surfaces S and $\tilde{S}$ such that $df(\boldsymbol{X})$ is non-zero whenever $\boldsymbol{X}$ is a non-zero tangent vector to S, and let $\boldsymbol{\alpha}$, $\boldsymbol{\beta}$ be regular curves on S intersecting at an angle θ. Then $f\boldsymbol{\alpha}$, $f\boldsymbol{\beta}$ are regular curves on $\tilde{S}$, but their angle of intersection will usually be different from θ. If f is such that the angle of intersection of $f\boldsymbol{\alpha}$, $f\boldsymbol{\beta}$ is the same as the angle of intersection of $\boldsymbol{\alpha}$ and $\boldsymbol{\beta}$ for all intersecting regular curves $\boldsymbol{\alpha}$ and $\boldsymbol{\beta}$ on S (Figure 4.8), then we say that f is a *conformal* map. Informally, conformal maps are angle-preserving. In this section, we find conditions on the derivative of f which enable us to decide whether f is conformal.

We have already seen many examples of conformal maps; if we identify $\mathbb{R}^2$ with the xy-plane in $\mathbb{R}^3$ in the usual way via $(u, v) \equiv (u, v, 0)$, then Lemma 4 of §3.4 shows that an isothermal local parametrisation is conformal. It is an isometry onto its image if and only if $E = G = 1$ (and $F = 0$).

We now find conditions on the derivative of f which enable us to decide whether a given map f is conformal. It turns out that all local isometries are conformal, but the converse is not true. It is clear from (4.6) that a smooth map f is conformal if and only if, for all $p \in S$, the derivative $df_p : T_pS \to T_{f(p)}\tilde{S}$ is angle preserving in that if $\boldsymbol{X}_1$, $\boldsymbol{X}_2$ are non-zero tangent vectors to S at p then $df_p(\boldsymbol{X}_1)$, $df_p(\boldsymbol{X}_2)$ are also non-zero and the angle between them is equal to the angle between $\boldsymbol{X}_1$ and $\boldsymbol{X}_2$.

Proposition 1 *A smooth map $f : S \to \tilde{S}$ is a conformal map if and only if there exists a strictly positive function $\lambda : S \to \mathbb{R}$ such that, for all vectors $\boldsymbol{X}_1$, $\boldsymbol{X}_2$, tangential to S at any point $p \in S$, we have*

$$df(\boldsymbol{X}_1).df(\boldsymbol{X}_2) = \lambda^2 \boldsymbol{X}_1.\boldsymbol{X}_2 \,. \tag{4.12}$$

Proof This is an exercise in linear algebra, which we include here for those interested. We begin by assuming that (4.12) holds at each $p \in S$, and, for brevity, we let ℓ denote the

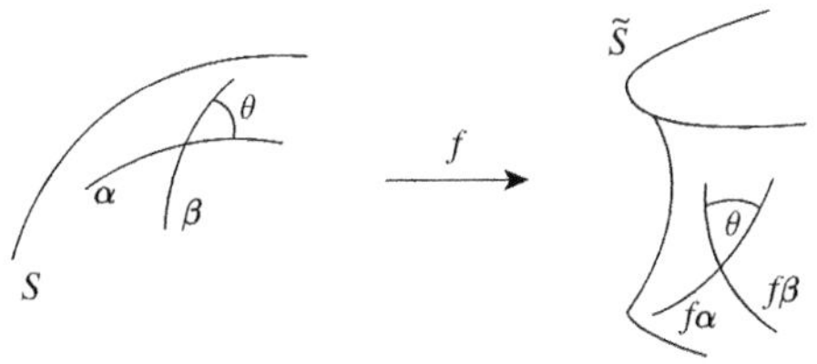

Figure 4.8 Conformal maps preserve angle of intersection

linear map df_p. We let θ (resp. ϕ) be the angle between non-zero vectors $\boldsymbol{X}_1, \boldsymbol{X}_2 \in T_pS$ (resp. $\ell(\boldsymbol{X}_1), \ell(\boldsymbol{X}_2) \in T_{f(p)}\tilde{S}$). Then

$$\cos\phi = \frac{\ell(\boldsymbol{X}_1).\ell(\boldsymbol{X}_2)}{|\ell(\boldsymbol{X}_1)|\,|\ell(\boldsymbol{X}_2)|} = \frac{\lambda^2\boldsymbol{X}_1.\boldsymbol{X}_2}{\lambda^2|\boldsymbol{X}_1|\,|\boldsymbol{X}_2|} = \cos\theta\,,$$

so that $\theta = \phi$ and f is conformal.

Conversely, assume that f is conformal and let $\{\boldsymbol{Y}_1, \boldsymbol{Y}_2\}$ be an orthonormal basis of T_pS. Then any non-zero tangent vector $\boldsymbol{X}$ at p may be written as

$$\boldsymbol{X} = |\boldsymbol{X}|(\cos\theta\boldsymbol{Y}_1 \pm \sin\theta\boldsymbol{Y}_2)\,,$$

where θ is the angle between $\boldsymbol{X}$ and $\boldsymbol{Y}_1$. Conformality implies that the angle between $\ell(\boldsymbol{X})$ and $\ell(\boldsymbol{Y}_1)$ is also θ, and that $\ell(\boldsymbol{Y}_1)$ and $\ell(\boldsymbol{Y}_2)$ are orthogonal. Hence, using linearity of ℓ,

$$\cos\theta = \frac{\ell(\boldsymbol{X}).\ell(\boldsymbol{Y}_1)}{|\ell(\boldsymbol{X})|\,|\ell(\boldsymbol{Y}_1)|} = \frac{\cos\theta|\boldsymbol{X}|\,|\ell(\boldsymbol{Y}_1)|^2}{|\ell(\boldsymbol{X})|\,|\ell(\boldsymbol{Y}_1)|} = \frac{\cos\theta|\boldsymbol{X}|\,|\ell(\boldsymbol{Y}_1)|}{|\ell(\boldsymbol{X})|}\,.$$

Hence, if we put $|\ell(\boldsymbol{Y}_1)| = \lambda$, we see that, for all tangent vectors $\boldsymbol{X}$,

$$|\ell(\boldsymbol{X})| = \lambda|\boldsymbol{X}|\,.$$

The proof of the proposition may now be completed as for Proposition 1 in §4.3 by applying the (square of the) above equation to tangent vectors $\boldsymbol{X}_1$, $\boldsymbol{X}_2$ and $\boldsymbol{X}_1 + \boldsymbol{X}_2$. □

The function λ is called the *conformal factor* of the conformal map f. The following corollary is immediate from Proposition 1 in this section and Proposition 1 in §4.3.

Corollary 2 *A map $f : S \to \tilde{S}$ is a local isometry if and only if f is a conformal map with conformal factor equal to 1.*

The next proposition follows from Proposition 1 for the same reason that Proposition 2 in §4.3 follows from Proposition 1 of that section. It provides a useful criterion in terms of partial derivatives for determining whether a given map is conformal.

Proposition 3 *Let $\boldsymbol{x} : U \to S$ be a local parametrisation of S. A map $f : S \to \tilde{S}$ is conformal on $\boldsymbol{x}(U)$ if and only if there is a strictly positive function $\lambda : U \to \mathbb{R}$ such that*

$$f_u.f_u = \lambda^2 E\,, \quad f_u.f_v = \lambda^2 F\,, \quad f_v.f_v = \lambda^2 G\,. \tag{4.13}$$

Moreover, f is a local isometry on $\boldsymbol{x}(U)$ if and only if $\lambda = 1$.

Example 4 (Gauss map of catenoid) We have seen (in Example 1 of §3.4) that the coefficients of the first fundamental form of the catenoid with the standard parametrisation given in Example 1 of §4.1 are

$$E = G = \cosh^2 v\,, \quad F = 0\,.$$

Easy calculations using the expressions for $\boldsymbol{N}_u$ and $\boldsymbol{N}_v$ obtained in Example 1 of §4.2 show that $\boldsymbol{N}_u.\boldsymbol{N}_u = \boldsymbol{N}_v.\boldsymbol{N}_v = \cosh^{-2} v$, while $\boldsymbol{N}_u.\boldsymbol{N}_v = 0$. It now follows from

Proposition 3 that $\boldsymbol{N}$ is a conformal map from the catenoid to the unit sphere $S^2(1)$ with conformal factor $\cosh^{-2} v$. In fact, as you are asked to prove in Exercise 4.4, $\boldsymbol{N}$ provides a conformal diffeomorphism from the catenoid to the 2-sphere minus the two poles.

Example 5 (The plane) We identify $\mathbb{C}$ with the xy-plane in $\mathbb{R}^3$ in the usual way. Recall from the theory of complex analysis that a complex differentiable function $f(z)$ is conformal at those points where the complex derivative $f'(z)$ is non-zero, and, in this case, the conformal factor is $|f'(z)|$. The conformal diffeomorphisms of $\mathbb{C}$ consist of complex functions of the form $z \mapsto az + b$, where $a \in \mathbb{C} \setminus \{0\}$ and $b \in \mathbb{C}$, together with the conjugates of such functions. The conformal factor λ in this case is just $|a|$, so it follows that the isometries of the plane form the *Euclidean group*, which is generated by rotations about the origin, translations, and reflection in the real axis.

It is clear from the definitions that the composite of two conformal maps is conformal, and the composite of two (local) isometries is a (local) isometry. It follows from Proposition 1 that the conformal factor of the composite of two conformal maps is equal to the product of the conformal factors of the two maps at the appropriate points.

4.5 Conformal maps and local parametrisations

Local parametrisations sometimes provide a useful way of constructing conformal diffeomorphisms and isometries between (open subsets of) surfaces. The process described in the following proposition is illustrated in Figure 4.9.

Proposition 1 *Suppose $\boldsymbol{x} : U \to S$, $\tilde{\boldsymbol{x}} : U \to \tilde{S}$ are local parametrisations of surfaces $S, \tilde{S}$ and that E, F, G and $\tilde{E}, \tilde{F}, \tilde{G}$ are the corresponding coefficients of the first fundamental forms. Then the bijective correspondence f from $\boldsymbol{x}(U)$ to $\tilde{\boldsymbol{x}}(U)$ given by*

$$f(\boldsymbol{x}(u, v)) = \tilde{\boldsymbol{x}}(u, v), \quad (u, v) \in U, \tag{4.14}$$

is a conformal diffeomorphism if and only if there exists a strictly positive function $\lambda(u, v)$ such that

$$\tilde{E} = \lambda^2 E, \quad \tilde{F} = \lambda^2 F, \quad \tilde{G} = \lambda^2 G.$$

Moreover, f is an isometry if and only if $\lambda = 1$.

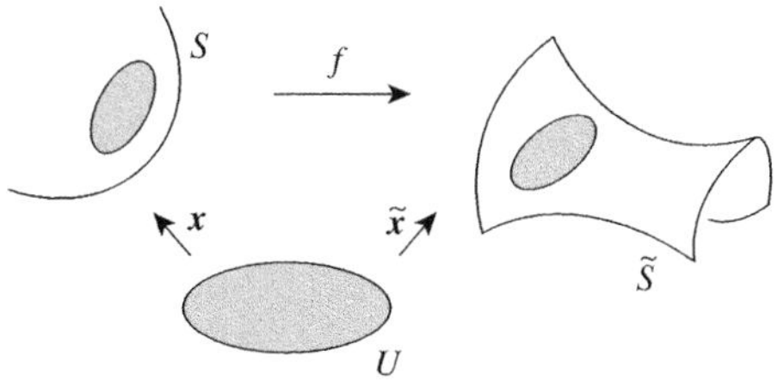

Figure 4.9 Using local parametrisations to construct a conformal map

Proof We first note that f is smooth since $f = \tilde{x}x^{-1}$ is the composite of two smooth maps. Since we are dealing with two parametrisations here, we do not use our usual abuse of notation; for instance, we do not write f_u since this could mean either $(f\boldsymbol{x})_u$ or $(f\tilde{\boldsymbol{x}})_u$.

It follows from Proposition 3 of §4.4 that f is conformal if and only if, for some positive function $\lambda(u, v)$, we have

$$(f\boldsymbol{x})_u.(f\boldsymbol{x})_u = \lambda^2 E\,, \quad (f\boldsymbol{x})_u.(f\boldsymbol{x})_v = \lambda^2 F\,, \quad (f\boldsymbol{x})_v.(f\boldsymbol{x})_v = \lambda^2 G\,, \tag{4.15}$$

and, by Proposition 2 of §4.3, f is a local isometry if (4.15) holds with $\lambda = 1$. But $f\boldsymbol{x} = \tilde{\boldsymbol{x}}$ so that $(f\boldsymbol{x})_u.(f\boldsymbol{x})_u = \tilde{\boldsymbol{x}}_u.\tilde{\boldsymbol{x}}_u = \tilde{E}$, $(f\boldsymbol{x})_u.(f\boldsymbol{x})_v = \tilde{F}$, and $(f\boldsymbol{x})_v.(f\boldsymbol{x})_v = \tilde{G}$. Since f is clearly a bijective map, the proof of the proposition now follows. □

We note that the local parametrisations $\boldsymbol{x}$ and $\tilde{\boldsymbol{x}}$ map a given curve in their common domain U to curves in S and $\tilde{S}$ which correspond under the bijective correspondence f. In particular, f maps the coordinate curves of $\boldsymbol{x}$ to those of $\tilde{\boldsymbol{x}}$.

Example 2 (Helicoid and catenoid) Let S be the helicoid in $\mathbb{R}^3$ defined by the equation $x \sin z = y \cos z$, and let $\tilde{S}$ be the catenoid with equation $x^2 + y^2 = \cosh^2 z$. Let $\boldsymbol{x} : \mathbb{R}^2 \to S$ and $\tilde{\boldsymbol{x}} : U \to \tilde{S}$, $U = (-\pi, \pi) \times \mathbb{R}$, be the local parametrisations of S, $\tilde{S}$ respectively given by

$$\boldsymbol{x}(u, v) = (\sinh v \cos u, \sinh v \sin u, u)\,, \quad (u, v) \in \mathbb{R}^2\,,$$

and

$$\tilde{\boldsymbol{x}}(u, v) = (\cosh v \cos u, \cosh v \sin u, v)\,, \quad (u, v) \in U\,.$$

Then, as we found in Example 1 of §3.4 and Example 1 of §3.6,

$$E = G = \cosh^2 v\,, \quad F = 0\,; \qquad \tilde{E} = \tilde{G} = \cosh^2 v\,, \quad \tilde{F} = 0\,,$$

and hence the map illustrated in Figure 4.10 and given by

$$f(\sinh v \cos u, \sinh v \sin u, u) = (\cosh v \cos u, \cosh v \sin u, v)\,, \quad (u, v) \in U\,, \tag{4.16}$$

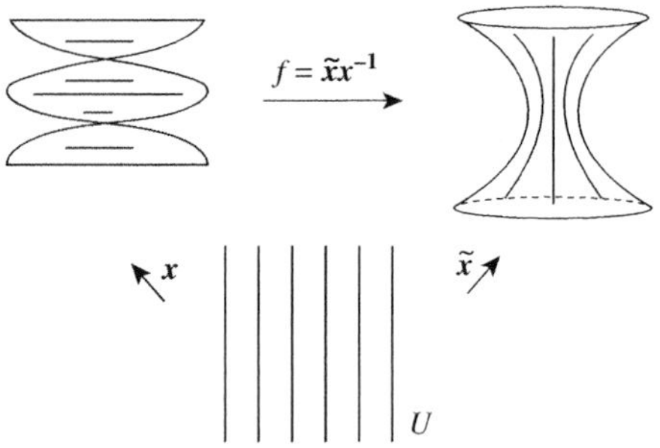

Figure 4.10 Isometry from one twist of a helicoid to a catenoid

is an isometry from one complete twist of the helicoid to the catenoid (with one meridian omitted). The coordinate curves u = constant on the helicoid give the rulings of S, and these map to the meridians u = constant on the catenoid. In a similar way, the helices v = constant on S map to the parallels on $\tilde{S}$.

In Exercise 4.15, you are asked to investigate a 1-parameter family of isometries which deforms one twist of a helicoid to give a catenoid (with one meridian omitted). The Gauss map stays constant throughout the deformation. This behaviour is characteristic of surfaces of a certain type, namely *minimal surfaces*; these form the main topic of Chapter 9. Animations of the deformation described here may be found on the internet.

We note that the formula given in (4.16) may be extended to the whole of $\mathbb{R}^2$ to give a local isometry from S onto $\tilde{S}$ that wraps the helicoid round the catenoid an infinite number of times.

We recall a remark made near the start of §3.3 to the effect that two surfaces having local parametrisations with the same coefficients of the first fundamental form have the same intrinsic metric geometry on the corresponding coordinate neighbourhoods. Proposition 1 shows that in this situation there is an isometry between the two coordinate neighbourhoods, and we will often say that the two surfaces are *metrically equivalent* on these coordinate neighbourhoods. So, Example 2 shows that, locally, the helicoid and the catenoid are metrically equivalent, although globally they are very different.

4.6 Appendix 1: Some substantial examples †

In the following two appendices, we shall present some rather more advanced examples of conformal maps and local isometries. These examples will not be needed in an essential way for the rest of the book, and so may be omitted if desired. The most accessible material, finding the conformal group and isometry group of the helicoid, may be covered by reading Appendix 2 up to the end of Example 2. This material does not depend on Appendix 1.

Example 1 (Conformal maps of the sphere) In Example 5 of §4.4, we mentioned the relation between complex differentiability and conformality for complex functions. In this example we use this, together with some particularly nice isothermal local parametrisations of the sphere $S^2(1)$ to construct smooth maps from $S^2(1)$ to itself which are conformal except perhaps at a finite number of points (where the derivative vanishes).

Let $\boldsymbol{x} : \mathbb{R}^2 \to S^2(1)$ be the local parametrisation discussed in Example 2 of §3.4. Specifically, $\boldsymbol{x}$ is a local parametrisation covering $S^2 \setminus \{(0,0,1)\}$, and, identifying $\mathbb{R}^2$ with $\mathbb{C}$ in the usual way,

$$\boldsymbol{x}(u+iv) = \frac{(2u, 2v, u^2+v^2-1)}{u^2+v^2+1}, \quad u+iv \in \mathbb{C}.$$

It is clear from the geometry of $\boldsymbol{x}$ as explained in Example 2 of §3.4 that the inverse map $\boldsymbol{x}^{-1}$ of $\boldsymbol{x}$ is given by *stereographic projection* π_N from the north pole $(0,0,1)$ of $S^2(1)$ onto the xy-plane. This latter map sends a point (x,y,z) of $S^2(1) \setminus \{(0,0,1)\}$ to the point of intersection of the line through (x,y,z) and $(0,0,1)$ with the xy-plane, and a short calculation shows that π_N (and hence $\boldsymbol{x}^{-1}$) is given by

$$\boldsymbol{x}^{-1}(x,y,z) = \pi_N(x,y,z) = \frac{x+iy}{1-z}\,, \quad (x,y,z) \in S^2(1) \setminus \{(0,0,1)\}\,.$$

If $f : \mathbb{C} \to \mathbb{C}$ is given by

$$f(w) = a_0 + \cdots + a_n w^n\,, \quad a_0, \ldots, a_n \in \mathbb{C}\,, \quad a_n \neq 0\,,$$

we define $\tilde{f} : S^2(1) \to S^2(1)$ (Figure 4.11) by

$$\tilde{f}(p) = \begin{cases} \boldsymbol{x} f \boldsymbol{x}^{-1}(p)\,, & p \neq (0,0,1)\,, \\ p\,, & p = (0,0,1)\,. \end{cases} \tag{4.17}$$

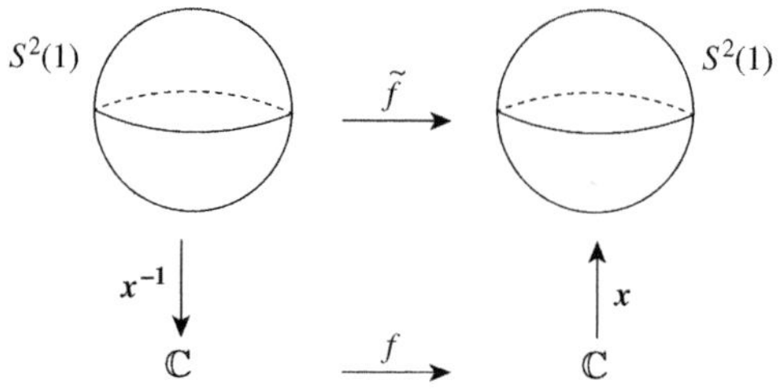

Figure 4.11 Definition of $\tilde{f}$

It is clear that $\tilde{f}$ is smooth on $S^2(1) \setminus \{(0,0,1)\}$, since its composite $\tilde{f}\boldsymbol{x}$ with $\boldsymbol{x}$ is a smooth map from the plane. We now show that $\tilde{f}$ is smooth at $(0,0,1)$ by considering a local parametrisation of $S^2(1)$ whose image contains $(0,0,1)$.

Specifically, let π_S denote stereographic projection from the south pole $(0,0,-1)$ of $S^2(1)$ onto the xy-plane. Then

$$\pi_S(x,y,z) = \frac{x+iy}{1+z}\,, \quad (x,y,z) \in S^2(1) \setminus \{(0,0,-1)\}\,,$$

and a short calculation shows that if $\boldsymbol{y} : \mathbb{C} \to S^2(1) \setminus \{(0,0,-1)\}$ is the smooth map given by

$$\boldsymbol{y}(u+iv) = \frac{(2u, -2v, -u^2 - v^2 + 1)}{u^2 + v^2 + 1}\,,$$

then $\pi_S \boldsymbol{y}$ is complex conjugation on $\mathbb{C}$.

It follows that $\boldsymbol{y}$ is a local parametrisation of $S^2(1)$ (whose image omits $(0,0,-1)$), and $\boldsymbol{y}^{-1}$ is given by π_S followed by complex conjugation. Another short calculation now shows that $\boldsymbol{y}^{-1}\boldsymbol{x} : \mathbb{C} \setminus \{0\} \to \mathbb{C} \setminus \{0\}$ is given by

$$\boldsymbol{y}^{-1}\boldsymbol{x}(w) = \frac{1}{w}\,, \quad w \in \mathbb{C} \setminus \{0\}\,. \tag{4.18}$$

This is the *transition function* between the two local parametrisations $\boldsymbol{x}$ and $\boldsymbol{y}$.

Equation (4.18) indicates the reason we didn't take $\boldsymbol{y}$ to be the inverse of stereographic projection from the south pole; without complex conjugation, the transition function would have been $w \mapsto 1/\bar{w}$, which is not as nice as (4.18) since it is not complex differentiable.

Notice that $\mathbf{y}(0) = (0, 0, 1)$, so if we show that $\tilde{f}\mathbf{y}$ is smooth at 0 it will follow from our definition of smoothness that $\tilde{f}$ is smooth at $(0, 0, 1)$. To do this, we first note that, for $w \neq 0$, we have, using (4.18),

$$\begin{aligned} \mathbf{y}^{-1}\tilde{f}\mathbf{y}(w) &= \mathbf{y}^{-1}\mathbf{x} f(\mathbf{y}^{-1}\mathbf{x})^{-1}(w) \\ &= \frac{1}{f(\frac{1}{w})} \\ &= \frac{1}{a_0 + \cdots + a_n(\frac{1}{w})^n} \\ &= \frac{w^n}{a_0 w^n + \cdots + a_n} \, . \end{aligned} \tag{4.19}$$

However, $\mathbf{y}^{-1}\tilde{f}\mathbf{y}(0) = 0$, so even when $w = 0$, $\mathbf{y}^{-1}\tilde{f}\mathbf{y}(w)$ is given by (4.19). It follows that $\mathbf{y}^{-1}\tilde{f}\mathbf{y}$ is smooth, indeed complex differentiable, at $w = 0$, so that $\tilde{f}\mathbf{y}$ is the composite of the smooth maps $\mathbf{y}$ and $\mathbf{y}^{-1}\tilde{f}\mathbf{y}$ and hence is smooth at 0. We may now conclude from our definition of smoothness that $\tilde{f}$ is smooth at $(0, 0, 1)$ and hence smooth on the whole of $S^2(1)$.

We now discuss conformality. As noted in Example 2 of §3.4, $\mathbf{x}$ is an isothermal local parametrisation, and hence is conformal. Since $f : \mathbb{C} \to \mathbb{C}$ is complex differentiable, f is also conformal except at the finite number of points where the complex derivative vanishes. It follows that $\tilde{f}$ is a conformal map of the sphere except at the corresponding points on the sphere and possibly at the north pole. In fact, the complex derivative of $\mathbf{y}^{-1}\tilde{f}\mathbf{y}$ is zero at $w = 0$ if and only if $n \geq 2$, so that $\tilde{f}$ is conformal at the north pole if and only if $n = 1$.

As a specific example, if $f(w) = w^2$ then the corresponding map $\tilde{f}$ may be described geometrically as follows. A point $p \in S^2(1)$ with z-coordinate z_0 may be written as $(\sqrt{1 - z_0{}^2}\, e^{i\theta}, z_0)$ and, when $f(w) = w^2$,

$$\tilde{f}(\sqrt{1 - z_0{}^2}\, e^{i\theta}, z_0) = \left(\frac{1 - z_0{}^2}{1 + z_0{}^2} e^{2i\theta}, \frac{2z_0}{1 + z_0{}^2} \right) .$$

Thus, points of $S^2(1)$ are moved around the sphere and towards to the poles by $\tilde{f}$.

We may extend the above ideas by considering rational functions rather than just polynomials on the complex plane. So, if $f(w) = g(w)/h(w)$, where $g(w)$ and $h(w)$ are polynomials with no common factors, we may use the above ideas to define a corresponding map $\tilde{f} : S^2(1) \to S^2(1)$. So, for instance, $\tilde{f}(\mathbf{x}(w)) = (0, 0, 1)$ whenever $h(w) = 0$. In fact, in complex analysis it is often convenient to consider complex functions as functions defined on the extended complex plane $\mathbb{C} \cup \{\infty\}$. We may use the parametrisation $\mathbf{x}$ to identify $\mathbb{C} \cup \{\infty\}$ with $S^2(1)$ (in this situation usually called the *Riemann* sphere) with ∞ being identified with the north pole. Under this identification, f and $\tilde{f}$ also become identified. The formula (4.18) for the transition function $\mathbf{y}^{-1}\mathbf{x}$ is the reason why the behaviour of complex functions at ∞ is studied by replacing w with $1/w$ and then seeing what happens when $w = 0$. For instance, if

$$f(w) = \frac{a_n w^n + \cdots + a_0}{b_n w^n + \cdots + b_0} \, ,$$

with at least one of a_n, b_n being non-zero, and the numerator and denominator having no common factors, then

$$f(1/w) = \frac{a_n + \cdots + a_0 w^n}{b_n + \cdots + b_0 w^n},$$

so that $f(\infty) = a_n/b_n$ (interpreted as ∞ if $b_n = 0$). We shall say a little more about the special case of *Möbius transformations* (for which $n = 1$) in Example 3 of the next appendix.

All the above examples are *orientation preserving* maps of $S^2(1)$ (except where the derivative vanishes) in the following sense. If $\boldsymbol{N}$ is a choice of orientation on an orientable surface $S \subset \mathbb{R}^3$ then a basis $\{\boldsymbol{X}, \boldsymbol{Y}\}$ of the tangent space T_pS at $p \in S$ is said to be *positively oriented* if $\boldsymbol{X} \times \boldsymbol{Y}$ is a positive scalar multiple of $\boldsymbol{N}(p)$. A smooth map $f : S \to S$ is *orientation preserving* if the derivative df maps one (and hence every) positively oriented basis at each point $p \in S$ to a positively oriented basis at $f(p)$. We note that this concept is independent of choice of orientation.

Orientation *reversing* conformal maps of $S^2(1)$ are obtained by considering rational functions of the complex conjugate $\bar{z}$.

Example 2 (Veronese surface) Let $f : S^2(1) \to \mathbb{R}^5$ be the map defined by

$$f(x, y, z) = \left(yz, zx, xy, \frac{1}{2}(x^2 - y^2), \frac{1}{2\sqrt{3}}(x^2 + y^2 - 2z^2)\right), \quad x^2 + y^2 + z^2 = 1. \tag{4.20}$$

It follows easily that $f(p) = f(q)$ if and only if $p = \pm q$, so that if we define the *real projective plane* $\mathbb{R}P^2$ to be the set of lines through the origin of $\mathbb{R}^3$ then f defines a bijective map from $\mathbb{R}P^2$ to $\mathbb{R}^5$. The image of f is a surface S in $\mathbb{R}^5$, and we now show that f is a local isometry from $S^2(1)$ onto S. The surface S is called the *Veronese* surface, and it has many interesting geometrical properties.

So, let $\boldsymbol{\alpha}_1(t) = (x_1(t), y_1(t), z_1(t))$, and $\boldsymbol{\alpha}_2(t) = (x_2(t), y_2(t), z_2(t))$ be curves on $S^2(1)$ with $\boldsymbol{\alpha}_1(0) = \boldsymbol{\alpha}_2(0) = (x, y, z) \in S^2(1)$.

Then, for each $i = 1, 2$, we have that ${x_i}^2 + {y_i}^2 + {z_i}^2 = 1$, so that

$$x_i x_i' + y_i y_i' + z_i z_i' = 0, \quad i = 1, 2. \tag{4.21}$$

We now note that

$$\begin{aligned} df(\boldsymbol{\alpha}_1') &= (f\boldsymbol{\alpha}_1)' \\ &= \Big(y_1' z_1 + y_1 z_1', z_1' x_1 + z_1 x_1', \\ &\qquad x_1' y_1 + x_1 y_1', x_1 x_1' - y_1 y_1', \frac{1}{\sqrt{3}}(x_1 x_1' + y_1 y_1' - 2z_1 z_1')\Big). \end{aligned}$$

Using the similar expression for $df(\boldsymbol{\alpha}_2')$, it follows that, evaluating all derivatives at $t = 0$,

$$\begin{aligned} df\left(\boldsymbol{\alpha}_i'(0)\right).df\left(\boldsymbol{\alpha}_j'(0)\right) &= (y_i' z + y z_i')(y_j' z + y z_j') + (z_i' x + z x_i')(z_j' x + z x_j') \\ &\quad + (x_i' y + x y_i')(x_j' y + x y_j') + (x x_i' - y y_i')(x x_j' - y y_j') \\ &\quad + \frac{1}{3}(x x_i' + y y_i' - 2z z_i')(x x_j' + y y_j' - 2z z_j') \\ &= (x_i' x_j' + y_i' y_j' + z_i' z_j')(x^2 + y^2 + z^2) \\ &\quad + \frac{1}{3}(x x_i' + y y_i' + z z_i')(x x_j' + y y_j' + z z_j'). \end{aligned}$$

Hence, using (4.21) and again evaluating all derivatives at $t = 0$,

$$df\left(\boldsymbol{\alpha}_i{}'(0)\right).df\left(\boldsymbol{\alpha}_j{}'(0)\right) = x_i{}'x_j{}' + y_i{}'y_j{}' + z_i{}'z_j{}' = \boldsymbol{\alpha}_i{}'(0).\boldsymbol{\alpha}_j{}'(0)\,.$$

It follows that, as claimed, f is a local isometry of $S^2(1)$ onto the Veronese surface. In a natural sense, the Veronese surface is obtained by identifying antipodal points of $S^2(1)$, so it follows that the Veronese surface has area 2π.

Example 3 (Models of the hyperbolic plane) The hyperbolic plane H was discussed in the optional Example 5 of §3.4. In that example, we described how to regard the hyperbolic plane as the upper half-plane $\{(u, v) \in \mathbb{R}^2 : v > 0\}$ which is equipped with a metric, the *hyperbolic metric*, which differs from the Euclidean metric by the conformal factor $1/v$. So, at a point $(u, v) \in H$, we take the inner product g given by

$$g\left((\lambda_1, \mu_1), (\lambda_2, \mu_2)\right) = \frac{1}{v^2}(\lambda_1\lambda_2 + \mu_1\mu_2)\,.$$

In the hyperbolic metric, the length of a curve is not its Euclidean length, but the angle of intersection of two curves is the same in both the Euclidean and hyperbolic metrics.

The coefficients of the first fundamental form of the hyperbolic metric are easy to work out; for instance, $E = g\left((1, 0), (1, 0)\right) = 1/v^2$. We find that

$$E = \frac{1}{v^2}\,, \quad F = 0\,, \quad G = \frac{1}{v^2}\,.$$

We now describe another way of putting a non-standard metric on a subset of the plane, and then show that this is isometric to H. We identify $\mathbb{C}$ with $\mathbb{R}^2$ in the usual way, and let $\tilde{H}$ denote the open unit disc $\{w \in \mathbb{C} : |w| < 1\}$, equipped with metric $\tilde{g}$ (again conformally equivalent to the standard Euclidean metric) having $\tilde{E} = \tilde{G} = 4/(1 - |w|^2)^2$, $\tilde{F} = 0$. We shall show that the Möbius transformation

$$f(z) = \frac{z - i}{z + i}$$

is an isometry from H onto $\tilde{H}$. To see this, we first note that f maps the upper half-plane onto the open unit disc. Then, differentiating with respect to z, we find that

$$f'(z) = \frac{2i}{(z + i)^2}\,.$$

We now recall that the *Cauchy–Riemann equations* from complex analysis give that

$$f_u = -if_v = f'(z)\,,$$

so, if $f'(z) = a + ib$ (which we identify with $(a, b) \in \mathbb{R}^2$), where a and b are real numbers, then

$$\begin{aligned}\tilde{g}(f_u, f_u) = \tilde{g}(f_v, f_v) = 4\frac{a^2 + b^2}{(1 - |f(z)|^2)^2} &= 4\frac{|f'(z)|^2}{(1 - |f(z)|^2)^2} \\ &= 4\left|\frac{2i}{(z + i)^2}\right|^2 \frac{1}{(1 - |\frac{z-i}{z+i}|^2)^2}\end{aligned}$$

$$= \frac{16}{(|z+i|^2 - |z-i|^2)^2}$$
$$= \frac{1}{(\text{Im } z)^2}$$
$$= E = G\,.$$

Also, $\tilde{g}(f_u, f_v) = 4(-ab+ab)/(1-|f(z)|^2)^2 = 0 = F$, so Proposition 2 of §4.3 shows that $f(z)$ is an isometry.

For obvious reasons, H is often referred to as the *upper half-plane model* of the hyperbolic plane, while $\tilde{H}$ is the *disc model.*

4.7 Appendix 2: Conformal and isometry groups †

As mentioned in the final paragraph of §4.4, the composite of two conformal maps is conformal, and the composite of two (local) isometries is a (local) isometry. It is clear that the conformal diffeomorphisms from a surface S to itself form a group under composition, the *conformal group* of S, and that the isometries form a subgroup of this.

In Example 5 of §4.4, we found these groups for the plane. In this optional appendix, we discuss the conformal group and the isometry group of a helicoid, the unit sphere, and the hyperbolic plane.

Example 1 (Conformal group of helicoid) As we saw in Example 1 of §3.6, the parametrisation

$$\boldsymbol{x}(u, v) = (\sinh v \cos u, \sinh v \sin u, u)\,, \quad (u, v) \in \mathbb{R}^2\,,$$

gives an isothermal parametrisation of the whole of a helicoid S. Thus (Figure 4.12) the map $f \mapsto \boldsymbol{x} f \boldsymbol{x}^{-1}$ gives a group isomorphism from the conformal group $C_{\mathbb{R}^2}$ of $\mathbb{R}^2$ to the conformal group C_S of S. (This is a similar idea to that used in Example 1 of Appendix 1, in that we have used an isothermal parametrisation to relate the conformal structure of the helicoid to that of the plane.)

We saw in Example 5 of §4.4 that $C_{\mathbb{R}^2}$ is made up of the (orientation preserving) maps

$$z \mapsto az + b\,, \quad a, b \in \mathbb{C}\,, \ a \neq 0\,,$$

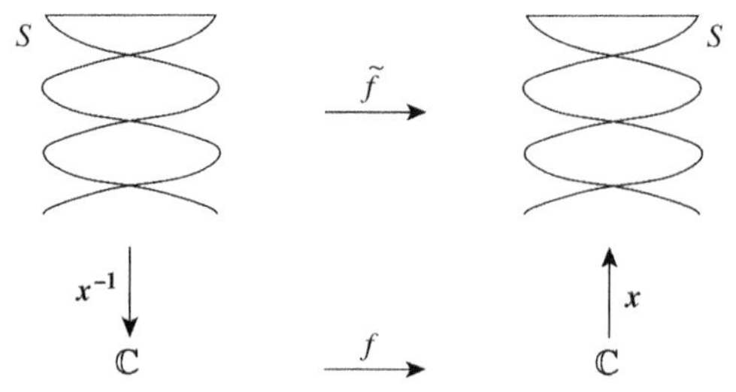

Figure 4.12 Conformal diffeomorphisms of the helicoid

and the (orientation reversing) maps

$$z \mapsto a\bar{z} + b\,, \quad a, b \in \mathbb{C}\,, \ a \neq 0\,,$$

where $\bar{}$ denotes complex conjugation. If $f(z) = az + b$ with $a = a_1 + ia_2$, $b = b_1 + ib_2$, then

$$f(u + iv) = a_1u - a_2v + b_1 + i(a_2u + a_1v + b_2)\,,$$

so that the corresponding orientation-preserving conformal diffeomorphism $\tilde{f} = \boldsymbol{x} f \boldsymbol{x}^{-1}$ of the helicoid is given by

$$\tilde{f}\,(\boldsymbol{x}(u, v)) = (\sinh \tilde{v} \cos \tilde{u}, \sinh \tilde{v} \sin \tilde{u}, \tilde{u})\,,$$

where

$$\tilde{u} = a_1u - a_2v + b_1\,, \quad \tilde{v} = a_2u + a_1v + b_2\,.$$

Since $|a|$ is the conformal factor λ of f at each point of $\mathbb{C}$, and $\cosh v$ is the conformal factor of $\boldsymbol{x}$ at (u, v), it follows that the conformal factor $\tilde{\lambda}$ of $\tilde{f}$ at $\boldsymbol{x}(u, v)$ is given by

$$\tilde{\lambda} = \frac{1}{\cosh v}|a| \cosh(a_2u + a_1v + b_2)\,. \tag{4.22}$$

Example 2 (Isometry group of helicoid) We already know many isometries of the helicoid S, namely the restriction to the helicoid of suitable screw motions of $\mathbb{R}^3$ about the z-axis. However, we may wonder if there are any more.

Since the previous example gives us the conformal group of the helicoid, we need only check which of these are isometries. To do this we must find those conformal maps f of the complex plane for which the corresponding map $\tilde{f}$ has conformal factor $\tilde{\lambda} = 1$. It thus follows from (4.22) that $\tilde{f}$ is an isometry if and only if

$$a_2 = b_2 = 0\,, \quad a_1 = \pm 1\,.$$

The orientation preserving isometries therefore come from the maps of the plane given by

$$(u, v) \mapsto (u + b_1, v) \quad \text{and} \quad (u, v) \mapsto (-u + b_1, -v),$$

where $b_1 \in \mathbb{R}$ is arbitrary. Notice that the second of these is the map $(u, v) \mapsto (-u, -v)$ followed by the first. Thus, the orientation preserving isometries of the helicoid are either of the form

$$(\sinh v \cos u, \sinh v \sin u, u) \mapsto (\sinh v \cos(u + b_1), \sinh v \sin(u + b_1), u + b_1)\,, \tag{4.23}$$

or the composite of a map of this type with the map

$$(\sinh v \cos u, \sinh v \sin u, u) \mapsto (-\sinh v \cos u, \sinh v \sin u, -u)\,. \tag{4.24}$$

We note that a map of the helicoid S of the form of (4.23) is the restriction to S of the screw motion of $\mathbb{R}^3$ given by

$$\begin{pmatrix} x \\ y \\ z \end{pmatrix} \mapsto R(b_1) \begin{pmatrix} x \\ y \\ z \end{pmatrix} + \begin{pmatrix} 0 \\ 0 \\ b_1 \end{pmatrix},$$

where

$$R(\theta) = \begin{pmatrix} \cos\theta & -\sin\theta & 0 \\ \sin\theta & \cos\theta & 0 \\ 0 & 0 & 1 \end{pmatrix}$$

is rotation of $\mathbb{R}^3$ about the z-axis through an angle θ. Similarly, the map (4.24) is the restriction to S of rotation of $\mathbb{R}^3$ about the y-axis through an angle π. This shows that each orientation preserving isometry of the helicoid is the restriction of a Euclidean motion of $\mathbb{R}^3$.

The case of orientation reversing isometries of the helicoid is left as an exercise.

Example 3 (Conformal and isometry groups of $S^2(1)$) This example follows on from Example 1 of the previous appendix, and uses the notation developed there. A *Möbius transformation* is a rational function of the form

$$f(w) = \frac{aw + b}{cw + d}\,, \quad a, b, c, d \in \mathbb{C}\,, \ ad - bc \neq 0\,,$$

and in Exercise 4.19 you are asked to prove that the corresponding maps $\tilde{f}$ are conformal diffeomorphisms of $S^2(1)$. If we assume without loss of generality that $ad - bc = 1$, then Exercise 4.19 also asks you to show that $\tilde{f}$ is an orientation preserving isometry (that is to say, a rotation) of $S^2(1)$ if and only if $d = \bar{a}$ and $c = -\bar{b}$. Although we shall not prove it, a standard result in complex analysis says that all orientation preserving conformal diffeomorphisms of $S^2(1)$ are induced by Möbius transformations as described in this example. The orientation reversing conformal diffeomorphisms are obtained by considering maps as above followed by reflection in, say, the xy-plane.

Example 4 (The hyperbolic plane) Let H denote the upper half-plane model of the hyperbolic plane discussed in Example 3 of Appendix 1. Identifying $\mathbb{R}^2$ with $\mathbb{C}$ as usual, the conformal diffeomorphisms of H are those Möbius transformations that map the upper half plane to itself, namely, $f : H \to H$ given by

$$f(z) = \frac{az + b}{cz + d}\,, \quad a, b, c, d \in \mathbb{R}\,, \ ad - bc > 0\,.$$

Rather surprisingly, it may be shown (see Exercise 4.20) that every conformal diffeomorphism of the hyperbolic plane is actually an isometry.

Exercises

4.1 Using the parametrisation of the helicoid $x \sin z = y \cos z$ given by

$$\boldsymbol{x}(u, v) = (\sinh v \cos u, \sinh v \sin u, u)\,, \quad u, v \in \mathbb{R}\,,$$

find the Gauss map $\boldsymbol{N}(u, v)$ of the helicoid, and show that it is not injective.

4.2 Show that the Gauss map of a surface of revolution S maps the parallels of S to the parallels of $S^2(1)$ and the meridians of S to the meridians of $S^2(1)$ (where $S^2(1)$ is considered as a surface of revolution with axis of rotation parallel to that of S).

4.3 Find the image of the Gauss map of the surface with equation $f(x, y, z) = 0$, where:

(i) $f(x, y, z) = x^2 + y^2 - z$ (paraboloid of revolution);
(ii) $f(x, y, z) = x^2 + y^2 - z^2 - 1$ (hyperboloid of 1 sheet);
(iii) $f(x, y, z) = x^2 + y^2 - z^2 + 1$ (hyperboloid of 2 sheets).

In each case use the orientation determined by grad f. (Note that the image of the Gauss map is a subset of $S^2(1)$. So, for example, the answer to (ii) is $\{(x, y, z) \in S^2(1) : |z| < 1/\sqrt{2}\}$.)

4.4 Show that the Gauss map of the catenoid $x^2 + y^2 = \cosh^2 z$ is an injective map onto $S^2(1) \setminus \{(0, 0, \pm 1)\}$.

4.5 (Height functions) Let S be a surface in $\mathbb{R}^n$, and let $\boldsymbol{v}$ be a unit vector in $\mathbb{R}^n$. Let $h : S \to \mathbb{R}$ be given by $h(p) = p.\boldsymbol{v}$. Show that h is a smooth function on S and that the derivative dh_p is zero if and only if $\boldsymbol{v}$ is orthogonal to S at p (Figure 4.13).

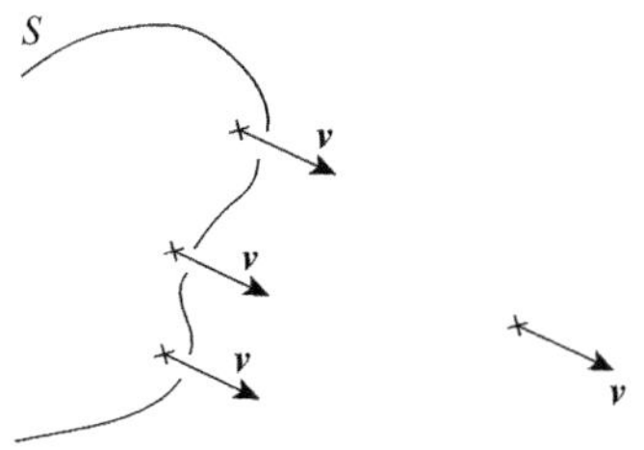

Figure 4.13 Height function

4.6 (Distance squared functions) Let S be a surface in $\mathbb{R}^n$ and let q be a point in $\mathbb{R}^n$. Let $f : S \to \mathbb{R}$ assign to each point $p \in S$ the square of the distance from p to q. Show that f is a smooth function on S and that the derivative df_p is zero if and only if either $p = q$ or $q - p$ is orthogonal to S at p (Figure 4.14).

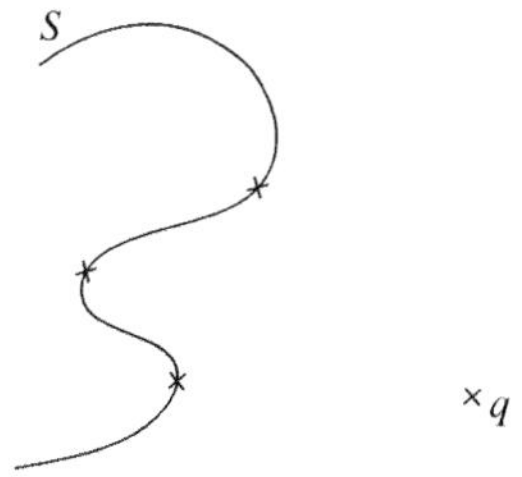

Figure 4.14 Square of the distance function

4.7 Let S be a connected surface in $\mathbb{R}^3$. If all the lines in $\mathbb{R}^3$ having orthogonal intersection with S pass through some fixed point of $\mathbb{R}^3$, show that S is an open subset of a sphere. (A surface S is *connected* if any two points of S may be joined by a smooth curve on S. You may use the fact that if the derivative of a differentiable function f on such a surface is everywhere zero, then f is constant.)

4.8 Complete Example 5 of §4.2 by showing directly that df_p maps $T_pS^2(1)$ to $T_{f(p)}\tilde{S}$.

4.9 For each positive real number a, find a local isometry from the xy-plane in $\mathbb{R}^3$ onto the cylinder in $\mathbb{R}^3$ with equation $x^2 + y^2 = a^2$.

4.10 Let b be a positive real number, and let $\tilde{S}$ be that part of the cone $z^2 = b^2(x^2 + y^2)$ for which $z > 0$. If S denotes the xy-plane in $\mathbb{R}^3$, and if n is a positive integer, show that the formula $f : S \setminus \{(0,0,0)\} \to \tilde{S}$ given by

$$f(r\cos\theta, r\sin\theta, 0) = \frac{1}{n}(r\cos n\theta, r\sin n\theta, br)\,, \quad r > 0\,,$$

gives a well-defined map onto $\tilde{S}$. Show also that if $b = \sqrt{n^2 - 1}$ then f is a local isometry. How would you model the effect of the map f by using a sheet of paper?

4.11 *(This exercise uses material in the optional Example 4 of §4.3.)* Let $\tilde{S}$ be the flat torus discussed in Example 4 of §4.3, and let S be the cylinder in $\mathbb{R}^3$ with equation $x^2 + y^2 = {r_1}^2$.

If $f : \mathbb{R}^3 \to \mathbb{R}^4$ is given by

$$f(x, y, z) = (x, y, r_2\cos(z/r_2), r_2\sin(z/r_2))\ ,$$

show that the restriction of f to S defines a surjective local isometry from S to $\tilde{S}$.

4.12 *(This exercise uses material in the optional Example 4 of §4.3.)* Show that any two points of the flat torus $\tilde{S}$ discussed in Example 4 of §4.3 may be joined by a curve in $\tilde{S}$ of length at most $\pi\sqrt{{r_1}^2 + {r_2}^2}$.

4.13 Show that the Gauss map of the helicoid $x\sin z = y\cos z$ is conformal.

4.14 *(This exercise uses material in the optional Example 4 of §4.3.)* Let S be the flat torus discussed in Example 4 of §4.3, and assume that ${r_1}^2 + {r_2}^2 = 1$. Let $T_{a,b}$ be the torus of revolution in $\mathbb{R}^3$ obtained by rotating the circle

$$(x - a)^2 + z^2 = b^2\,, \quad y = 0\,,$$

about the z-axis, where $a = 1/r_1$ and $b = r_2/r_1$.

Let $X = \{(x_1, x_2, x_3, x_4) : x_4 \neq 1\}$ denote $\mathbb{R}^4$ with the plane $x_4 = 1$ omitted, and let $f : X \to \mathbb{R}^3$ be stereographic projection from $(0,0,0,1)$ onto the plane $x_4 = 0$ (so that, if $p \in X$, then $f(p)$ is the point of intersection with the plane $x_4 = 0$ of the line through p and $(0,0,0,1)$). Show (or assume that) f is given by

$$f(x_1, x_2, x_3, x_4) = \left(\frac{x_1}{1 - x_4}, \frac{x_2}{1 - x_4}, \frac{x_3}{1 - x_4}\right)\,.$$

Show that f defines a conformal diffeomorphism of the flat torus S onto the torus of revolution $T_{a,b}$.

4.15 Let $\boldsymbol{x}$ and $\tilde{\boldsymbol{x}}$ be the local parametrisations of (one twist of) the helicoid S and the catenoid $\tilde{S}$ given in Exercise 3.10, namely

$$\boldsymbol{x}(u, v) = (\sinh v\sin u, -\sinh v\cos u, u)\,, \quad -\pi < u < \pi\,,\ v \in \mathbb{R}\,,$$

$$\tilde{\boldsymbol{x}}(u, v) = (\cosh v\cos u, \cosh v\sin u, v)\,, \quad -\pi < u < \pi\,,\ v \in \mathbb{R}\,.$$

In Exercise 3.10 we found that $\boldsymbol{x}_u = \tilde{\boldsymbol{x}}_v$ and $\boldsymbol{x}_v = -\tilde{\boldsymbol{x}}_u$ and, for each $\theta \in \mathbb{R}$, we showed that the coefficients of the first fundamental form of the surface S_θ parametrised by

$$\boldsymbol{x}_\theta(u,v) = \cos\theta\ \boldsymbol{x}(u,v) + \sin\theta\ \tilde{\boldsymbol{x}}(u,v)\,, \qquad -\pi < u < \pi\,,\ v \in \mathbb{R}\,,$$

are independent of θ. Use this to show that the correspondence $f_\theta\,(\boldsymbol{x}(u,v)) = \boldsymbol{x}_\theta(u,v)$ is an isometry from one twist of the helicoid S onto S_θ. Show also that the corresponding Gauss maps $\boldsymbol{N}_\theta$ are independent of θ. This provides a 1-parameter family of isometries which deforms one twist of the helicoid to form the catenoid (with one meridian removed) in such a way that the Gauss map remains constant throughout the deformation. As mentioned in §4.5, this behaviour is characteristic of surfaces of a certain type, namely *minimal surfaces*; these form the main topic of Chapter 9. Animations of the deformation described in this exercise may be found on the internet.

The following exercises use material in the optional appendices.

4.16 Let $\tilde{f} : S^2(1) \to S^2(1)$ be the map defined as in Example 1 of Appendix 1, with $f(w) = w + 1$. Draw sketches of $S^2(1)$ showing the curves of intersection of $S^2(1)$ with the coordinate planes, and their images under $\tilde{f}$. Provide justification for your sketches.

4.17 (The hyperbolic plane for relativity theorists!) Let B be the symmetric bilinear form defined on $\mathbb{R}^3 \times \mathbb{R}^3$ by

$$B\left((x_1,x_2,x_3),(y_1,y_2,y_3)\right) = x_1y_1 + x_2y_2 - x_3y_3$$

(this is an example of an *indefinite metric* on $\mathbb{R}^3$), and let S be the upper sheet of the hyperboloid of two sheets given by

$$S = \{(x_1,x_2,x_3) \in \mathbb{R}^3 : B\left((x_1,x_2,x_3),(x_1,x_2,x_3)\right) = -1,\ x_3 > 0\}\,,$$

(so that S is a " sphere of radius $\sqrt{-1}$" in terms of the indefinite metric).

(i) Show that, if $p \in S$, then $B(\boldsymbol{X}, p) = 0$ for all $\boldsymbol{X} \in T_pS$. (It now follows from a result in linear algebra called **Sylvester's law of inertia** that the restriction of B to the tangent spaces of S defines a positive definite inner product $\langle\ ,\ \rangle$ on each tangent space of S.)

(ii) Let $(\tilde{H}, \tilde{g})$ denote the disc model of the hyperbolic plane (see Example 3 of Appendix 1) equipped with the metric $\tilde{g}$ described in that example. For each $(u,v) \in \tilde{H}$, show that the line through $(0,0,-1)$ and $(u,v,0)$ intersects S at the unique point

$$f(u,v) = \frac{(2u, 2v, 1+u^2+v^2)}{1-u^2-v^2}\,.$$

(iii) Show that f maps $(\tilde{H}, \tilde{g})$ isometrically onto $(S, \langle\ ,\ \rangle)$. (This gives an alternative way of showing that, as noted above, $\langle\ ,\ \rangle$ is positive definite on each tangent space of S.)

4.18 Let $H = \{(u, v) \in \mathbb{R}^2 : v > 0\}$ be the upper half-plane model of the hyperbolic plane discussed in Example 5 of §3.4 and in Example 3 of Appendix 1. Let S be the pseudosphere obtained by rotating the tractrix

$$\boldsymbol{\alpha}(v) = \left(\frac{1}{v}\,,\ 0\,,\ \operatorname{arccosh} v - \frac{(v^2-1)^{1/2}}{v}\right), \quad v > 1\,,$$

around the z-axis, where $\operatorname{arccosh} v$ is taken to be the positive number w with $\cosh w = v$. Show that the map

$$f(u, v) = \left(\frac{1}{v}\cos u,\ \frac{1}{v}\sin u,\ \operatorname{arccosh} v - \frac{(v^2-1)^{1/2}}{v}\right), \quad v > 1\,,$$

is a local isometry of the open subset $\tilde{H} = \{(u, v) \in H : v > 1\}$ of H onto S. This local isometry wraps $\tilde{H}$ round the pseudosphere an infinite number of times, rather like the local isometry considered in Example 3 of §4.3 wraps the plane round the cylinder an infinite number of times.

4.19 This exercise follows on from Example 1 of Appendix 1 and Example 3 of Appendix 2, and uses the notation developed there.

(i) Using the fact that the conformal factor of $\boldsymbol{x}$ at w is $2/(1 + |w|^2)$, show that, if $f(w)$ is a complex differentiable function with non-zero derivative at w, then the conformal factor of $\tilde{f}$ at $\boldsymbol{x}(w)$ is

$$\frac{(1 + |w|^2)|f'(w)|}{1 + |f(w)|^2}\,.$$

(ii) If $f(w)$ is the Möbius transformation given by

$$f(w) = \frac{aw + b}{cw + d}\,,$$

show that the conformal factor of the corresponding map $\tilde{f}$ at $\boldsymbol{x}(w)$ is equal to

$$\frac{(1 + |w|^2)\,|ad - bc|}{|cw + d|^2 + |aw + b|^2}\,.$$

(iii) Assuming that $\tilde{f}$ is smooth at $(0, 0, 1)$, show that $\tilde{f}$ is conformal at $(0, 0, 1)$ and hence is a conformal diffeomorphism of $S^2(1)$.

(iv) Assuming without loss of generality that $ad - bc = 1$, show that $\tilde{f}$ is an isometry of $S^2(1)$ if and only if $d = \bar{a}$ and $c = -\bar{b}$.

4.20 Use the upper half-plane model of the hyperbolic plane H described in Example 3 of Appendix 1, and the description of the conformal diffeomorphisms of the hyperbolic plane given in Example 4 of Appendix 2, to show that every conformal diffeomorphism of the hyperbolic plane is actually an isometry.

5 Measuring how surfaces curve

In Chapters 3 and 4 we studied intrinsic properties of surfaces; those which depend on only the inner product on each tangent space. In this chapter we study *extrinsic* properties of a surface in $\mathbb{R}^3$. These consider the measurement and consequences of the curvature of the surface in the containing Euclidean space.

We saw in Chapter 1 that the bending of a regular curve $\boldsymbol{\alpha}$ in $\mathbb{R}^2$ is measured by the rate of change of its unit normal vector $\mathbf{n}$. In a similar manner, the way in which a surface S is curving in $\mathbb{R}^3$ at a point $p \in S$ may be measured by the rate of change at p of its unit normal vector $\boldsymbol{N}$. This is quite complicated, since it is given by the derivative $d\boldsymbol{N}_p$ which is a linear map from the tangent space T_pS to $\mathbb{R}^3$. However, we shall see that this linear map may be used to define scalar quantities, the *Gaussian curvature* K and the *mean curvature* H, which turn out to be of fundamental importance in describing the geometry of S.

In this chapter, we begin the study of these two measures of curvature, and relate them to other quantities determined by the rate of change of $\boldsymbol{N}$. This involves a discussion of the *second fundamental form* II of S, and its role in determining the *normal curvature* κ_n of a regular curve $\boldsymbol{\alpha}$ on S; this latter quantity may be thought of as a measure of the rate at which S is curving in $\mathbb{R}^3$ as we travel along $\boldsymbol{\alpha}$, or, as an alternative interpretation, the minimum amount of bending $\boldsymbol{\alpha}$ must do in order to stay on S.

The situation for surfaces in higher dimensional Euclidean spaces is rather more complicated than for surfaces in $\mathbb{R}^3$, since here the normal space at a point is more than 1-dimensional so it is more difficult to measure the rate of change. However, although beyond the scope of this book, much can be done and many of the results in this chapter may be generalised.

In §5.2 to §5.6 we define various quantities determined by the rate of change of $\boldsymbol{N}$, and give several examples of how to calculate them. We then begin an investigation of the geometric information carried by these quantities.

5.1 The Weingarten map

Let S be a surface in $\mathbb{R}^3$ and let $\boldsymbol{N} : S \to S^2(1)$ be the corresponding Gauss map, which gives a smooth choice of unit normal vector on S (and so is locally defined up to sign) as described in §4.1, where some examples were given.

The rate of change of $\boldsymbol{N}$ at a point $p \in S$ is measured by the derivative

$$d\boldsymbol{N}_p : T_pS \to T_{\boldsymbol{N}(p)}S^2(1) ,$$

and it is this map which captures the way in which S is curving at p.

In this section, we give some examples and first properties of $d\boldsymbol{N}_p$.

Example 1 (Plane) Let $f(x, y, z) = ax + by + cz$, where at least one of a, b, c is non-zero. Then, for each real number k, the equation $f(x, y, z) = k$ gives a plane in $\mathbb{R}^3$ and, as proved in Proposition 5 of §3.1,

$$\boldsymbol{N} = \frac{\text{grad } f}{|\text{grad } f|} = \frac{(a, b, c)}{\sqrt{a^2 + b^2 + c^2}} .$$

Thus $\boldsymbol{N}$ is constant and its rate of change is the zero map at each point p of S, reflecting the fact that the plane doesn't curve at all.

Example 2 (Unit sphere) The outward unit normal to $S^2(1)$ at a point p on $S^2(1)$ is equal to the position vector of p (Figure 5.1); the corresponding Gauss map of $S^2(1)$ is simply the identity map. Since this is the restriction to $S^2(1)$ of a linear map of $\mathbb{R}^3$, namely the identity map, the derivative $d\boldsymbol{N}_p$ is the inclusion map $T_pS^2(1) \hookrightarrow \mathbb{R}^3$. In this book we shall always use the orientation of $S^2(1)$ given by the outward unit normal.

We note that, for a general surface S in $\mathbb{R}^3$, both T_pS and $T_{N(p)}S^2(1)$ have the same unit normal, namely $\boldsymbol{N}(p)$. It follows that, as illustrated in Figure 5.2, $T_{N(p)}S^2(1) = T_pS$. Hence $d\boldsymbol{N}_p$ is actually a linear map from T_pS to itself. This may also be seen by using a local parametrisation $\boldsymbol{x}(u, v)$; since $\boldsymbol{N}.\boldsymbol{N} = 1$ we have that $\boldsymbol{N}_u.\boldsymbol{N} = 0 = \boldsymbol{N}_v.\boldsymbol{N}$, so that $\boldsymbol{N}_u$ and $\boldsymbol{N}_v$, which span the image of $d\boldsymbol{N}$, are both in T_pS.

For each $p \in S$ the map

$$-d\boldsymbol{N}_p : T_pS \to T_pS$$

is called the *Weingarten map* of S at p. The reason for the minus sign will become apparent when we discuss normal curvature in §5.7.

A linear map is determined by its effect on a basis of tangent vectors, and, in terms of a local parametrisation $\boldsymbol{x}(u, v)$, equation (4.4) shows that

$$-\,d\boldsymbol{N}(\boldsymbol{x}_u) = -\boldsymbol{N}_u\,, \quad -d\boldsymbol{N}(\boldsymbol{x}_v) = -\boldsymbol{N}_v. \tag{5.1}$$

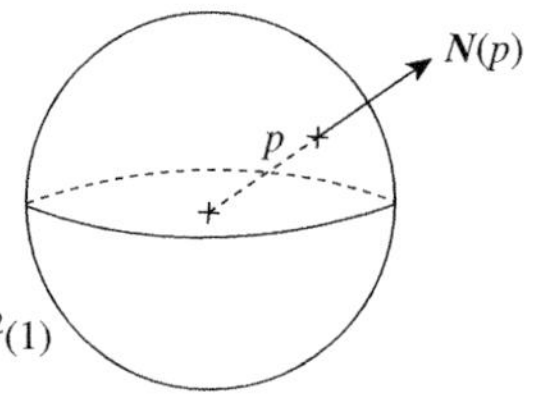

Figure 5.1 Gauss map of sphere

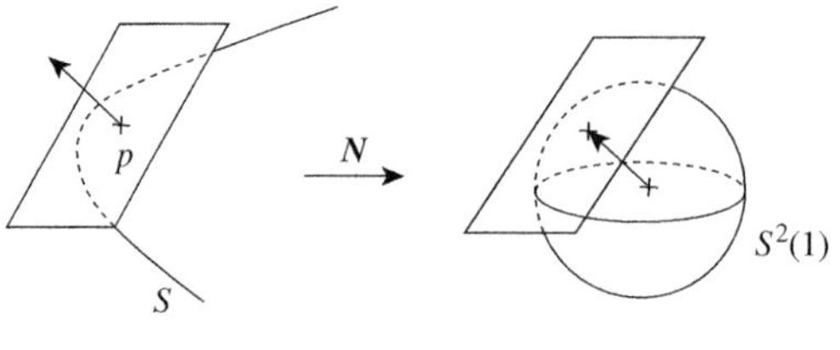

Figure 5.2 Gauss map of general surface

Example 3 (Surface of revolution) Let S be the surface generated by rotating the curve $(f(v), 0, g(v))$, $f(v) > 0\ \forall v$, about the z-axis. Then S has a local parametrisation

$$\boldsymbol{x}(u, v) = (f(v)\cos u, f(v)\sin u, g(v)), \quad u \in (-\pi, \pi),$$

and we saw in Example 2 of §4.1 that

$$\boldsymbol{N} = \frac{\boldsymbol{x}_u \times \boldsymbol{x}_v}{|\boldsymbol{x}_u \times \boldsymbol{x}_v|} = \frac{(g'\cos u, g'\sin u, -f')}{(f'^2 + g'^2)^{1/2}}\,.$$

If the generating curve is parametrised by arc length then

$$f'^2 + g'^2 = 1 \tag{5.2}$$

and

$$\begin{aligned} -d\boldsymbol{N}(\boldsymbol{x}_u) &= -\boldsymbol{N}_u = (g'\sin u, -g'\cos u, 0)\,, \\ -d\boldsymbol{N}(\boldsymbol{x}_v) &= -\boldsymbol{N}_v = (-g''\cos u, -g''\sin u, f'')\,. \end{aligned}$$

Differentiating (5.2), we obtain

$$f'f'' + g'g'' = 0\,, \tag{5.3}$$

from which it follows that, when the generating curve is parametrised by arc length,

$$-\,d\boldsymbol{N}(\boldsymbol{x}_u) = -\frac{g'}{f}\boldsymbol{x}_u\,, \quad -d\boldsymbol{N}(\boldsymbol{x}_v) = -\frac{g''}{f'}\boldsymbol{x}_v = \frac{f''}{g'}\boldsymbol{x}_v\,. \tag{5.4}$$

(We give two expressions for $-d\boldsymbol{N}(\boldsymbol{x}_v)$, so we can evaluate this at points where either f' or g' is zero.)

As already noted, the Gauss map of a surface in $\mathbb{R}^3$ is only defined up to sign. When, as in the previous example, a local parametrisation has been chosen, we shall always take $\boldsymbol{N}$ to be a positive scalar multiple of $\boldsymbol{x}_u \times \boldsymbol{x}_v$.

We now discuss a very important property of the Weingarten map $-d\boldsymbol{N}_p$; it is *self-adjoint* at each point $p \in S$, or, in symbols,

$$d\boldsymbol{N}_p(\boldsymbol{X}).\boldsymbol{Y} = \boldsymbol{X}.d\boldsymbol{N}_p(\boldsymbol{Y})\,, \quad \forall\, \boldsymbol{X}, \boldsymbol{Y} \in T_pS\,. \tag{5.5}$$

Theorem 4 *For each point p on a surface S in $\mathbb{R}^3$, the Weingarten map $-d\boldsymbol{N}_p : T_pS \to T_pS$ is a self-adjoint linear map.*

Proof It suffices to check (5.5) in the case in which $\{\boldsymbol{X}, \boldsymbol{Y}\}$ is a basis of T_pS. So, if $\boldsymbol{x}(u, v)$ is a local parametrisation whose image contains p, we need to show that

$$d\boldsymbol{N}(\boldsymbol{x}_u).\boldsymbol{x}_v = \boldsymbol{x}_u.d\boldsymbol{N}(\boldsymbol{x}_v)\,.$$

To do this, we note that $\boldsymbol{N}.\boldsymbol{x}_u = 0$ and $\boldsymbol{N}.\boldsymbol{x}_v = 0$, so by differentiation,

$$\boldsymbol{N}_v.\boldsymbol{x}_u + \boldsymbol{N}.\boldsymbol{x}_{uv} = 0\,, \qquad \boldsymbol{N}_u.\boldsymbol{x}_v + \boldsymbol{N}.\boldsymbol{x}_{vu} = 0\,. \tag{5.6}$$

Subtracting and using the fact that $\boldsymbol{x}_{uv} = \boldsymbol{x}_{vu}$, we find that

$$\boldsymbol{N}_v.\boldsymbol{x}_u - \boldsymbol{N}_u.\boldsymbol{x}_v = 0\,, \tag{5.7}$$

and the result follows from (5.1). □

It will be recalled from a first course in linear algebra that if V is an n-dimensional vector space equipped with an inner product, and if ℓ is a self-adjoint linear map from V to V, then it is always possible to find an orthonormal basis $\{\boldsymbol{w}_1, \dots, \boldsymbol{w}_n\}$ of V consisting of *eigenvectors* of ℓ, that is to say, there are real numbers $\lambda_1, \dots, \lambda_n$, the corresponding *eigenvalues*, such that $\ell(\boldsymbol{w}_i) = \lambda_i \boldsymbol{w}_i$ for each $i = 1, \dots, n$. This is often very useful; for instance it is the key fact which leads to the classification of conics in $\mathbb{R}^2$ and, more generally, quadrics in $\mathbb{R}^n$. In our situation, for each $p \in S$ the self-adjoint map $-d\boldsymbol{N}_p$ maps T_pS to itself, and so T_pS admits an orthonormal basis $\{\boldsymbol{e}_1, \boldsymbol{e}_2\}$ of eigenvectors.

Example 5 (Surface of revolution) If we consider the standard parametrisation of a surface of revolution when the generating curve is parametrised by arc length, then (5.4) shows that $\{\boldsymbol{x}_u/f,\ \boldsymbol{x}_v\}$ is an orthonormal basis of eigenvectors of $-d\boldsymbol{N}$, and the corresponding eigenvalues are $-g'/f$ and $-g''/f'$.

It is clear that the eigenvectors and eigenvalues of the Weingarten map are going to be important in describing how S curves in $\mathbb{R}^3$; we discuss these, and related quantities, later in this chapter.

5.2 Second fundamental form

As was mentioned at the beginning of the chapter, the Weingarten map $-d\boldsymbol{N}_p$ is crucial in describing the way in which a surface S in $\mathbb{R}^3$ is curving at a point $p \in S$. Since the Weingarten map is self-adjoint it may be studied by using the associated quadratic form, the *second fundamental form*, which is defined for vectors $\boldsymbol{X}$ tangential to S using the inner product by

$$II(\boldsymbol{X}) = -\boldsymbol{X}.d\boldsymbol{N}(\boldsymbol{X})\,. \tag{5.8}$$

In a similar way as for the first fundamental form, when a local parametrisation $\boldsymbol{x}(u, v)$ has been chosen, the *coefficients of the second fundamental form* are given by

$$L = -\boldsymbol{x}_u.\boldsymbol{N}_u\,, \quad M = -\boldsymbol{x}_u.\boldsymbol{N}_v = -\boldsymbol{x}_v.\boldsymbol{N}_u\,, \quad N = -\boldsymbol{x}_v.\boldsymbol{N}_v\,, \tag{5.9}$$

so that

$$\begin{aligned} II(a\boldsymbol{x}_u + b\boldsymbol{x}_v) &= -(a\boldsymbol{x}_u + b\boldsymbol{x}_v).d\boldsymbol{N}(a\boldsymbol{x}_u + b\boldsymbol{x}_v) \\ &= a^2L + 2abM + b^2N\,. \end{aligned} \tag{5.10}$$

We obtain alternative expressions for the coefficients of the second fundamental form by differentiating $\boldsymbol{x}_u.\boldsymbol{N} = 0$ and $\boldsymbol{x}_v.\boldsymbol{N} = 0$. We find that

$$L = \boldsymbol{x}_{uu}.\boldsymbol{N}\,, \quad M = \boldsymbol{x}_{uv}.\boldsymbol{N}\,, \quad N = \boldsymbol{x}_{vv}.\boldsymbol{N}\,. \tag{5.11}$$

Example 1 (Surface of revolution) Let $\boldsymbol{x}(u, v)$ be the standard parametrisation of a surface of revolution as considered, for example, in Example 3 of §5.1. Then, as stated in that example,

$$N(u,v) = \frac{(g'\cos u, g'\sin u, -f')}{(f'^2 + g'^2)^{1/2}}\,.$$

We also have that

$$\begin{aligned} x_{uu} &= (-f(v)\cos u, -f(v)\sin u, 0)\,, \\ x_{uv} &= \big(-f'(v)\sin u, f'(v)\cos u, 0\big)\,, \\ x_{vv} &= \big(f''(v)\cos u, f''(v)\sin u, g''(v)\big)\,. \end{aligned}$$

Hence

$$L = x_{uu}.N = -\frac{fg'}{(f'^2 + g'^2)^{1/2}}\,,$$

and in a similar way,

$$M = 0\,, \quad N = \frac{f''g' - f'g''}{(f'^2 + g'^2)^{1/2}}\,.$$

5.3 Matrix of the Weingarten map

A linear map is often studied by considering its matrix with respect to some suitable basis. A local parametrisation $x(u,v)$ of a surface S in $\mathbb{R}^3$ provides us with a basis $\{x_u, x_v\}$ of each tangent space, and in this section we show how to use the coefficients E, F, G of the first fundamental form and L, M, N of the second fundamental form to calculate the matrix of the Weingarten map with respect to this basis. To aid the use of matrix notation we shall write u_1, u_2 in place of u, v; x_1, x_2 in place of x_u, x_v; and N_1, N_2 in place of N_u, N_v. In a similar spirit we replace E, F, G by g_{11}, g_{12}, g_{22}, so that $g_{ij} = x_i.x_j$ for $i,j = 1,2$. We also replace L, M, N by h_{11}, h_{12}, h_{22}, so that, using (5.9) and (5.11),

$$h_{ij} = x_{ij}.N = -x_j.N_i = -x_i.N_j\,, \quad i,j = 1,2\,.$$

The matrix

$$(a_{ij}) = \begin{pmatrix} a_{11} & a_{12} \\ a_{21} & a_{22} \end{pmatrix}$$

of the Weingarten map $-dN$ with respect to the basis $\{x_1,\ x_2\}$ of the tangent space of S is defined by setting

$$-N_1 = -dN(x_1) = a_{11}x_1 + a_{21}x_2, \qquad (5.12)$$

$$-N_2 = -dN(x_2) = a_{12}x_1 + a_{22}x_2, \qquad (5.13)$$

or, more compactly,

$$-N_k = -dN(x_k) = \sum_{j=1}^{2} a_{jk}x_j\,, \quad k = 1,2\,. \qquad (5.14)$$

Example 1 (Surface of revolution) We see from (5.4) that the matrix of the Weingarten map for the standard parametrisation of a surface of revolution when the generating curve is parametrised by arc length is given by

$$\begin{pmatrix} -g'/f & 0 \\ 0 & -g''/f' \end{pmatrix}.$$

The above example is rather simple. If the matrix of the Weingarten map is not diagonal then the entries are more difficult to find directly. To help with this, we obtain an expression for the matrix (a_{ij}) in terms of the coefficients E, F, G and L, M, N of the first and second fundamental forms.

We first note that, using (5.14),

$$h_{ik} = -\boldsymbol{x}_i . \boldsymbol{N}_k = \sum_{j=1}^{2} g_{ij} a_{jk},$$

which gives the matrix equation

$$\begin{pmatrix} h_{11} & h_{12} \\ h_{21} & h_{22} \end{pmatrix} = \begin{pmatrix} g_{11} & g_{12} \\ g_{21} & g_{22} \end{pmatrix} \begin{pmatrix} a_{11} & a_{12} \\ a_{21} & a_{22} \end{pmatrix}.$$

The first matrix on the right hand side of the above equation is non-singular since it has determinant $EG - F^2$, which, by Lemma 3 of §3.2, is non-zero. Hence

$$\begin{pmatrix} a_{11} & a_{12} \\ a_{21} & a_{22} \end{pmatrix} = \begin{pmatrix} g_{11} & g_{12} \\ g_{21} & g_{22} \end{pmatrix}^{-1} \begin{pmatrix} h_{11} & h_{12} \\ h_{21} & h_{22} \end{pmatrix}.$$

If we now replace the g_{ij} and the h_{ij} by the coefficients of the first and second fundamental forms, the above equation gives the matrix (a_{ij}) of the Weingarten map $-d\boldsymbol{N}$ in terms of these coefficients as

$$\begin{pmatrix} a_{11} & a_{12} \\ a_{21} & a_{22} \end{pmatrix} = \begin{pmatrix} E & F \\ F & G \end{pmatrix}^{-1} \begin{pmatrix} L & M \\ M & N \end{pmatrix}, \tag{5.15}$$

which leads to the following expression for the matrix of the Weingarten map.

Proposition 2 *Let $\boldsymbol{x}(u, v)$ be a local parametrisation of a surface S in $\mathbb{R}^3$. Then the matrix (a_{ij}) of the Weingarten map is given in terms of the coefficients of the first and second fundamental forms of $\boldsymbol{x}$ by*

$$\begin{pmatrix} a_{11} & a_{12} \\ a_{21} & a_{22} \end{pmatrix} = \frac{1}{EG - F^2} \begin{pmatrix} GL - FM & GM - FN \\ -FL + EM & -FM + EN \end{pmatrix}. \tag{5.16}$$

Example 3 (Hyperbolic paraboloid) We consider the hyperbolic paraboloid (Figure 5.3) with equation $z = xy$, and parametrise it as a graph,

$$\boldsymbol{x}(u, v) = (u, v, uv)\,, \quad u, v \in \mathbb{R}\,.$$

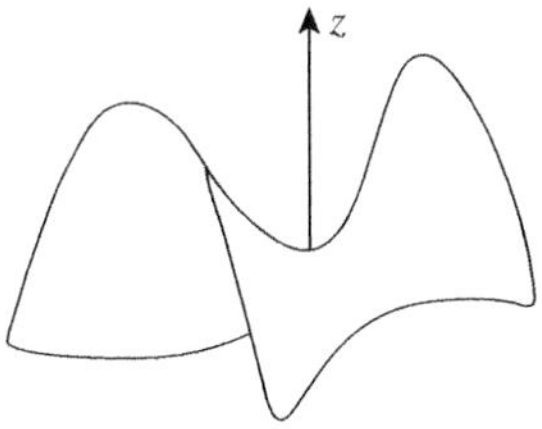

Figure 5.3 Hyperbolic paraboloid

Routine calculations then show that the coefficients of the first fundamental form are given by

$$E = 1 + v^2\,, \quad F = uv\,, \quad G = 1 + u^2\,.$$

We may calculate the coefficients of the second fundamental form by showing that

$$N(u, v) = \frac{\boldsymbol{x}_u \times \boldsymbol{x}_v}{|\boldsymbol{x}_u \times \boldsymbol{x}_v|} = \frac{(-v, -u, 1)}{D^{1/2}}\,,$$

where

$$D = 1 + u^2 + v^2\,,$$

while

$$\boldsymbol{x}_{uu} = (0, 0, 0)\,, \quad \boldsymbol{x}_{uv} = (0, 0, 1)\,, \quad \boldsymbol{x}_{vv} = (0, 0, 0)\,,$$

so that

$$\begin{aligned} L &= \boldsymbol{x}_{uu}.\boldsymbol{N} = 0\,, \\ M &= \boldsymbol{x}_{uv}.\boldsymbol{N} = D^{-1/2}\,, \\ N &= \boldsymbol{x}_{vv}.\boldsymbol{N} = 0\,. \end{aligned}$$

A straightforward substitution now shows that the matrix of the Weingarten map is given by

$$\begin{pmatrix} a_{11} & a_{12} \\ a_{21} & a_{22} \end{pmatrix} = D^{-3/2} \begin{pmatrix} -uv & 1 + u^2 \\ 1 + v^2 & -uv \end{pmatrix}.$$

5.4 Gaussian and mean curvature

The individual entries a_{ij} of the matrix of the Weingarten map are not, in themselves, of geometrical significance since they depend on the choice of local parametrisation $\boldsymbol{x}$. However, the *trace* and *determinant* of this matrix **are** geometrically important quantities because, as for any linear transformation of a finite dimensional vector space, they do not depend on the choice of basis; they are quantities which depend on only the map itself. The determinant has a geometrical interpretation as the scale factor by which area in the tangent space is multiplied under the Weingarten map; its sign indicates whether $\boldsymbol{N}_u \times \boldsymbol{N}_v$ is a positive or a negative scalar multiple of $\boldsymbol{x}_u \times \boldsymbol{x}_v$.

The determinant of the Weingarten map $-d\boldsymbol{N}_p$ at a point $p \in S$ is called the *Gaussian curvature* $K(p)$ of S at p, while the *mean curvature* $H(p)$ is defined to be half the trace. So, if (a_{ij}) is the matrix of the Weingarten map with respect to any basis of the tangent space, then

$$K = a_{11}a_{22} - a_{12}a_{21}, \quad H = \frac{1}{2}(a_{11} + a_{22}). \tag{5.17}$$

Just to reiterate, since it is so important; K and H are functions **defined on the surface** S, and are independent of choice of local parametrisation. However, as for all functions on a surface, once a local parametrisation $\boldsymbol{x}(u, v)$ has been chosen, K and H may be considered as functions of u and v, and we now obtain expressions for these functions in terms of the coefficients of the fundamental forms of $\boldsymbol{x}$.

Since the determinant of the product of two matrices is the product of the determinants, equation (5.15) shows that the Gaussian curvature K is given in terms of the coefficients of the fundamental forms of a local parametrisation by

$$K = \det(-d\boldsymbol{N}) = \frac{LN - M^2}{EG - F^2}, \tag{5.18}$$

while (5.16) shows that the mean curvature H is given by

$$H = \frac{1}{2}\mathrm{tr}(-d\boldsymbol{N}) = \frac{1}{2}\frac{EN - 2FM + GL}{EG - F^2}. \tag{5.19}$$

Example 1 (Hyperbolic paraboloid) Following on from Example 3 of §5.3, the hyperbolic paraboloid in that example has Gaussian curvature K and mean curvature H given by

$$K = \det(-d\boldsymbol{N}) = \frac{LN - M^2}{EG - F^2} = -1/D^2,$$
$$H = \frac{1}{2}\mathrm{tr}(-d\boldsymbol{N}) = \frac{1}{2}\frac{EN - 2FM + GL}{EG - F^2} = -uv/D^{3/2}.$$

In particular, we note that the hyperbolic paraboloid has negative Gaussian curvature at all points. The geometrical significance of this will be explored later in the chapter.

We note that, for an isothermal parametrisation, the expression for H takes a particularly simple form.

Lemma 2 *If $\boldsymbol{x}$ is an isothermal local parametrisation of S, with $E = G = \lambda^2$ (and $F = 0$), then the mean curvature H is given by*

$$H = \frac{L + N}{2\lambda^2}.$$

Note that, for a general surface in $\mathbb{R}^3$, if $\boldsymbol{N}$ is replaced by $-\boldsymbol{N}$ then K remains unchanged, but H changes sign. However the *mean curvature vector* $\boldsymbol{H}$ given by $\boldsymbol{H} = H\boldsymbol{N}$ does not change. For example, the mean curvature vector $\boldsymbol{H}$ for the sphere $S^2(r)$ is the inward normal of length $1/r$.

Much of the rest of this book describes some of the geometry associated with the Gaussian curvature K and the mean curvature H. For instance, we shall see that:

(i) S is a "soap film" if $H \equiv 0$, and is a "soap bubble" if H is a non-zero constant (these surfaces are discussed in Chapter 9);
(ii) the sign of K at a point $p \in S$ determines whether a sufficiently small open neighbourhood of p in S lies on one side of its tangent plane (think of a sphere, where the answer is "yes", and a hyperbolic paraboloid where the answer is "no").

In fact, we shall see in the next chapter that K is **much** more important than you may currently think; so important, indeed, that its study instigated and motivated a major branch of modern mathematics, called Riemannian geometry.

5.5 Principal curvatures and directions

We saw in §5.1 that at each point p of a surface S in $\mathbb{R}^3$, the Weingarten map $-d\boldsymbol{N}_p$ is self-adjoint, so that the tangent space T_pS has an orthonormal basis of eigenvectors. The eigenvalues of $-d\boldsymbol{N}_p$ are called the *principal curvatures* κ_1, κ_2, of S at p, and the eigenvectors of $-d\boldsymbol{N}_p$ are called the *principal vectors*. The directions determined by the principal vectors are the *principal directions*. The following lemma is a direct consequence of the definitions.

Lemma 1 *A non-zero tangent vector $\boldsymbol{X}$ to a surface S in $\mathbb{R}^3$ is in a principal direction if and only if*

$$d\boldsymbol{N}(\boldsymbol{X}) = \lambda \boldsymbol{X}$$

for some real number λ. In this case, $-\lambda$ is the corresponding principal curvature. In particular, if $\boldsymbol{x}$ is a local parametrisation of S, then $\boldsymbol{x}_u$ is in a principal direction if and only if

$$\boldsymbol{N}_u = \lambda \boldsymbol{x}_u \,,$$

where $-\lambda$ is the corresponding principal curvature (with a similar result, of course, for $\boldsymbol{x}_v$).

In this section we investigate the relation between the principal curvatures and the Gaussian and mean curvatures.

Since we already have the expressions (5.18) and (5.19) for K and H in terms of the coefficients of the fundamental forms of a local parametrisation, we begin by finding the principal curvatures κ_1 and κ_2 in terms of K and H. If A is the matrix of the Weingarten map with respect to some basis of the tangent space, and if I denotes the identity matrix, then the principal curvatures are the roots of the *characteristic equation* $\det(A - \kappa I) = 0$ of the Weingarten map, which, using the fact that K and H are, respectively, the determinant and half the trace of the Weingarten map, may be written as

$$\kappa^2 - 2H\kappa + K = 0 \,. \tag{5.20}$$

Since the Weingarten map is self-adjoint, this quadratic equation has two real roots (allowing the possibility of one repeated root), namely the principal curvatures. The result of the next lemma follows from the well-known formula for the roots of a quadratic equation.

Lemma 2 *$H^2 - K \geq 0$ and the principal curvatures are given by $H \pm \sqrt{H^2 - K}$.*

As is the case for K and H, the principal curvatures and directions are properties of the surface itself. However, once a local parametrisation $\boldsymbol{x}(u, v)$ has been chosen, we may regard them as functions of u and v.

Example 3 (Hyperbolic paraboloid) Using the expressions for K and H found in Example 1 in §5.4, we quickly find that the principal curvatures of the hyperbolic paraboloid parametrised by

$$\boldsymbol{x}(u, v) = (u, v, uv)\,, \quad u, v \in \mathbb{R}\,,$$

are given by

$$\kappa_1, \kappa_2 = H \pm \sqrt{H^2 - K} = -\frac{uv \pm \sqrt{(1+u^2)(1+v^2)}}{D^{3/2}}\,,$$

where $D = 1 + u^2 + v^2$.

We have just seen how the principal curvatures κ_1 and κ_2 may be found in terms of K and H. Conversely, since the left hand side of the characteristic equation (5.20) is equal to $(\kappa - \kappa_1)(\kappa - \kappa_2)$, we find the following.

Lemma 4 *The mean curvature H is the average of the two principal curvatures, and the Gaussian curvature K is their product. In symbols,*

$$H = \frac{1}{2}(\kappa_1 + \kappa_2), \quad K = \kappa_1 \kappa_2\,.$$

We note that Lemma 4 also follows immediately from the fact that the matrix of the Weingarten map with respect to a basis of eigenvectors is diagonal, and the corresponding eigenvalues κ_1 and κ_2 are the entries down the diagonal.

The following result is now immediate from (5.4).

Corollary 5 *Let S be a surface of revolution whose generating curve $(f(v), 0, g(v))$, $f(v) > 0\ \forall v$, is parametrised by arc length. Then the principal directions are given by the coordinate vectors, the principal curvatures are given by*

$$\kappa_1 = -g'/f\,, \quad \kappa_2 = -g''/f' = f''/g',$$

and the mean and Gaussian curvatures are given by

$$H = -\frac{1}{2}\left(\frac{g'}{f} + \frac{g''}{f'}\right), \quad K = -\frac{f''}{f}\,.$$

In the next section, we generalise much of Corollary 5 to the case of a surface of revolution whose generating curve is not necessarily parametrised by arc length.

5.6 Examples: surfaces of revolution

In this section we give several examples of explicit calculations of the principal curvatures and directions for a surface S in $\mathbb{R}^3$ using a particular type of local parametrisation $\boldsymbol{x}(u, v)$. As previously mentioned, we take the Gauss map $\boldsymbol{N}$ to be in the direction of $\boldsymbol{x}_u \times \boldsymbol{x}_v$.

As we have seen, we may use the coefficients of the first and second fundamental forms to find K and H, and then use Lemma 2 of §5.5 (or factorise the characteristic equation) to find the principal curvatures κ_1 and κ_2. If we have a local parametrisation for which the coordinate vectors $\boldsymbol{x}_u$ and $\boldsymbol{x}_v$ are both in principal directions at each point, then κ_1 and κ_2 may be found more directly. For one of the statements of the following lemma, we have to assume that we are not at an *umbilic* of S, that is to say, not at a point of S where the characteristic equation has just one (repeated) root. We consider umbilics in §5.8.

Lemma 1 *If a local parametrisation $\boldsymbol{x}(u, v)$ of a surface S in $\mathbb{R}^3$ has $F = M = 0$, then the coordinate vectors $\boldsymbol{x}_u$ and $\boldsymbol{x}_v$ are in principal directions. Conversely, at a non-umbilic point, if $\boldsymbol{x}_u$ and $\boldsymbol{x}_v$ are in principal directions then $F = M = 0$.*

If $F = M = 0$ then the principal curvatures are L/E and N/G, and the matrix of the Weingarten map with respect to $\boldsymbol{x}_u$ and $\boldsymbol{x}_v$ is given by

$$\begin{pmatrix} L/E & 0 \\ 0 & N/G \end{pmatrix}.$$

Proof A proof may be easily given using the expression (5.16) for the matrix of the Weingarten map in terms of the coefficients of the fundamental forms of $\boldsymbol{x}(u, v)$. However, we prefer to present an alternative proof which reinforces several of the ideas we have discussed.

First assume that $F = M = 0$. Then $\boldsymbol{x}_v.\boldsymbol{N}_u = -M = 0$, so that $\boldsymbol{N}_u$ is orthogonal to $\boldsymbol{x}_v$, and hence, since $F = 0$, is a scalar multiple of $\boldsymbol{x}_u$. A similar argument shows that $\boldsymbol{N}_v$ is a scalar multiple of $\boldsymbol{x}_v$. Hence $\boldsymbol{x}_u$ and $\boldsymbol{x}_v$ are in principal directions, and, if κ_1 and κ_2 are the principal curvatures, then

$$-\boldsymbol{N}_u = \kappa_1 \boldsymbol{x}_u\,, \quad -\boldsymbol{N}_v = \kappa_2 \boldsymbol{x}_v\,. \tag{5.21}$$

Conversely, if the coordinate vectors $\boldsymbol{x}_u$ and $\boldsymbol{x}_v$ are in principal directions at a non-umbilic point then, being eigenvectors corresponding to different eigenvalues of a self-adjoint operator, they are orthogonal. Also, if κ_1 and κ_2 are the principal curvatures, then

$$M = -\boldsymbol{N}_u.\boldsymbol{x}_v = \kappa_1 \boldsymbol{x}_u.\boldsymbol{x}_v = 0\,.$$

Assume now that $F = M = 0$. Then, using (5.21),

$$L = -\boldsymbol{N}_u.\boldsymbol{x}_u = \kappa_1 \boldsymbol{x}_u.\boldsymbol{x}_u = \kappa_1 E\,,$$

so that $\kappa_1 = L/E$. The expression for κ_2 follows in a similar manner, and the lemma is proved. □

Example 2 (Surface of revolution) We found in Example 2 of §3.2 and Example 1 of §5.2 that, for the standard parametrisation of a surface of revolution

$$\boldsymbol{x}(u,v) = (f(v)\cos u, f(v)\sin u, g(v)), \quad u \in (-\pi,\pi), \ f(v) > 0 \,\forall v,$$

the coefficients of the first and second fundamental forms are given by

$$\begin{aligned} E &= f^2, \qquad F = 0, \qquad G = f'^2 + g'^2, \\ L &= -\frac{fg'}{(f'^2+g'^2)^{1/2}}, \quad M = 0, \quad N = \frac{f''g' - f'g''}{(f'^2+g'^2)^{1/2}}. \end{aligned} \tag{5.22}$$

Hence, by Lemma 1, the coordinate vectors are in principal directions and the principal curvatures are given by

$$\kappa_1 = -\frac{g'}{f(f'^2+g'^2)^{1/2}}, \qquad \kappa_2 = \frac{f''g' - f'g''}{(f'^2+g'^2)^{3/2}}. \tag{5.23}$$

It may be easily checked that if $g'(v) > 0$ then the following give alternative ways of writing the formulae in (5.23) for κ_1 and κ_2:

$$\kappa_1 = -\frac{1}{f(1+(f'/g')^2)^{1/2}}, \quad \kappa_2 = \left(\frac{f'}{g'}\right)' \frac{1}{g'(1+(f'/g')^2)^{3/2}}. \tag{5.24}$$

If $g'(v) < 0$ then the formulae for κ_1 and κ_2 are the negative of those given in (5.24).

We now obtain an expression for the Gaussian curvature of a surface of revolution. This generalises the formula obtained in Corollary 5 in §5.5 for the case in which the generating curve is parametrised by arc length. The proposition may be proved by expanding formula (5.25) in the statement of the proposition to obtain the product of κ_1 and κ_2 as given in (5.23).

Proposition 3 *Let S be the surface of revolution parametrised by*

$$\boldsymbol{x}(u,v) = (f(v)\cos u, f(v)\sin u, g(v)), \quad f(v) > 0 \,\forall v.$$

Then the Gaussian curvature K is given by

$$K = \frac{1}{2ff'}\left\{\frac{g'^2}{f'^2+g'^2}\right\}'. \tag{5.25}$$

The following proposition is easily proved using Lemma 4 of §5.5, and taking $g(v) = v$ in formulae (5.23) for the principal curvatures κ_1 and κ_2.

Proposition 4 *Let S be the surface of revolution parametrised by*

$$\boldsymbol{x}(u,v) = (f(v)\cos u, f(v)\sin u, v), \quad f(v) > 0 \,\forall v.$$

Then the mean curvature H and the Gaussian curvature K are given by

$$H = \frac{ff'' - f'^2 - 1}{2f(1+f'^2)^{3/2}}, \qquad K = -\frac{f''}{f(1+f'^2)^2}.$$

We now consider some particular examples of surfaces of revolution.

Example 5 (Catenoid) We consider the catenoid obtained by rotating the catenary ($\cosh v$, $0, v$) about the z-axis. It follows quickly from Proposition 4 that this catenoid has constant mean curvature $H = 0$ (as, indeed, do all the other catenoids – see Exercise 5.10). This has great geometrical significance; it means that catenoids are *minimal* or *soap-film* surfaces. Each catenoid is the shape taken up by a soap film suspended between two circular loops of wire. The mathematics associated with minimal surfaces is very elegant, and involves complex analysis in a crucial way. We discuss minimal surfaces in Chapter 9.

Example 6 (Torus of revolution) This is generated by rotating the curve given by

$$f(v) = a + b\cos v\,, \quad g(v) = b\sin v\,, \quad u, v \in (-\pi, \pi)\,, \quad a > b > 0\,,$$

about the z-axis.

A short calculation using (5.23) shows that

$$\kappa_1 = -\frac{\cos v}{a + b\cos v}, \qquad \kappa_2 = -\frac{1}{b}.$$

Example 7 (Pseudosphere) This is generated by rotating the tractrix, which may be parametrised by taking

$$f(v) = \operatorname{sech} v\,, \quad g(v) = v - \tanh v\,, \quad v > 0\,,$$

about the z-axis (see Figure 2.10 for a picture of the pseudosphere).

In this case,

$$f' = -\operatorname{sech} v \tanh v\,, \quad g' = \tanh^2 v\,,$$

so that

$$\frac{f'}{g'} = -\frac{\operatorname{sech} v}{\tanh v} = -\frac{1}{\sinh v}.$$

Hence, using (5.24),

$$\kappa_1 = -\frac{\cosh v}{(1 + \sinh^{-2} v)^{1/2}} = -\sinh v\,,$$

while

$$\kappa_2 = -\left(\frac{1}{\sinh v}\right)' \frac{1}{\tanh^2 v \coth^3 v} = \frac{1}{\sinh v}.$$

In contrast to the unit sphere, which has constant Gaussian curvature $K = 1$, we see that the pseudosphere has constant Gaussian curvature $K = -1$. This provides the motivation for the name of this surface.

You may recall that we discussed the pseudosphere in (the optional) Example 5 of §3.4, where we considered the hyperbolic plane. We described an isometry from part of the hyperbolic plane to the pseudosphere, and we shall indicate the significance of this in the next chapter, where we concentrate on the geometric information carried by the Gaussian curvature.

5.7 Normal curvature

We now begin our investigation of the geometrical information which is contained in the quantities we have defined.

In this section we define the *normal curvature* κ_n of a regular curve $\boldsymbol{\alpha}$ on a surface S in $\mathbb{R}^3$, and show how it may be measured using the second fundamental form of S. As usual, we let d/ds denote differentiation with respect to an arc-length parameter s along $\boldsymbol{\alpha}$.

We first consider the orthonormal moving frame $\{\boldsymbol{t}, \boldsymbol{N}\times\boldsymbol{t}, \boldsymbol{N}\}$ along $\boldsymbol{\alpha}$, where $\boldsymbol{t} = d\boldsymbol{\alpha}/ds$ is the unit tangent vector to $\boldsymbol{\alpha}$, and, as usual, $\boldsymbol{N}$ is the unit normal to S. This moving frame, shown in Figure 5.4, reflects both the geometry of the curve and the geometry of the surface on which it lies, whereas the orthonormal moving frame $\{\boldsymbol{t}, \boldsymbol{n}, \boldsymbol{b}\}$ along $\boldsymbol{\alpha}$ described in Chapter 1 depends on only the geometry of the curve itself.

Since $d\boldsymbol{t}/ds$ is orthogonal to $\boldsymbol{t}$, we have the decomposition

$$\frac{d\boldsymbol{t}}{ds} = \kappa_g \boldsymbol{N} \times \boldsymbol{t} + \kappa_n \boldsymbol{N} \tag{5.26}$$

of $d\boldsymbol{t}/ds$ into components, the first of which is tangential to S and the second is orthogonal to S. Then κ_g is called the *geodesic curvature* of $\boldsymbol{\alpha}$, while κ_n is the *normal curvature* of $\boldsymbol{\alpha}$. We note that both κ_g and κ_n change sign when $\boldsymbol{N}$ is replaced by $-\boldsymbol{N}$.

In this chapter we shall consider normal curvature κ_n, and we begin by showing that κ_n gives a measure of how the surface S curves in $\mathbb{R}^3$ as we travel along $\boldsymbol{\alpha}$.

Proposition 1 *Let $\boldsymbol{\alpha}$ be a regular curve on a surface S in $\mathbb{R}^3$ and let $\boldsymbol{t} = d\boldsymbol{\alpha}/ds$ be the unit tangent vector to $\boldsymbol{\alpha}$. Then the normal curvature κ_n of $\boldsymbol{\alpha}$ is given by*

$$\kappa_n = II(\boldsymbol{t})\,. \tag{5.27}$$

Proof Since $\boldsymbol{t}.\boldsymbol{N} = 0$, we have

$$\kappa_n = \frac{d\boldsymbol{t}}{ds}.\boldsymbol{N} = -\boldsymbol{t}.\frac{d\boldsymbol{N}}{ds}\,. \tag{5.28}$$

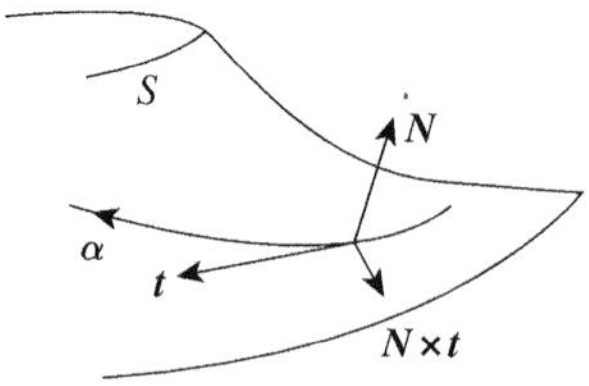

Figure 5.4 An orthonormal frame

However, from (4.7), $dN/ds = dN(t)$, so the result follows from the definition of the second fundamental form. □

Proposition 1 is perhaps rather surprising; dt/ds is the **rate of change** of the unit tangent vector of $\boldsymbol{\alpha}$ as we travel along $\boldsymbol{\alpha}$ at unit speed, but (5.27) shows that the normal component of this rate of change at a point $p \in S$ depends only on the tangent vector to $\boldsymbol{\alpha}$ **at the point** p **itself**. This gives a theorem due to Meusnier.

Theorem 2 (Meusnier) *All regular curves on a surface S in $\mathbb{R}^3$ through a point p on S having the same tangent line at p have the same normal curvature at p.*

We now see one reason why the Weingarten map is traditionally defined to be *minus* the derivative of the Gauss map. Without this sign, the right hand side of (5.27) would have a minus sign in it.

We recall that if κ is the curvature of $\boldsymbol{\alpha}$ as a regular curve in $\mathbb{R}^3$, then $\kappa = |dt/ds|$, so it follows from (5.26) that

$$\kappa^2 = \kappa_g^2 + \kappa_n^2 . \tag{5.29}$$

This leads to another interpretation of the normal curvature; it gives a measure of the minimum amount of bending a curve $\boldsymbol{\alpha}$ on a surface S must do in order to stay on S. We shall see in Chapter 7 that geodesic curvature κ_g may be interpreted as a measure of the extra bending that $\boldsymbol{\alpha}$ does within S.

Finding the normal curvature of a curve $\boldsymbol{\alpha}$ not necessarily parametrised by arc length is straightforward since, in this case,

$$\kappa_n = \frac{1}{|\boldsymbol{\alpha}'|^2}\boldsymbol{\alpha}''.\boldsymbol{N} , \tag{5.30}$$

or, in terms of the second fundamental form,

$$\begin{aligned}\kappa_n &= II(\boldsymbol{\alpha}'/|\boldsymbol{\alpha}'|)\\ &= \frac{II(\boldsymbol{\alpha}')}{|\boldsymbol{\alpha}'|^2} .\end{aligned} \tag{5.31}$$

If we have a local parametrisation $\boldsymbol{x}(u, v)$ of our surface S, we may use the coefficients of the first and second fundamental forms to find the normal curvature κ_n of a curve $\boldsymbol{\alpha}(t) = \boldsymbol{x}\,(u(t), v(t))$. In fact, using (5.10),

$$II(\boldsymbol{\alpha}') = u'^2 L + 2u'v'M + v'^2 N , \tag{5.32}$$

so that

$$\kappa_n = \frac{u'^2 L + 2u'v'M + v'^2 N}{u'^2 E + 2u'v'F + v'^2 G} . \tag{5.33}$$

Hence, in some sense, κ_n gives the ratio of the second and first fundamental forms.

Example 3 (Hyperbolic paraboloid) Continuing with Example 3 of §5.3; for each fixed θ, we shall find the normal curvature at $t = 0$ of the curve $\boldsymbol{\alpha}(t) = \boldsymbol{x}(t\cos\theta, t\sin\theta)$. This curve is the image under $\boldsymbol{x}$ of the line through the origin in the uv-plane making an angle θ with the u-axis (Figure 5.5).

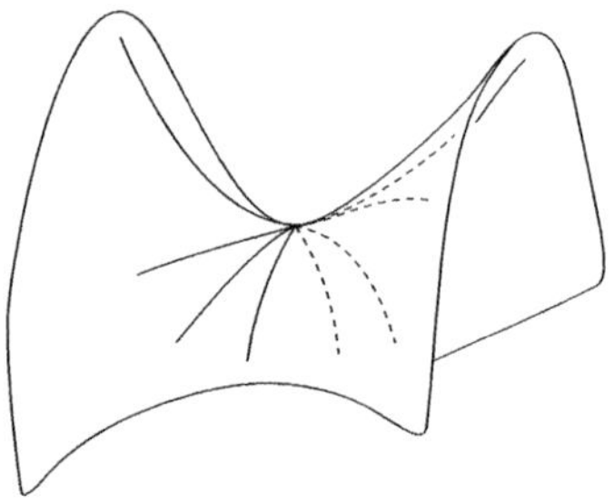

Figure 5.5 Curves through $(0, 0, 0)$ on a hyperbolic paraboloid

We first note that, at $\boldsymbol{\alpha}(0)$,

$$E = G = 1\,, \quad F = 0\,; \qquad L = N = 0\,, \quad M = 1\,,$$

so the normal curvature of $\boldsymbol{\alpha}(t)$ at $t = 0$ is given by

$$\begin{aligned}\kappa_n &= \frac{\cos^2\theta\ L + 2\cos\theta\sin\theta\ M + \sin^2\theta\ N}{\cos^2\theta\ E + 2\cos\theta\sin\theta\ F + \sin^2\theta\ G} \\ &= 2\cos\theta\sin\theta = \sin 2\theta\,.\end{aligned}$$

If we have a surface S defined by an equation, then we do not necessarily have a convenient local parametrisation. However, it follows from (5.30) that if grad f is never zero on the surface with equation $f(x, y, z) = c$, then the normal curvature κ_n of a regular curve $\boldsymbol{\alpha}(t)$ on S is given by

$$\kappa_n = \frac{\boldsymbol{\alpha}''}{|\boldsymbol{\alpha}'|^2} \cdot \frac{\operatorname{grad} f}{|\operatorname{grad} f|}\,. \tag{5.34}$$

Returning to the general situation, if $p \in S$ and $\boldsymbol{X} \in T_pS$ is a non-zero vector, then, as we saw in Meusnier's Theorem, all curves on S through p in direction $\boldsymbol{X}$ have the same normal curvature, namely $II(\boldsymbol{X}/|\boldsymbol{X}|)$. For this reason, we say that $II(\boldsymbol{X}/|\boldsymbol{X}|))$ is the *normal curvature* $\kappa_n(\boldsymbol{X})$ of S at p in direction $\boldsymbol{X}$.

The following proposition illustrates the geometrical significance of the principal curvatures and principal directions (which we defined in §5.5).

Proposition 4 *The principal curvatures of S at a point $p \in S$ give the extremal values of the normal curvatures of S at p; these values being taken in the principal directions.*

Proof This is an exercise in linear algebra, which we include for completeness. Since $-d\boldsymbol{N}_p$ is self-adjoint, we may choose an orthonormal basis $\{\boldsymbol{e}_1, \boldsymbol{e}_2\}$ of T_pS consisting of eigenvectors of $-d\boldsymbol{N}_p$. We let κ_1, κ_2 be the corresponding eigenvalues. Then, for any unit vector $\boldsymbol{e}(\theta) = \cos\theta\ \boldsymbol{e}_1 + \sin\theta\ \boldsymbol{e}_2$, we have

$$-\,dN_p(e) = \kappa_1 \cos\theta \; e_1 + \kappa_2 \sin\theta \; e_2 \,. \tag{5.35}$$

Hence

$$\begin{aligned} II(e) &= (\kappa_1 \cos\theta \; e_1 + \kappa_2 \sin\theta \; e_2).(\cos\theta \; e_1 + \sin\theta \; e_2) \\ &= \kappa_1 \cos^2\theta + \kappa_2 \sin^2\theta \\ &= (\kappa_1 - \kappa_2)\cos^2\theta + \kappa_2 \,. \end{aligned} \tag{5.36}$$

If $\kappa_1 \neq \kappa_2$, then the extremal values of $II(e)$ are κ_1 and κ_2, and they occur when $\cos^2\theta = 1$ and 0 respectively, which correspond to $e = \pm e_1$ and $e = \pm e_2$ respectively. If $\kappa_1 = \kappa_2$ then (5.35) shows that every non-zero vector is an eigenvector of the Weingarten map, and (5.36) shows that the normal curvatures are the same in all directions. A point at which $\kappa_1 = \kappa_2$ is called an *umbilic*, and these will be discussed in the next section. □

Example 5 (Hyperbolic paraboloid) Referring back to Example 3, $T_{(0,0,0)}S$ is the xy-plane, and the normal curvature of S at $(0, 0, 0)$ in direction $(\cos\theta, \sin\theta, 0)$ is $\sin 2\theta$. This attains its extremal values, namely ± 1, when $\theta = \pm\pi/4$.

This is in accord with Proposition 4, since, from Example 3 of §5.3, the matrix of the Weingarten map at $(0, 0)$ is $\begin{pmatrix} 0 & 1 \\ 1 & 0 \end{pmatrix}$, which has unit eigenvectors $(1/\sqrt{2}, \pm 1/\sqrt{2})$ with corresponding eigenvalues ± 1.

In this example, the directions for the extremal values of the normal curvature are also clear geometrically from Figure 5.5; travelling from the saddle point at the origin, the direction of maximal upward curvature is along the ridge while the direction of minimal upward curvature (that is to say, maximal downward curvature) is down the valley floor. We note these two directions are mutually orthogonal, as must be the case since they are eigenvectors corresponding to different eigenvalues of a self-adjoint linear map.

Returning to the general situation, formula (5.36) for the normal curvature in the direction of a unit vector e enables us to give further justification to the term mean curvature. We have already seen that H is the average of the eigenvalues of the Weingarten map, and we now show it is the average normal curvature over all directions on the surface at the point in question.

Proposition 6 *Let $p \in S$ and for $0 \leq \theta < 2\pi$, let $e(\theta)$ be the unit vector in T_pS making an angle θ with some fixed direction in T_pS. Then*

$$H = \frac{1}{2\pi} \int_0^{2\pi} II\,(e(\theta))\, d\theta \,.$$

Proof Choose the fixed direction to be a principal direction. Then from (5.36),

$$\begin{aligned} \int_0^{2\pi} II\,(\mathbf{e}(\theta))\, d\theta &= (\kappa_1 - \kappa_2) \int_0^{2\pi} \cos^2\theta \, d\theta + 2\pi\kappa_2 \\ &= \pi(\kappa_1 - \kappa_2) + 2\pi\kappa_2 \\ &= \pi(\kappa_1 + \kappa_2) = 2\pi H. \end{aligned}$$

□

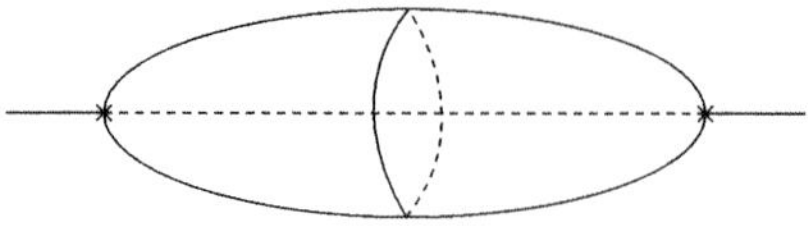

Figure 5.6 Umbilics on a rugby ball

5.8 Umbilics

An *umbilic* on a surface S in $\mathbb{R}^3$ is a point at which the characteristic equation (5.20) of the Weingarten map has just one (repeated) root, or, alternatively, where $H^2 = K$. At an umbilic we have that $-d\boldsymbol{N}_p = \kappa \operatorname{Id}$, where $\kappa = \kappa_1 = \kappa_2$ and $\operatorname{Id} : T_pS \to T_pS$ is the identity map (which follows, for instance from (5.35)). Hence every direction is a principal direction and all normal curvatures are equal. This explains the name; umbilic comes from the Latin *umbilicus* (which means navel), since the surface curves equally in all directions at such points. For instance, every point of a sphere is an umbilic, whereas on an American football or a rugby ball there are just two umbilics, namely the points where the axis of rotation cuts the ball (Figure 5.6).

The question of the existence of umbilics on surfaces is a very interesting one. It is also very important since, for instance, at such points the principal directions do not provide two distinguished directions on the surface. It follows quickly from Example 6 in §5.6 that the standard torus of revolution has no umbilics, whereas a consequence of the *Hairy Ball Theorem* is that any surface in $\mathbb{R}^3$ that is diffeomorphic to a 2-sphere must have at least one umbilic. So, what can we say if **every** point on a surface in $\mathbb{R}^3$ is an umbilic?

For the next theorem, we need to assume that our surface S is *connected*, that is to say any two points on S can be joined by a smooth curve on S. It may be shown that any surface in $\mathbb{R}^n$ is a disjoint union of connected surfaces, so the condition is not particularly restrictive, but it is clearly necessary for Theorem 1.

Here is how we use the connectedness assumption in the proof of Theorem 1. Firstly, it follows easily from the chain rule (and the Mean Value Theorem) that a smooth function with everywhere zero derivative on a connected surface is constant. Secondly, some elementary topology shows that if S is connected then S is not the disjoint union of two non-empty open subsets.

Theorem 1 *If every point of a connected surface S in $\mathbb{R}^3$ is an umbilic then S is an open subset of a plane or a sphere.*

Proof It is clear that the image of a connected set under a smooth map is connected, so, in particular, every point of any surface is contained in a connected coordinate neighbourhood. So, let $\boldsymbol{x}(u, v)$ be a local parametrisation of S whose image V is connected, and let $\boldsymbol{N} = \boldsymbol{x}_u \times \boldsymbol{x}_v / |\boldsymbol{x}_u \times \boldsymbol{x}_v|$. The hypothesis that every point of S is an umbilic implies that there is a smooth real-valued function $\lambda(u, v)$ such that

$$d\boldsymbol{N} = \lambda \operatorname{Id}, \tag{5.37}$$

where Id is the identity map of the appropriate tangent space. We show that λ is constant by showing that the derivative of λ is everywhere zero. In fact, (5.37) is equivalent to

$$N_u = dN(x_u) = \lambda x_u , \quad N_v = dN(x_v) = \lambda x_v , \qquad (5.38)$$

and, by differentiating, we find that

$$N_{uv} = \lambda_v x_u + \lambda x_{uv} , \quad N_{vu} = \lambda_u x_v + \lambda x_{vu} .$$

Subtracting these equations, we obtain

$$\lambda_v x_u = \lambda_u x_v ,$$

so, since x_u and x_v are linearly independent, we see that $\lambda_u = \lambda_v = 0$. Hence λ has everywhere zero derivative and so is constant.

Suppose first that $\lambda = 0$. Then $dN = 0$ at all points in the image V of x. Hence N is constant and V is contained in a plane (see Exercise 5.14).

Now consider the case in which λ is not zero, and consider the map from V to $\mathbb{R}^3$ given by $p \mapsto p - \lambda^{-1}N$. The derivative of this is given by Id $- \lambda^{-1}\lambda$ Id, and, since this is zero, $p - \lambda^{-1}N$ is constant, equal to a, say. Then V is an open subset of the sphere centre a radius $1/\lambda$.

The above shows that every point of S has an open neighbourhood which is a subset of either a plane or a sphere. We now use the connectivity of S to show that all points of S lie on the same plane or sphere. So, let P denote either a plane or a sphere in $\mathbb{R}^3$, and let S_P be the subset of S consisting of those points having an open neighbourhood which is a subset of P. It is clear that S_P is open, and that if Q is a plane or sphere different from P then S_P and S_Q are disjoint. Since S is assumed connected, it follows that S_P is non-empty for exactly one plane or sphere, and the result follows. □

The above proof is just what a proof in differential geometry should be! We wrote the assumption in the form of an equation, then we differentiated the equation, then we deduced a local conclusion, and then we globalised it.

Remark 2 In Example 4 of §4.4, we proved that the Gauss map of the catenoid is conformal. In fact, using an orthonormal basis of principal vectors, it follows quickly from condition (4.12) (which need only be checked when X_1, X_2 are members of that basis) that the Gauss map of a surface S in $\mathbb{R}^3$ is conformal if and only if the principal curvatures satisfy $\kappa_1 = \pm\kappa_2 \neq 0$. If $\kappa_1 = -\kappa_2$ then (as is the case for the catenoid) the mean curvature $H = 0$, while if $\kappa_1 = \kappa_2$ then every point of S is an umbilic, so that if S is connected then S is an open subset of a sphere.

5.9 Special families of curves

A regular curve $\alpha(t)$ on a surface S in $\mathbb{R}^3$ is a *line of curvature* if its tangent vector $\alpha'(t)$ is always in a principal direction. It follows from Proposition 4 of §5.7 that, at each point, a line of curvature has maximal (or minimal) normal curvature of all curves on S through that point.

There are two families of lines of curvature; they form an orthogonal net at all non-umbilic points of S. For instance, the discussion in Example 2 of §5.6 shows that the the parallels and meridians on a surface of revolution are also the lines of curvature.

The following theorem follows directly from our definitions and from (4.7), which says that $d\boldsymbol{N}(\boldsymbol{\alpha}')$ is the rate of change $\boldsymbol{N}'$ of $\boldsymbol{N}$ along $\boldsymbol{\alpha}$.

Theorem 1 (Rodrigues) *A regular curve $\boldsymbol{\alpha}(t)$ on a surface S in $\mathbb{R}^3$ is a line of curvature if and only if*

$$\boldsymbol{N}'(t) = \lambda(t)\boldsymbol{\alpha}'(t) \tag{5.39}$$

for some real-valued function $\lambda(t)$. In this case, $-\lambda(t)$ is the principal curvature of S in the principal direction $\boldsymbol{\alpha}'(t)$.

We now show how the lines of curvature on a surface S may be investigated using a local parametrisation $\boldsymbol{x}(u, v)$ of S. The condition (5.39) that a regular curve $\boldsymbol{\alpha}(t) = \boldsymbol{x}\,(u(t), v(t))$ be a line of curvature becomes

$$d\boldsymbol{N}(u'\boldsymbol{x}_u + v'\boldsymbol{x}_v) = \lambda(u'\boldsymbol{x}_u + v'\boldsymbol{x}_v)\,, \tag{5.40}$$

and, using the formula (5.16) for the matrix of the Weingarten map $-d\boldsymbol{N}$ in terms of the coefficients of the first and second fundamental forms, we may write condition (5.40) as

$$-\frac{1}{EG - F^2}\begin{pmatrix} GL - FM & GM - FN \\ -FL + EM & -FM + EN \end{pmatrix}\begin{pmatrix} u' \\ v' \end{pmatrix} = \lambda \begin{pmatrix} u' \\ v' \end{pmatrix},$$

or, equivalently,

$$\frac{(GL - FM)u' + (GM - FN)v'}{(-FL + EM)u' + (-FM + EN)v'} = \frac{u'}{v'}\,.$$

Cross-multiplying and simplifying, we obtain

$$(EM - FL)u'^2 + (EN - GL)u'v' + (FN - GM)v'^2 = 0\,,$$

from which we obtain the following lemma.

Lemma 2 *The regular curve $\boldsymbol{\alpha}(t) = \boldsymbol{x}\,(u(t), v(t))$ is a line of curvature on S if and only if*

$$\begin{vmatrix} v'^2 & -u'v' & u'^2 \\ E & F & G \\ L & M & N \end{vmatrix} = 0\,. \tag{5.41}$$

Example 3 (Hyperbolic paraboloid) We shall again consider the hyperbolic paraboloid S with equation $z = xy$, parametrised by $\boldsymbol{x}(u, v) = (u, v, uv)$.

As we saw in Example 3 of §5.3,

$$E = 1 + v^2\,, \quad F = uv\,, \quad G = 1 + u^2\,,$$
$$L = 0\,, \quad M = (1 + u^2 + v^2)^{-1/2}\,, \quad N = 0\,,$$

so $\boldsymbol{x}\,(u(t), v(t))$ is a line of curvature if and only if

$$\frac{u'^2}{1+u^2} - \frac{v'^2}{1+v^2} = 0\,.$$

Taking square roots and integrating, we obtain

$$\int \frac{1}{(1+u^2)^{1/2}} \frac{du}{dt} dt = \pm \int \frac{1}{(1+v^2)^{1/2}} \frac{dv}{dt} dt\,,$$

or, using the substitution rule for integration,

$$\int \frac{du}{(1+u^2)^{1/2}} = \pm \int \frac{dv}{(1+v^2)^{1/2}}\,. \tag{5.42}$$

We saw in §3.5 how to describe, in the form $\phi(u, v) = \text{constant}$, families of curves on a surface, and it follows from (5.42) that the two families of lines of curvature on S are given by

$$\operatorname{arcsinh} u \pm \operatorname{arcsinh} v \;=\; \text{const.}$$

So, for instance, if we wish to find the two lines of curvature through $(0, 0, 0)$, we should take $u = v = 0$, so that the lines of curvature are given by

$$\operatorname{arcsinh} u \pm \operatorname{arcsinh} v = 0,$$

which immediately simplifies to give $u = \pm v$. Thus the lines of curvature through $(0, 0, 0)$ are $\boldsymbol{\alpha}(t) = \boldsymbol{x}(t, t) = (t, t, t^2)$ and $\boldsymbol{\beta}(t) = \boldsymbol{x}(t, -t) = (t, -t, -t^2)$. One of these curves travels up the ridge of the hyperbolic paraboloid, and the other travels down the valley.

We now consider other geometrically significant families of curves. A regular curve $\boldsymbol{\alpha}(t)$ on a surface S in $\mathbb{R}^3$ is called an *asymptotic curve* on S if its normal curvature is identically zero. The following proposition is immediate from (5.31).

Proposition 4 *A regular curve $\boldsymbol{\alpha}(t)$ on a surface S in $\mathbb{R}^3$ is an asymptotic curve if and only if*

$$II\left(\boldsymbol{\alpha}'(t)\right) = 0\ \forall t\,.$$

Asymptotic curves may be found using a local parametrisation in a similar way to the lines of curvature. The following lemma is immediate from (5.32).

Lemma 5 *The regular curve $\boldsymbol{\alpha}(t) = \boldsymbol{x}\,(u(t), v(t))$ is an asymptotic curve on S if and only if*

$$u'^2 L + 2u'v'M + v'^2 N = 0\,. \tag{5.43}$$

Example 6 (Catenoid) We consider the catenoid parametrised as a surface of revolution by

$$\boldsymbol{x}(u, v) = (\cosh v \cos u, \cosh v \sin u, v)\,.$$

It is easy to check that $L = -1$, $M = 0$ and $N = 1$, so it follows from (5.43) that the asymptotic curves are given by $\boldsymbol{\alpha}(t) = \boldsymbol{x}\,(u(t), v(t))$ where

$$-u'^2 + v'^2 = 0\,.$$

Therefore

$$(v' - u')(v' + u') = 0,$$

so that

$$u' \pm v' = 0\,.$$

Integrating, we see that the asymptotic curves are given by

$$u \pm v = \text{constant},$$

and so may be parametrised by taking $\boldsymbol{\alpha}(t) = \boldsymbol{x}(t, c \pm t)$.

For instance, we might wonder if one or both of the asymptotic curves through $(1, 0, 0)$ intersect the xz-plane again, and, if so, at what height above the equatorial circle. To decide this, we note that the asymptotic curves through $(1, 0, 0)$ satisfy $u \pm v = 0$, and so may be parametrised by

$$t \mapsto (\cosh t \cos t, \pm \cosh t \sin t, t)\,.$$

After $t = 0$, these curves next intersect the xz-plane when $t = \pi$, which is at height π above the equatorial circle.

As we saw in Rodrigues' Theorem, a regular curve $\boldsymbol{\alpha}(t)$ on a surface S is a line of curvature if and only if $\boldsymbol{N}'$ is a scalar multiple of $\boldsymbol{\alpha}'$. In contrast, we may use (5.28) to obtain the following characterisation of asymptotic curves.

Proposition 7 *A regular curve $\boldsymbol{\alpha}(t)$ is an asymptotic curve on S if and only if, for all t, $\boldsymbol{N}'(t)$ is orthogonal $\boldsymbol{\alpha}'(t)$.*

Remark 8 In Exercise 5.23 you are invited to obtain a characterisation of asymptotic curves in terms of the geometry of the curves as space curves in $\mathbb{R}^3$.

Returning to the general situation, we note that there are no asymptotic curves through a point where $K > 0$. This follows from Proposition 4 of §5.7, but may also be seen using the fact that, at such points, $LN - M^2 > 0$ so that (5.43) has only $u' = v' = 0$ as a solution. For similar reasons, there are exactly two asymptotic curves through a point where $K < 0$, and, in this case (see Exercise 5.22), the lines of curvature through the point bisect the angles between the asymptotic curves.

In summary, in this section we have investigated curves on a surface S in $\mathbb{R}^3$ whose tangent vectors are in geometrically significant directions at each point, namely those directions which maximise and minimise normal curvature, and those directions in which the normal curvature is zero (these last directions are called *asymptotic directions*; they exist at those points where the Gaussian curvature is non-positive).

In §5.11 we illustrate the significance of the asymptotic directions in terms of the intersection of a surface with its tangent plane.

5.10 Elliptic, hyperbolic, parabolic and planar points

We now begin our investigation of the geometry associated with the Gaussian curvature K of a surface S in $\mathbb{R}^3$. Of course, since K is determined by the rate of change of the unit normal, its value at a point will only reflect the behaviour of a surface S in $\mathbb{R}^3$ in a sufficiently small open neighbourhood of that point (and a similar comment holds for the mean curvature H).

A point $p \in S$ is said to be an *elliptic point* if $K(p) > 0$, a *hyperbolic point* if $K(p) < 0$, a *parabolic point* if $K(p) = 0$ but $d\boldsymbol{N}_p \neq 0$, and a *planar point* if $d\boldsymbol{N}_p = 0$. Since K is the product of the principal curvatures, at an elliptic point the principal curvatures κ_1, κ_2 have the same sign, while at a hyperbolic point the principal curvatures have opposite signs. Exactly one principal curvature is zero at a parabolic point, while both are zero at a planar point.

Example 1 (Hyperbolic paraboloid) We saw in Example 1 of §5.4 that each point of the hyperbolic paraboloid S with equation $z = xy$ is a hyperbolic point.

Example 2 (Torus of revolution) In terms of the parametrisation

$$\boldsymbol{x}(u,v) = ((a + b\cos v)\cos u, (a + b\cos v)\sin u, b\sin v)\,, \quad u, v \in (-\pi, \pi), \quad a > b > 0\,,$$

we found in Example 6 of §5.6 that the principal curvatures are given by

$$\kappa_1 = -\frac{\cos v}{a + b\cos v}\,, \qquad \kappa_2 = -\frac{1}{b}\,.$$

Hence (Figure 5.7), the elliptic points correspond to $-\frac{\pi}{2} < v < \frac{\pi}{2}$, the parabolic points to $v = \pm\frac{\pi}{2}$, with the remaining points being hyperbolic points.

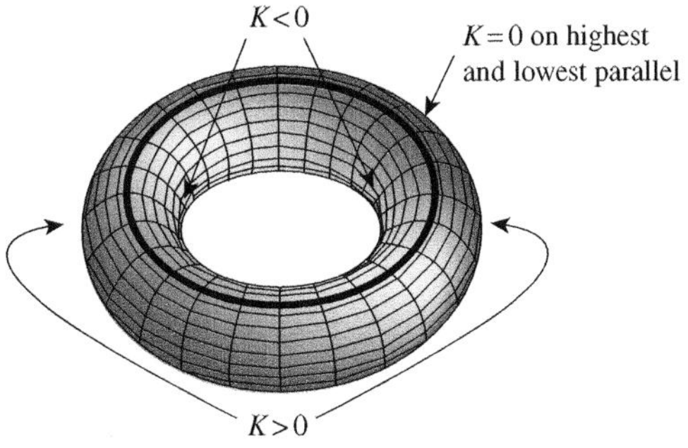

Figure 5.7 Torus of revolution

In terms of the coefficients of the second fundamental form, it follows from (5.18) that $LN - M^2$ is greater than zero at an elliptic point, is less than zero at a hyperbolic point, and equal to zero at a planar or parabolic point.

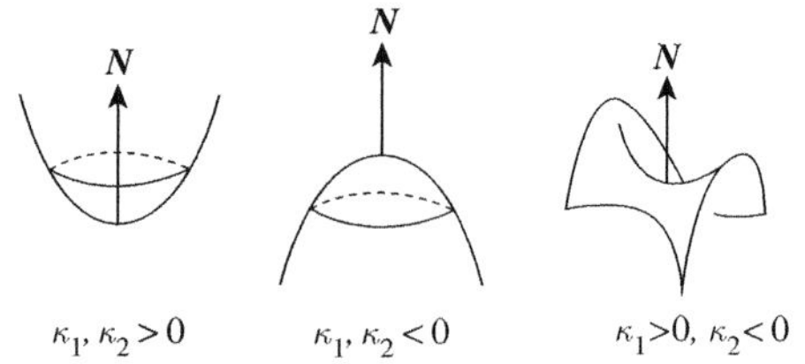

Figure 5.8 Elliptic and hyperbolic points

We saw in Proposition 4 of §5.7 that the principal curvatures at a point p on S are the extremal values of the normal curvatures of the curves in S through p. Hence, at a hyperbolic point the normal curvatures take both positive and negative values, while at an elliptic point they all have the same (non-zero) sign. Since $\kappa_n \boldsymbol{N}$ is the component orthogonal to S of the acceleration vector of a curve parametrised by arc length on S, at an elliptic point these acceleration vectors all point to the same side of S. It would seem reasonable, therefore, that the points of S sufficiently close to an elliptic point should all be on one side of the tangent plane at that point. Similar reasoning would indicate that, in any open neighbourhood of a hyperbolic point there would be points of S on both sides of the tangent plane.

Figures 5.3 and 5.7 add weight to these conjectures, and we now state and prove the theorem which has been suggested by the above discussions.

Theorem 3 *Let S be a surface in $\mathbb{R}^3$ and let $p \in S$.*

(i) If p is an elliptic point then there is an open neighbourhood of p in S which lies entirely on one side of the tangent plane T_pS.
(ii) If p is a hyperbolic point then every open neighbourhood of p in S contains points on both sides of T_pS.

Proof We shall need Taylor's Theorem for functions of two variables, which gives

$$\begin{aligned}\boldsymbol{x}(u,v) &= \boldsymbol{x}(0,0) + u\boldsymbol{x}_u(0,0) + v\boldsymbol{x}_v(0,0)\\ &\quad + \frac{1}{2}\Big(u^2\boldsymbol{x}_{uu}(0,0) + 2uv\boldsymbol{x}_{uv}(0,0) + v^2\boldsymbol{x}_{vv}(0,0)\Big) + o(u^2+v^2)\,, \qquad (5.44)\end{aligned}$$

where $o(u^2+v^2)$ stands for a remainder term $R(u,v)$ with the property that

$$\lim_{(u,v)\to(0,0)} \frac{R(u,v)}{u^2+v^2} = 0\,.$$

For ease of discussion and notation, we translate S if necessary so that $p = (0,0,0)$. We also choose (without loss of generality) a local parametrisation $\boldsymbol{x}(u,v)$ such that $\boldsymbol{x}(0,0) = p$, and such that $\{\boldsymbol{x}_u(0,0), \boldsymbol{x}_v(0,0)\}$ is an orthonormal basis of principal vectors of T_pS. This assumption implies that

$$\boldsymbol{x}_{uu}(0,0).\boldsymbol{N}(p) = \kappa_1, \quad \boldsymbol{x}_{uv}(0,0).\boldsymbol{N}(p) = 0, \quad \boldsymbol{x}_{vv}(0,0).\boldsymbol{N}(p) = \kappa_2\,, \qquad (5.45)$$

where κ_1 and κ_2 are the corresponding principal curvatures at $(0,0)$.

The sign of the inner product $\boldsymbol{x}(u, v).\boldsymbol{N}(p)$ tells us which side of T_pS the point $\boldsymbol{x}(u, v)$ lies, and, with the above assumptions on $\boldsymbol{x}$, (5.44) and (5.45) imply that

$$\boldsymbol{x}(u, v).\boldsymbol{N}(p) = \frac{1}{2}(\kappa_1 u^2 + \kappa_2 v^2) + o(u^2 + v^2) . \tag{5.46}$$

At an elliptic point, the principal curvatures κ_1 and κ_2 have the same sign, and, in this case, an elementary limiting argument may now be used to prove that $\boldsymbol{x}(u, v).\boldsymbol{N}(p)$ always has the same sign on some open neighbourhood of p in S. On the other hand, if κ_1 and κ_2 have opposite signs, then $\boldsymbol{x}(u, v).\boldsymbol{N}(p)$ takes both positive and negative values on any open neighbourhood of p in S. The theorem now follows. □

We now discuss a very nice theorem which will have important consequences in later chapters. We need a little notation before stating it. A *closed surface* is a surface that is a closed subset of its containing Euclidean space. So, for instance, ellipsoids, tori, catenoids and helicoids are all closed surfaces, but the disc $D = \{(x, y, 0) : x^2 + y^2 < 1\}$ is not. Intuitively speaking, a closed surface has no edges to fall off. Also, a subset W of $\mathbb{R}^n$ is said to be *compact* if it is both closed and bounded. So, ellipsoids and tori are compact surfaces, but catenoids and helicoids are not.

The crucial property of compact sets for the proof of the following theorem is that if $f : W \to \mathbb{R}$ is a continuous function defined on a compact subset W of $\mathbb{R}^n$ then there are points $p_0, p_1 \in W$ such that $f(p_0) \leq f(p) \leq f(p_1)$ for all $p \in W$. This result, Weierstrass's Extremal Value Theorem, is usually stated as: any real-valued continuous function on a compact set is bounded and attains its bounds.

Theorem 4 *Every compact surface S in $\mathbb{R}^3$ has at least one elliptic point.*

Proof We first describe the geometrical idea of the proof, which is illustrated in Figure 5.9. Since S is compact, it is, in particular, bounded. This means that there is a sphere centred on the origin of $\mathbb{R}^3$ which has S inside it. If we shrink this sphere until it first touches S at the point p_1, say, then S and the sphere are tangential at p_1 and S is completely on one side of their common tangent plane. This would lead us to hope that p_1 would be an elliptic point.

The details of the proof are as follows. Let $f : \mathbb{R}^3 \to \mathbb{R}$ be given by $f(x) = |x|^2$. Then f is continuous (in fact, differentiable), so, since S is compact, there is some point $p_1 \in S$ with the property that, for all $p \in S$, $f(p) \leq f(p_1)$. Let $\boldsymbol{\alpha} : (-\epsilon, \epsilon) \to S$ be a curve on S, parametrised by arc length, with $\boldsymbol{\alpha}(0) = p_1$. Then $f\boldsymbol{\alpha}(s) = \boldsymbol{\alpha}(s).\boldsymbol{\alpha}(s)$ has a maximum at $s = 0$ so that

$$\boldsymbol{\alpha}(0).\boldsymbol{\alpha}'(0) = 0 , \quad \boldsymbol{\alpha}(0).\boldsymbol{\alpha}''(0) + \boldsymbol{\alpha}'(0).\boldsymbol{\alpha}'(0) \leq 0 .$$

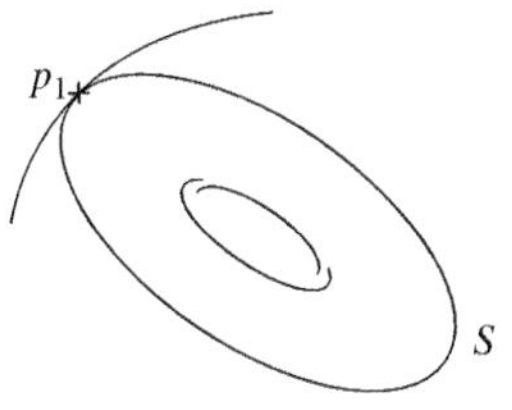

Figure 5.9 Sphere tangential to S at p_1

Since these equations hold for all curves on S through p_1, the first equation shows that, as suggested by Figure 5.9, $\boldsymbol{N}(p_1) = p_1/|p_1|$. The second equation now implies that $\boldsymbol{\alpha}''(0).\boldsymbol{N}(p_1) \leq -1/|p_1|$. However, from (5.26), $\boldsymbol{\alpha}''(0).\boldsymbol{N}(p_1)$ is the normal curvature $\kappa_n(\boldsymbol{\alpha}'(0))$, from which it follows that all the normal curvatures of S at p_1 are strictly negative. In particular, the extremal values are both negative, so that their product, the Gaussian curvature, is positive. □

5.11 Approximating a surface by a quadric †

In this section, we extend Theorem 3 of §5.10 by giving rather more quantitative information about the way the principal curvatures influence the local behaviour of a surface. This material is optional and could be omitted if time is short.

We begin by showing that any surface S in $\mathbb{R}^3$ may be locally parametrised near a point $p \in S$ as the graph of a function.

Proposition 1 *Let S be a surface in $\mathbb{R}^3$ and assume that the unit normal $\boldsymbol{N}(p)$ at a point $p \in S$ is not parallel to the xy-plane. Then there is a local parametrisation of an open neighbourhood of p in S of the form*

$$\boldsymbol{x}(u, v) = (u, v, g(u, v)),$$

for some smooth function $g(u, v)$.

Proof Let $\boldsymbol{y} : U \to S$ be a local parametrisation of an open neighbourhood of p in S, and let $\pi : \mathbb{R}^3 \to \mathbb{R}^2$ be given by $\pi(x, y, z) = (x, y)$. The derivative $d(\pi\, \boldsymbol{y})$ is non-singular at p, so the Inverse Function Theorem shows that, by taking U smaller if necessary, there exists a smooth map h such that $\pi\, \boldsymbol{y} h$ is the identity map. It then follows that $\boldsymbol{x} = \boldsymbol{y} h$ is a local parametrisation of the required form. □

The following proposition gives a good description of a surface near any point p since it shows that, after applying a suitable rigid motion of $\mathbb{R}^3$, the surface may be approximated up to second order near p by the quadric with equation $2z = \kappa_1 x^2 + \kappa_2 y^2$. Here, as before, κ_1 and κ_2 are the principal curvatures at p, and, in the statement of the proposition, $o(u^2 + v^2)$ stands for a remainder term $R(u, v)$ with the property that

$$\lim_{(u,v)\to(0,0)} \frac{R(u, v)}{u^2 + v^2} = 0.$$

Proposition 2 *Let S be a surface in $\mathbb{R}^3$ and let κ_1 and κ_2 be the principal curvatures at $p \in S$. By applying a suitable rigid motion of $\mathbb{R}^3$ we may assume that $p = (0, 0, 0)$, that $\boldsymbol{N}(p) = (0, 0, 1)$ and that $(1, 0, 0)$, $(0, 1, 0)$ are principal directions at p.*

In this case, if κ_1 and κ_2 are the principal curvatures at p, then there is a local parametrisation of an open neighbourhood of p in S such that

$$\boldsymbol{x}(u, v) = (u, v, g(u, v)),$$

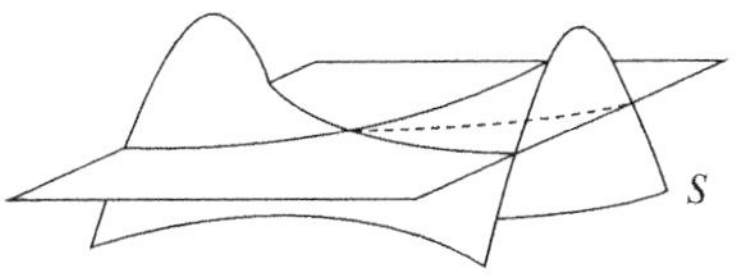

Figure 5.10 Intersection of a surface and its tangent plane

where

$$g(u,v) = \frac{1}{2}(\kappa_1 u^2 + \kappa_2 v^2) + o(u^2 + v^2)\,. \tag{5.47}$$

Proof Proposition 1 shows that if $p = (0,0,0)$ and $\boldsymbol{N}(p) = (0,0,1)$, then S has a local parametrisation of the form

$$\boldsymbol{x}(u,v) = (u,v,g(u,v)) \tag{5.48}$$

for some smooth function $g(u,v)$ defined on an open neighbourhood of $(0,0)$ with $g(0,0) = g_u(0,0) = g_v(0,0) = 0$. Under the remaining assumptions of the proposition, equation (5.46) reduces to (5.47), and the proposition is proved. □

Remark 3 In Exercise 5.27, you are invited to prove that if a surface S is parametrised as a graph $\boldsymbol{x}(u,v) = (u,v,g(u,v))$ with $g(0,0) = g_u(0,0) = g_v(0,0) = 0$, then the second fundamental form of S at $(0,0,0)$ is equal to the *Hessian* $\mathcal{H}$ of g at $(0,0)$. Here, the *Hessian* of g is the quadratic form used in the calculus of functions of two variables given by

$$\mathcal{H}(u,v) = u^2 g_{uu}(0,0) + 2uv g_{uv}(0,0) + v^2 g_{vv}(0,0)\,.$$

We end this section by giving an illustration of the significance of the asymptotic directions in terms of the intersection of a surface with its tangent plane. It is intuitively clear (and may be proved using Proposition 2) that if p is a hyperbolic point on S then, in an open neighbourhood of p, the intersection of S and its tangent plane T_pS is the union of two regular curves through p. We now show (Figure 5.10) that the tangent vectors at p of these curves give the asymptotic directions at p.

So, let $\boldsymbol{\alpha}(t)$ be a regular curve lying on the intersection of S and T_pS and having $\boldsymbol{\alpha}(0) = p$. Then, for all t, $\boldsymbol{\alpha}'(t).\boldsymbol{N}(t) = 0$ and $\boldsymbol{\alpha}'(t).\boldsymbol{N}(0) = 0$. Differentiating these two equations, we find that

$$\boldsymbol{\alpha}''(t).\boldsymbol{N}(t) + \boldsymbol{\alpha}'(t).\boldsymbol{N}'(t) = 0 \quad \text{and} \quad \boldsymbol{\alpha}''(t).\boldsymbol{N}(0) = 0\,.$$

Hence $II\left(\boldsymbol{\alpha}'(0)\right) = -\boldsymbol{\alpha}'(0).\boldsymbol{N}'(0) = 0$, and $\boldsymbol{\alpha}'(0)$ is in an asymptotic direction as claimed.

5.12 Gaussian curvature and the area of the image of the Gauss map †

In this section, we investigate further the geometrical information carried by the Gaussian curvature. Although geometrically interesting, this section may be omitted since the material is not used elsewhere in the book.

We remarked at the beginning of §5.4 that, since the Gaussian curvature is the determinant of the Weingarten map, then $|K|$ is the scale factor by which area in the tangent space is multiplied under the Weingarten map. In this section we justify this comment (see equation (5.50)), and then "integrate it up" to obtain expression (5.51) for the area of the image of a region R of a surface S in $\mathbb{R}^3$ under the Gauss map. We then indicate that, in a sense made a little more precise below, $|K(p)|$ gives the ratio of the area of the image under the Gauss map $\boldsymbol{N}$ of a region R of S compared to the area of R itself, as the region contracts down towards p (see equation (5.52)).

Remark 1 We recall that in Remark 5 of §1.3, we interpreted the modulus of the curvature κ of a plane curve $\boldsymbol{\alpha}$ as the ratio of $|\boldsymbol{n}'|$ and $|\boldsymbol{\alpha}'|$. Formula (5.50) is the analogue for surfaces.

Area and integration on a surface were discussed in §3.7, and we use the notation of that section. Let R be the image under a local parametrisation $\boldsymbol{x}(u, v)$ of a suitable region Q in the (u, v)-plane. Then, quoting (3.24), the area $A(R)$ of R is given by

$$A(R) = \iint_R dA = \iint_Q |\boldsymbol{x}_u \times \boldsymbol{x}_v| \, du \, dv \, . \tag{5.49}$$

In a similar way, the image $\boldsymbol{N}(R)$ of R under $\boldsymbol{N}$, being the image of Q under $\boldsymbol{N}\boldsymbol{x}$, has area

$$A\left(\boldsymbol{N}(R)\right) = \iint_Q |\boldsymbol{N}_u \times \boldsymbol{N}_v| \, du \, dv$$

(areas being counted with multiplicity, that is to say the number of times they are covered by $\boldsymbol{N}(R)$). However, from (5.12) and (5.13),

$$\begin{aligned} |\boldsymbol{N}_u \times \boldsymbol{N}_v| &= |(a_{11}\boldsymbol{x}_u + a_{21}\boldsymbol{x}_v) \times (a_{12}\boldsymbol{x}_u + a_{22}\boldsymbol{x}_v)| \\ &= |a_{11}a_{22} - a_{12}a_{21}| \, |\boldsymbol{x}_u \times \boldsymbol{x}_v| \\ &= |K| \, |\boldsymbol{x}_u \times \boldsymbol{x}_v| \, , \end{aligned} \tag{5.50}$$

where we have used (5.17) for the last equality.

Hence, using (3.25),

$$A\left(\boldsymbol{N}(R)\right) = \iint_Q |K| \, |\boldsymbol{x}_u \times \boldsymbol{x}_v| \, du \, dv \; = \iint_R |K| \, dA \, . \tag{5.51}$$

As mentioned in §3.7 (where integration on surfaces is discussed), we may extend (5.51) to more general subsets of S; all we need is that the subset may be broken up into the types of piece we have considered above. In particular, we may integrate $|K|$ over the whole of the surface (although the result is not necessarily finite if S is not compact).

The following proposition may be proved using (5.49) and (5.51), together with some standard analysis involving double integrals.

Proposition 2 *If we consider a sequence of contracting regions R of S containing a point $p \in S$ then, subject to suitable mathematical assumptions on the way the contraction is made,*

$$|K(p)| = \lim_{A(R) \to 0} \frac{A\left(\boldsymbol{N}(R)\right)}{A(R)} \, . \tag{5.52}$$

In order to prove the final theorem of this chapter, we must first quote a deep result concerning compact surfaces (without self-intersections) in $\mathbb{R}^3$. The result is a 2-dimensional analogue of (part of) a famous theorem concerning simple closed curves in the plane called the Jordan Curve Theorem (a statement of which may be found in §8.6).

Theorem 3 *Let S be a compact connected surface (without self-intersections) in $\mathbb{R}^3$. Then the complement $\mathbb{R}^3 \setminus S$ of S in $\mathbb{R}^3$ is the disjoint union of two connected sets; one of these (the outside) is unbounded, while the other one (the inside) is bounded. Such a surface is orientable, since a unit normal may be assigned smoothly over the whole of the surface (either the outward unit normal or the inward unit normal).*

Using this theorem, we may now prove the following.

Theorem 4 *Let S be a compact surface without self-intersections in $\mathbb{R}^3$. Then*

$$\iint_S |K| dA \geq 4\pi \, .$$

Proof The proof is similar to the first part of the proof of Theorem 4 of §5.10. We show that every point of the unit sphere is in the image of the Gauss map $\boldsymbol{N}$ of S so that the area of the image (counted with multiplicity) is at least 4π. To do this, let $q_0 \in S^2(1)$, and let $h : S \to \mathbb{R}$ be given by $h(p) = p.q_0$. (This map was considered in Exercise 4.5.) Since S is compact, h has a maximum value on S taken at p_0, say. Then, arguing as in the proof of Theorem 4 of §5.10, it follows that, if we take $\boldsymbol{N}$ to be the outward unit normal, then $\boldsymbol{N}(p_0)$ is equal to q_0. Thus the image of $\boldsymbol{N}$ covers every point of $S^2(1)$ at least once, and the theorem now follows. □

Exercises

5.1 Find the coefficients L, M and N of the second fundamental form of the graph of a smooth function $g(u, v)$, when the graph is parametrised in the usual way by

$$\boldsymbol{x}(u, v) = (u, v, g(u, v)) \, .$$

Hence show that the graph of the function $g(u, v) = u^2 + v^2$ has everywhere positive Gaussian curvature.

5.2 **Enneper's surface** is the image of the map $\boldsymbol{x} : \mathbb{R}^2 \to \mathbb{R}^3$ given by

$$\boldsymbol{x}(u, v) = \left(u - \frac{u^3}{3} + uv^2, \; v - \frac{v^3}{3} + u^2 v, \; u^2 - v^2 \right) .$$

Show that the coefficients of the first and second fundamental forms are given by

$$E = G = (1 + u^2 + v^2)^2 \, , \quad F = 0 \, , \quad \text{and} \quad L = 2 \, , \quad M = 0 \, , \quad N = -2 \, ,$$

and deduce that Enneper's surface has constant mean curvature $H = 0$. (In fact, although sufficiently small pieces of Enneper's surface really are surfaces, the whole of the image of $\boldsymbol{x}$ is not actually a surface as discussed in this book. This is because $\boldsymbol{x}$ is not injective, so that Enneper's surface, which is illustrated in Figure 9.2, has self-intersections.)

5.3 Find the coefficients L, M and N of the second fundamental form of the helicoid parametrised by

$$\boldsymbol{x}(u, v) = (\sinh v \cos u, \sinh v \sin u, u), \quad (u, v) \in \mathbb{R}^2.$$

Hence find the mean and Gaussian curvatures, and the principal curvatures.

5.4 Find the Gaussian curvature K and the mean curvature H for a graph as parametrised in Exercise 5.1.

5.5 **(Developable surfaces)** Show that each of the following types of ruled surface in $\mathbb{R}^3$ has the property that the unit normal $\boldsymbol{N}$ is constant along each line of the ruling. Such ruled surfaces are called *developable* surfaces.

(a) **(Tangent surfaces)** Let $\boldsymbol{\alpha}(u)$ be a regular curve in $\mathbb{R}^3$ with nowhere zero curvature κ, and let S be the image of the map

$$\boldsymbol{x}(u, v) = \boldsymbol{\alpha}(u) + v\boldsymbol{\alpha}'(u), \quad v > 0\,.$$

(See Figure 2.17 for a picture of a tangent surface.)

(b) **(Generalised cones)** Let $\boldsymbol{\alpha}(u)$ be a regular curve in $\mathbb{R}^3$ not passing through the origin 0 of $\mathbb{R}^3$, and assume that $\boldsymbol{\alpha}'(u)$ is never a scalar multiple of $\boldsymbol{\alpha}(u)$. Let S be the image of the map

$$\boldsymbol{x}(u, v) = v\boldsymbol{\alpha}(u)\,, \quad v > 0\,.$$

(c) **(Generalised cylinders)** Let $\boldsymbol{\alpha}(u)$ be a regular curve in $\mathbb{R}^3$ and let $\boldsymbol{e}$ be a non-zero vector such that $\boldsymbol{\alpha}'(u)$ is never a scalar multiple of $\boldsymbol{e}$. Let S be the image of the map

$$\boldsymbol{x}(u, v) = \boldsymbol{\alpha}(u) + v\boldsymbol{e}, \quad v \in \mathbb{R}\,.$$

5.6 Let S be a ruled surface. Show that the Gaussian curvature of S is identically zero if and only if the unit normal $\boldsymbol{N}$ is constant along each line of the ruling. (So that S is a developable surface, as defined in the previous exercise.) In Exercise 5.16 we see that, conversely, every surface (not even assuming it is ruled) with $K = 0$ is locally a developable surface away from the umbilics.

5.7 **(Parallel surfaces)** Let S be a surface in $\mathbb{R}^3$ with Gauss map $\boldsymbol{N}$, and for each real number λ let $f^\lambda : S \to \mathbb{R}^3$ be given by

$$f^\lambda(p) = p + \lambda \boldsymbol{N}\,.$$

The image S^λ of f^λ is a *parallel surface* of S. At those points where S^λ is a surface, show that:

(i) $\boldsymbol{N}^\lambda f^\lambda = \boldsymbol{N}$, where $\boldsymbol{N}^\lambda$ is the Gauss map of S^λ;

(ii) if $\boldsymbol{X}$ is a principal vector of S with corresponding principal curvature κ, then $\boldsymbol{X}$ is also a principal vector of S^λ, but with corresponding principal curvature $\kappa/(1 - \lambda\kappa)$;

(iii) the Gaussian curvature K^λ and the mean curvature H^λ of S^λ are given by

$$K^\lambda = \frac{K}{1 - 2\lambda H + \lambda^2 K}, \quad H^\lambda = \frac{H - \lambda K}{1 - 2\lambda H + \lambda^2 K};$$

(iv) if S has constant non-zero mean curvature H, then the parallel surface obtained by taking $\lambda = 1/(2H)$ has constant Gaussian curvature $4H^2$.

5.8 **(Surfaces of revolution with constant Gaussian curvature)** In this exercise we find all surfaces of revolution S in $\mathbb{R}^3$ with constant Gaussian curvature K. If $K \neq 0$ we may assume by re-scaling that $K = \pm 1$. We also assume that S is generated by rotating the curve $\boldsymbol{\alpha}(v) = (f(v), 0, g(v))$, $f(v) > 0\ \forall v$, about the z-axis, with $\boldsymbol{\alpha}$ being parametrised by arc length.

(i) If $K = 1$ show that $\boldsymbol{\alpha}$ may be parametrised by arc length in such a way that $f(v) = A\cos v$, for some positive constant A. By noting that $|f'(v)| < |\boldsymbol{\alpha}'(v)| = 1$, show that the domain of $\boldsymbol{\alpha}$ is $(-\pi/2, \pi/2) \cap (-v_0, v_0)$, where $0 < v_0 \leq \pi/2$ and $A\sin v_0 = 1$ (this condition giving no restriction if $0 < A \leq 1$).
On the same set of axes, sketch the generating curve $\boldsymbol{\alpha}(v)$ for $A = 1/2$, $A = 1$, and $A = 3/2$.

(ii) If $K = 0$ show that S is an open subset of a cylinder, a cone or a plane.

(iii) If $K = -1$ show that $\boldsymbol{\alpha}$ may be parametrised by arc length in such a way that one of the following three cases occurs:

(a) $f(v) = A\cosh v$, with $A > 0$ and $-v_0 < v < v_0$ for some $v_0 > 0$;
(b) $f(v) = e^{-v}$, $v > 0$;
(c) $f(v) = B\sinh v$, with $0 < B < 1$ and $0 < v < v_0$ for some $v_0 > 0$.

In each of cases (a) and (c), determine the value of v_0.
On the same set of axes, sketch the generating curve for $A = 1$ in Case (a), the generating curve in Case (b), and the generating curve for $B = \sqrt{3}/2$ in Case (c). In each case assume that 0 is the infimum of the values taken by g.

5.9 Let S be a surface of revolution with a parametrisation of the form

$$\boldsymbol{x}(u, v) = (v\cos u, v\sin u, g(v)), \quad v > 0.$$

Show that the Gaussian curvature K is given by

$$K = \frac{g'g''}{v(1 + g'^2)^2}.$$

Hence find the regions of the surface

$$z = \frac{1}{1 + x^2 + y^2}$$

where $K > 0$ and where $K < 0$. Indicate these regions on a sketch of the surface.

5.10 Up to rigid motions of $\mathbb{R}^3$, all catenoids are obtained by rotating a catenary

$$\boldsymbol{\alpha}(v) = \left(a\cosh\frac{v}{a}, 0, v\right), \quad v \in \mathbb{R}, \quad a > 0,$$

about the z-axis. Show that all catenoids have constant mean curvature $H = 0$.

5.11 Find conditions on real numbers $a > b > 0$ such that the torus of revolution defined in Example 6 of §5.6 has points where the mean curvature H is equal to zero.

5.12 Let a, b be positive real numbers, and let $f(z)$ be a positive function of z. Show that the surface S in $\mathbb{R}^3$ with equation

$$\frac{x^2}{a^2} + \frac{y^2}{b^2} = (f(z))^2$$

has Gaussian curvature $K > 0$, $K = 0$, $K < 0$ respectively, at those points of S with $f'' < 0$, $f'' = 0$, $f'' > 0$ respectively.

5.13 Let $\boldsymbol{\alpha}(t)$ be the curve on the cone $x^2 + y^2 = z^2$, $z > 0$, given by

$$\boldsymbol{\alpha}(t) = e^t(\cos t, \sin t, 1), \quad t \in \mathbb{R}\,.$$

Show that the normal curvature of $\boldsymbol{\alpha}$ is inversely proportional to e^t by either or both of the following methods.

(i) Parametrise the cone as a surface of revolution and use formula (5.33).

(ii) Use the definition of κ_n, namely (in the usual notation),

$$\kappa_n = \frac{d\boldsymbol{t}}{ds}.\boldsymbol{N}\,.$$

5.14 Let V be a connected coordinate neighbourhood on a surface S in $\mathbb{R}^3$, and assume that the unit normal $\boldsymbol{N}$ is constant on V. Show that V is contained in a plane. (This is used in the proof of Theorem 1 of §5.8.)

5.15 Use the parametrisation given in Exercise 5.1 of the surface S with equation $z = x^2 + y^2$ to show that the curve

$$\boldsymbol{\alpha}(t) = \Big(u(t), v(t), u^2(t) + v^2(t)\Big)$$

is a line of curvature on S if and only if

$$(uu' + vv')(u'v - uv') = 0.$$

Hence find functions $\phi(u, v)$, $\psi(u, v)$ so that the two families of lines of curvature on S are given by $\phi(u, v) =$ constant and $\psi(u, v) =$ constant. Give a sketch of S illustrating the lines of curvature, and then say why you knew before you started that these were indeed the lines of curvature.

5.16 Let S be a surface in $\mathbb{R}^3$ with zero Gaussian curvature. Show that every non-umbilic point of S has an open neighbourhood which is a developable surface; that is to say, is a ruled surface with the property that the unit normal $\boldsymbol{N}$ is constant along each line of the ruling. (See Exercises 5.5 and 5.6.) **You may assume (correctly) that there is a local parametrisation around any non-umbilic point of a surface S in $\mathbb{R}^3$ such that the coordinate curves are lines of curvature.**

5.17 Show that, for Enneper's surface as defined in Exercise 5.2:

(i) the principal curvatures are given by

$$\kappa_1 = \frac{2}{(1 + u^2 + v^2)^2}\,, \qquad \kappa_2 = -\frac{2}{(1 + u^2 + v^2)^2}\,;$$

(ii) the lines of curvature are the coordinate curves;
(iii) the asymptotic curves are given by $u \pm v = \text{const}$.

Prove that any pair of asymptotic curves, one from each family, have non-empty intersection.

5.18 Use the parametrisation of the helicoid given in Exercise 5.3 to find the asymptotic curves on the helicoid. Sketch the helicoid, indicating the asymptotic curves.

5.19 Use the parametrisation

$$\boldsymbol{x}(u, v) = (\operatorname{sech} v \cos u, \operatorname{sech} v \sin u, v - \tanh v), \qquad -\pi < u < \pi\ ,\ \ v > 0,$$

of the pseudosphere S as a surface of revolution to show that the asymptotic curves on S are given by $u \pm v = \text{constant}$. Hence show that the angle of intersection θ of the two asymptotic curves through $\boldsymbol{x}(u, v)$ is given by

$$\cos\theta = 2\operatorname{sech}^2 v - 1.$$

5.20 Show that any straight line lying on a surface is an asymptotic curve on the surface. Deduce that the Gaussian curvature of a ruled surface in $\mathbb{R}^3$ is everywhere non-positive.

5.21 **(Theorem of Beltrami–Enneper)** Let S be a surface in $\mathbb{R}^3$ and let K denote its Gaussian curvature. If $\boldsymbol{\alpha}$ is an asymptotic curve on S whose curvature is never zero, prove that the modulus $|\tau|$ of the torsion τ of $\boldsymbol{\alpha}$ is given by

$$|\tau| = \sqrt{-K}\ .$$

5.22 Let p be a point on a surface S in $\mathbb{R}^3$ with $K(p) < 0$. Show that the lines of curvature through p bisect the angles between the asymptotic curves through p. Show also that the following three conditions are equivalent:

(i) the asymptotic curves through p bisect the angle between the lines of curvature through p;
(ii) the asymptotic curves intersect orthogonally at p;
(iii) $H(p) = 0$.

5.23 Let $\boldsymbol{\alpha}$ be a regular curve on a surface S in $\mathbb{R}^3$, and let $\boldsymbol{n}$ be the principal normal of $\boldsymbol{\alpha}$ as a space curve (as described in Chapter 1). Show that $\boldsymbol{\alpha}$ is an asymptotic curve on S if and only if $\boldsymbol{n}$ is everywhere tangential to S. (An equivalent condition is that the osculating planes of an asymptotic curve on S coincide with the tangent planes of S.)

5.24 **(Theorem of Joachimsthal)** Let S_1 and S_2 be two surfaces in $\mathbb{R}^3$ which intersect along a regular curve C in such a way that, for each point $p \in C$, the angle $\theta(p)$ between their normals at p is never zero or π (S_1 and S_2 are then said to intersect *transversally*). If C is a line of curvature on S_1, prove that θ is constant if and only if C is also a line of curvature on S_2.

5.25 **(Monkey saddle)** The graph of $g(u, v) = u^3 - 3uv^2$ is called a *monkey saddle* (Figure 5.11). Show that the origin is a planar point, and every other point is hyperbolic.

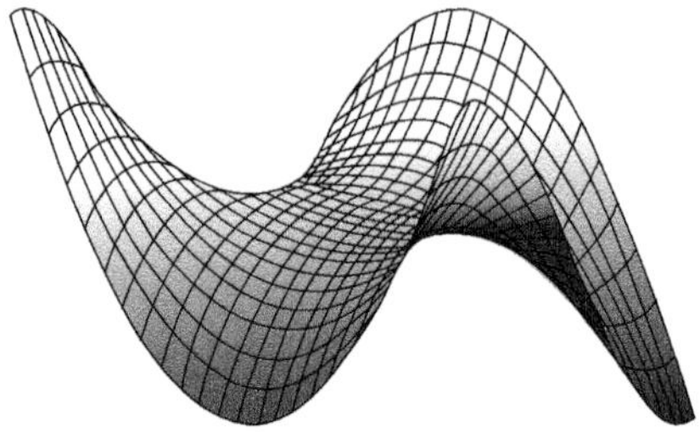

Figure 5.11 A monkey saddle

5.26 If S is the surface in $\mathbb{R}^3$ with equation $x^4+y^4+z^4=1$, use (5.34) to show that each of the six points $(\pm 1,0,0)$, $(0,\pm 1,0)$, $(0,0,\pm 1)$ is a planar point. Show also that all other points of intersection of S with the coordinate planes $x=0$, $y=0$, $z=0$ are parabolic points, and all other points of S are elliptic points.

5.27 *(This exercise uses material in the optional §5.11)* Let S be a surface in $\mathbb{R}^3$ which is parametrised as a graph

$$\boldsymbol{x}(u,v)=(u,v,g(u,v)),$$

with $g(0,0)=g_u(0,0)=g_v(0,0)=0$.

Show that the second fundamental form of S at $(0,0,0)$ is equal to the *Hessian* $\mathcal{H}$ of g at $(0,0)$. Here, the *Hessian* of g is the quadratic form used in the calculus of functions of two variables given by

$$\mathcal{H}(u,v)=u^2g_{uu}(0,0)+2uvg_{uv}(0,0)+v^2g_{vv}(0,0).$$

5.28 *(This exercise uses material in the optional §5.12)* Find the image of the Gauss map of the paraboloid of revolution S with equation $z=x^2+y^2$. Show that, in accordance with (5.51),

$$\iint_S |K|dA = \text{area of the image of the Gauss map.}$$

5.29 *(This exercise uses material in the optional §5.12)* Let S be a compact surface without self-intersections in $\mathbb{R}^3$. Show that the Gauss map $\boldsymbol{N}$ of S maps the union of the elliptic, parabolic and planar points of S surjectively onto the unit sphere.

6 The Theorema Egregium

In this chapter we consider one of the most important theorems in differential geometry, the Theorema Egregium of Gauss. Gauss's name for the theorem is well chosen; the word "egregium" comes from the Latin for "remarkable" (literally, "standing out from the flock"), and the theorem has had a profound effect on the development of not only geometry but other areas of mathematics, particularly relativity theory. There is no doubt that Gaussian curvature is the most important and interesting notion, both historically and mathematically, discussed in this book.

We recall from §5.4 that the Gaussian curvature K of a surface S in $\mathbb{R}^3$ is defined to be the determinant of $-d\boldsymbol{N}$, where $\boldsymbol{N}$ is the Gauss map of S; in terms of the coefficients of the first and second fundamental forms of a local parametrisation of S,

$$K = \frac{LN - M^2}{EG - F^2} .$$

It would appear that K is an extrinsic property of the surface, in that it seems to depend on the coefficients of both the first and the second fundamental forms. However, the Theorema Egregium states that K is actually intrinsic; two surfaces with local parametrisations having the same E, F and G must have the same Gaussian curvature at corresponding points. The theorem is proved by finding an explicit formula, the *Gauss formula*, for K solely in terms of the coefficients of the first fundamental form and their derivatives. For the case of an isothermal parametrisation, for example, we shall see that K is given by

$$K = -\frac{1}{2E}\left(\frac{\partial^2}{\partial u^2} + \frac{\partial^2}{\partial v^2}\right) \log E .$$

It is clear that the functions E, F and G do not determine L, M and N; consider the standard parametrisations of the plane and the cylinder, which have the same coefficients of the first fundamental form but different coefficients of the second fundamental form. However, as we shall see, the expression $LN - M^2$ **is** determined by E, F and G, which gives the proof of Gauss's Theorem.

This shows that the coefficients E, F, G of the first fundamental form and the coefficients L, M, N of the second fundamental form of a local parametrisation of a surface in $\mathbb{R}^3$ are related in a rather subtle way. Relations between the coefficients of the two fundamental forms are explored in §6.2 and §6.3.

We have already investigated some of the geometrical information carried by the Gaussian curvature, and the fact that K is intrinsic makes it an even more important function on a surface. It is therefore interesting to find as much information as we can about surfaces of constant Gaussian curvature, and we consider this in (optional) §6.4.

The Theorema Egregium is very important historically since it suggests (correctly!) that Gaussian curvature can be defined for surfaces in $\mathbb{R}^n$ and, more generally, abstract surfaces with metric with no reference to any containing Euclidean space. This observation instigated and motivated the study of a major branch of modern mathematics, Riemannian geometry. In (optional) §6.5, we indicate how to generalise the notion of Gaussian curvature to surfaces in $\mathbb{R}^n$ for $n > 3$, and show that all the results in §6.1 and §6.2, and all the formulae except (6.1)–(6.3) and (6.6) (each of which has to be slightly modified) hold in this more general setting. This implies that

the Theorema Egregium and its corollaries also hold for surfaces S in $\mathbb{R}^n$ and $\tilde{S}$ in $\mathbb{R}^m$.

Readers whose primary interest is surfaces in $\mathbb{R}^3$ may choose to omit §6.5.

In its simplest form, the Theorema Egregium may be stated as follows.

Theorem 1 (Theorema Egregium of Gauss) *The Gaussian curvature K at a point p of a surface S in $\mathbb{R}^3$ may be expressed solely in terms of the coefficients of the first fundamental form (and their derivatives) of any local parametrisation of S whose image contains p.*

The following corollary is an immediate consequence of the Theorema Egregium.

Corollary 2 *Assume that $\boldsymbol{x} : U \to S$, $\tilde{\boldsymbol{x}} : U \to \tilde{S}$ are local parametrisations of surfaces S, $\tilde{S}$ in $\mathbb{R}^3$ with coefficients of the first fundamental form satisfying $E = \tilde{E}$, $F = \tilde{F}$, $G = \tilde{G}$. Then, for each $q \in U$, the Gaussian curvature of $\tilde{S}$ at $\tilde{\boldsymbol{x}}(q)$ is equal to that of S at $\boldsymbol{x}(q)$.*

For instance, the corollary implies that the helicoid and the catenoid have the same Gaussian curvature at points which correspond under the local parametrisations $\boldsymbol{x}$ and $\tilde{\boldsymbol{x}}$ described in Example 2 of §4.5; this is not at all obvious from the actual shapes of the two surfaces.

The next corollary is also important and useful.

Corollary 3 *Let S and $\tilde{S}$ be surfaces in $\mathbb{R}^3$. If there is a local isometry f from an open neighbourhood of a point $p \in S$ to $\tilde{S}$, then the Gaussian curvature of $\tilde{S}$ at $f(p)$ is equal to that of S at p.*

Proof of Corollary 3 Let $\boldsymbol{x}(u, v)$ be a local parametrisation of an open neighbourhood U of p in S. It follows from Theorem 3 of §2.5 that, choosing U smaller if necessary, $\tilde{\boldsymbol{x}} = f\boldsymbol{x}$ is a local parametrisation of an open neighbourhood of $f(p)$ in $\tilde{S}$. Proposition 3 of §4.4 shows that the coefficients of the first fundamental form of $\tilde{\boldsymbol{x}}$ are equal to those of $\boldsymbol{x}$, and the result follows from Corollary 2. □

6.1 The Christoffel symbols

In this section we define the Christoffel symbols determined by a local parametrisation of a surface in $\mathbb{R}^3$, and show that they can be expressed in terms of the coefficents of the first fundamental form and their derivatives. The Christoffel symbols are of fundamental importance in their own right, as well as helping us to prove the Theorema Egregium.

Let $\boldsymbol{x}(u, v)$ be a local parametrisation of a surface S in $\mathbb{R}^3$. Then any vector may be written as a linear combination of $\boldsymbol{x}_u$, $\boldsymbol{x}_v$ and $\boldsymbol{N}$ so, in particular, we may write

$$\boldsymbol{x}_{uu} = \Gamma^1_{11}\boldsymbol{x}_u + \Gamma^2_{11}\boldsymbol{x}_v + L\boldsymbol{N}, \tag{6.1}$$
$$\boldsymbol{x}_{uv} = \Gamma^1_{12}\boldsymbol{x}_u + \Gamma^2_{12}\boldsymbol{x}_v + M\boldsymbol{N}, \tag{6.2}$$
$$\boldsymbol{x}_{vu} = \Gamma^1_{21}\boldsymbol{x}_u + \Gamma^2_{21}\boldsymbol{x}_v + M\boldsymbol{N},$$
$$\boldsymbol{x}_{vv} = \Gamma^1_{22}\boldsymbol{x}_u + \Gamma^2_{22}\boldsymbol{x}_v + N\boldsymbol{N}, \tag{6.3}$$

for suitable functions $\{\Gamma^k_{ij}\}$. These functions are called the *Christoffel symbols* of S with respect to the parametrisation $\boldsymbol{x}$. Since $\boldsymbol{x}_{uv} = \boldsymbol{x}_{vu}$, it is clear that $\Gamma^i_{12} = \Gamma^i_{21}$ for $i = 1, 2$.

We shall be taking the inner product of the above equations with $\boldsymbol{x}_u$ and $\boldsymbol{x}_v$, and it will be useful to note that

$$\boldsymbol{x}_{uu}.\boldsymbol{x}_u = \frac{1}{2}\frac{\partial}{\partial u}(\boldsymbol{x}_u.\boldsymbol{x}_u) = \frac{1}{2}E_u ,$$
$$\boldsymbol{x}_{uu}.\boldsymbol{x}_v = \frac{\partial}{\partial u}(\boldsymbol{x}_u.\boldsymbol{x}_v) - \boldsymbol{x}_u.\boldsymbol{x}_{uv} = F_u - \frac{1}{2}E_v,$$

with similar expressions for the other inner products that occur. So, taking the inner product of each of equations (6.1), (6.2), (6.3) with $\boldsymbol{x}_u$ and $\boldsymbol{x}_v$, we obtain the following three pairs of linear equations.

$$E\Gamma^1_{11} + F\Gamma^2_{11} = \frac{1}{2}E_u ,$$
$$F\Gamma^1_{11} + G\Gamma^2_{11} = F_u - \frac{1}{2}E_v ;$$

$$E\Gamma^1_{12} + F\Gamma^2_{12} = \frac{1}{2}E_v ,$$
$$F\Gamma^1_{12} + G\Gamma^2_{12} = \frac{1}{2}G_u ;$$

$$E\Gamma^1_{22} + F\Gamma^2_{22} = F_v - \frac{1}{2}G_u ,$$
$$F\Gamma^1_{22} + G\Gamma^2_{22} = \frac{1}{2}G_v. \tag{6.4}$$

The determinant of each of the three pairs of linear equations for the Γ^k_{ij} is $EG - F^2$ which, being non-zero, gives the following lemma.

Lemma 1 *The Christoffel symbols $\{\Gamma^k_{ij}\}$ are determined by the coefficients of the first fundamental form and their derivatives.*

This lemma is very important, and is not immediately clear from the definition of the Christoffel symbols given at the start of this section.

Example 2 (Orthogonal parametrisations) For a local parametrisation with $F = 0$, the Christoffel symbols are given by:

$$\Gamma^1_{11} = \frac{1}{2}\frac{E_u}{E}, \qquad \Gamma^1_{12} = \Gamma^1_{21} = \frac{1}{2}\frac{E_v}{E}, \qquad \Gamma^1_{22} = -\frac{1}{2}\frac{G_u}{E},$$
$$\Gamma^2_{11} = -\frac{1}{2}\frac{E_v}{G}, \qquad \Gamma^2_{12} = \Gamma^2_{21} = \frac{1}{2}\frac{G_u}{G}, \qquad \Gamma^2_{22} = \frac{1}{2}\frac{G_v}{G}.$$

Example 3 (Tchebycheff parametrisations) A local parametrisation $\boldsymbol{x}(u, v)$ with coefficients of the first fundamental form satisfying $E = G = 1$ is called a *Tchebycheff parametrisation*. In this case, the coordinate curves are parametrised by arc length, and $F = \cos\theta$ where θ is the angle of intersection of the coordinate curves. Intuitively, a Tchebycheff parametrisation may be thought of as moulding a piece of fabric over the surface without stretching the fibres but changing the angle θ at which the two sets of fibres (the weft and the warp) meet.

When using a Tchebycheff parametrisation, equations (6.4) for the Γ^i_{jk} reduce to

$$\begin{aligned}
\Gamma^1_{11} \quad &+ \Gamma^2_{11}\cos\theta = 0\,, \\
\Gamma^1_{11}\cos\theta + \Gamma^2_{11} \quad &= -\theta_u \sin\theta\,; \\[1em]
\Gamma^1_{12} \quad &+ \Gamma^2_{12}\cos\theta = 0\,, \\
\Gamma^1_{12}\cos\theta + \Gamma^2_{12} \quad &= 0\,; \\[1em]
\Gamma^1_{22} \quad &+ \Gamma^2_{22}\cos\theta = -\theta_v \sin\theta\,, \\
\Gamma^1_{22}\cos\theta + \Gamma^2_{22} \quad &= 0.
\end{aligned} \tag{6.5}$$

These three pairs of equations are easily solved. In particular, the second pair of equations imply that, for a Tchebycheff parametrisation,

$$\Gamma^1_{12} = \Gamma^2_{12} = 0.$$

6.2 Proof of the theorem

We prove the Theorema Egregium by showing that, having chosen a local parametrisation $\boldsymbol{x}$ of a surface S in $\mathbb{R}^3$, then $LN - M^2$ is expressible in terms of the Christoffel symbols $\{\Gamma^i_{jk}\}$ and the coefficients E, F, G of the first fundamental form and their derivatives. This, together with Lemma 1 of §6.1, will show that we may write $LN - M^2$, and hence the Gaussian curvature K, in terms of E, F, G and their derivatives.

Using (6.1), (6.2) and (6.3) we have

$$LN - M^2 = LN.NN - MN.MN \tag{6.6}$$

$$\begin{aligned}
&= (\boldsymbol{x}_{uu} - \Gamma^1_{11}\boldsymbol{x}_u - \Gamma^2_{11}\boldsymbol{x}_v).(\boldsymbol{x}_{vv} - \Gamma^1_{22}\boldsymbol{x}_u - \Gamma^2_{22}\boldsymbol{x}_v) \\
&\quad - (\boldsymbol{x}_{uv} - \Gamma^1_{12}\boldsymbol{x}_u - \Gamma^2_{12}\boldsymbol{x}_v).(\boldsymbol{x}_{uv} - \Gamma^1_{12}\boldsymbol{x}_u - \Gamma^2_{12}\boldsymbol{x}_v)\,,
\end{aligned} \tag{6.7}$$

so that

$$\begin{aligned}
LN - M^2 &= \boldsymbol{x}_{uu}.\boldsymbol{x}_{vv} - \boldsymbol{x}_{uv}.\boldsymbol{x}_{uv} \\
&\quad + \text{terms involving the } \Gamma^k_{ij}, E, F, G \text{ and their derivatives.}
\end{aligned}$$

However,

$$
\begin{aligned}
\boldsymbol{x}_{uu}.\boldsymbol{x}_{vv} - \boldsymbol{x}_{uv}.\boldsymbol{x}_{uv} &= \frac{\partial}{\partial u}(\boldsymbol{x}_u.\boldsymbol{x}_{vv}) - \frac{\partial}{\partial v}(\boldsymbol{x}_u.\boldsymbol{x}_{uv}) \\
&= \frac{\partial}{\partial u}\left(F_v - \frac{1}{2}G_u\right) - \frac{\partial}{\partial v}\left(\frac{1}{2}E_v\right) \\
&= -\frac{1}{2}(E_{vv} - 2F_{uv} + G_{uu}), \qquad (6.8)
\end{aligned}
$$

which completes the proof of the theorem.

An explicit formula for K may be deduced quite quickly from (6.7) if one remembers that, for instance, $\Gamma^1_{11}\boldsymbol{x}_u + \Gamma^2_{11}\boldsymbol{x}_v$ is the component of $\boldsymbol{x}_{uu}$ tangential to S, whereas $\boldsymbol{x}_{vv} - \Gamma^1_{22}\boldsymbol{x}_u - \Gamma^2_{22}\boldsymbol{x}_v$ is orthogonal to S. Hence,

$$
\begin{aligned}
&(\boldsymbol{x}_{uu} - \Gamma^1_{11}\boldsymbol{x}_u - \Gamma^2_{11}\boldsymbol{x}_v).(\boldsymbol{x}_{vv} - \Gamma^1_{22}\boldsymbol{x}_u - \Gamma^2_{22}\boldsymbol{x}_v) \\
&\quad = \boldsymbol{x}_{uu}.(\boldsymbol{x}_{vv} - \Gamma^1_{22}\boldsymbol{x}_u - \Gamma^2_{22}\boldsymbol{x}_v) \\
&\quad = \boldsymbol{x}_{uu}.\boldsymbol{x}_{vv} - \boldsymbol{x}_{uu}.(\Gamma^1_{22}\boldsymbol{x}_u + \Gamma^2_{22}\boldsymbol{x}_v) \\
&\quad = \boldsymbol{x}_{uu}.\boldsymbol{x}_{vv} - (\Gamma^1_{11}\boldsymbol{x}_u + \Gamma^2_{11}\boldsymbol{x}_v).(\Gamma^1_{22}\boldsymbol{x}_u + \Gamma^2_{22}\boldsymbol{x}_v) \\
&\quad = \boldsymbol{x}_{uu}.\boldsymbol{x}_{vv} - E\Gamma^1_{11}\Gamma^1_{22} - F(\Gamma^1_{11}\Gamma^2_{22} + \Gamma^2_{11}\Gamma^1_{22}) - G\Gamma^2_{11}\Gamma^2_{22}.
\end{aligned}
$$

A similar expression may be obtained for $(\boldsymbol{x}_{uv} - \Gamma^1_{12}\boldsymbol{x}_u - \Gamma^2_{12}\boldsymbol{x}_v).(\boldsymbol{x}_{uv} - \Gamma^1_{12}\boldsymbol{x}_u - \Gamma^2_{12}\boldsymbol{x}_v)$, and, putting these together, we find that K is given by

$$
\begin{aligned}
(EG - F^2)K = &-\frac{1}{2}(E_{vv} - 2F_{uv} + G_{uu}) - E\left(\Gamma^1_{11}\Gamma^1_{22} - (\Gamma^1_{12})^2\right) \\
&- F\left(\Gamma^1_{11}\Gamma^2_{22} - 2\Gamma^1_{12}\Gamma^2_{12} + \Gamma^1_{22}\Gamma^2_{11}\right) - G\left(\Gamma^2_{11}\Gamma^2_{22} - (\Gamma^2_{12})^2\right). \qquad (6.9)
\end{aligned}
$$

This is the *Gauss formula* for K.

Since the right hand side of (6.9) is determined by the coefficients of the first fundamental form, while the left hand side is equal to $LN - M^2$, the Gauss formula gives a relation between the coefficients of the two fundamental forms. We investigate further relations of this type in §6.3.

The Gauss formula is perhaps a little complicated in the general case, but for some types of local parametrisation the formula for K takes a simpler form.

Example 1 (Orthogonal parametrisations) For an orthogonal local parametrisation, the expressions for the Christoffel symbols given in Example 2 of §6.1 may be used to show that the Gaussian curvature is given by

$$K = -\frac{1}{2\sqrt{EG}}\left\{\left(\frac{E_v}{\sqrt{EG}}\right)_v + \left(\frac{G_u}{\sqrt{EG}}\right)_u\right\}. \qquad (6.10)$$

In particular, if the local parametrisation is isothermal, so that $E = G = \lambda^2$, $F = 0$, then

$$K = -\frac{1}{\lambda^2}\Delta \log \lambda, \qquad (6.11)$$

where $\Delta = \frac{\partial^2}{\partial u^2} + \frac{\partial^2}{\partial v^2}$ is the *Laplacian*. We note that the surface has **constant** Gaussian curvature if and only if λ satisfies *Liouville's equation*,

$$\frac{1}{\lambda^2}\Delta \log \lambda = \text{constant}.$$

Formula (6.10) for K in terms of an orthogonal parametrisation is rather nice, in that the Christoffel symbols do not occur explicitly.

Example 2 (Tchebycheff parametrisations) This continues Example 3 of §6.1, where we wrote down the equations satisfied by the Christoffel symbols when $E = G = 1$ and $F = \cos\theta$. As already noted, the second pair of equations in (6.5) give that $\Gamma^1_{12} = \Gamma^2_{12} = 0$, and, using this, we find that

$$\begin{aligned}(EG - F^2)K &= F_{uv} - \Gamma^1_{11}\Gamma^1_{22} - (\Gamma^1_{11}\Gamma^2_{22} + \Gamma^1_{22}\Gamma^2_{11})\cos\theta - \Gamma^2_{11}\Gamma^2_{22}\\ &= F_{uv} - \Gamma^1_{22}(\Gamma^1_{11} + \Gamma^2_{11}\cos\theta) - \Gamma^2_{22}(\Gamma^1_{11}\cos\theta + \Gamma^2_{11}).\end{aligned}$$

Hence, using the first pair of equations in (6.5),

$$(EG - F^2)K = F_{uv} + \Gamma^2_{22}\theta_u \sin\theta.$$

However, the third pair of equations in (6.5) shows that $\Gamma^2_{22}\sin\theta = \theta_v\cos\theta$, so it follows that

$$(EG - F^2)K = -\theta_{uv}\sin\theta, \tag{6.12}$$

so that, for a local parametrisation with $E = G = 1$, $F = \cos\theta$,

$$K = -\frac{\theta_{uv}\sin\theta}{\sin^2\theta} = -\frac{\theta_{uv}}{\sin\theta}.$$

In particular, we note that $K = -1$ if and only if θ satisfies the *sine-Gordon* equation,

$$\theta_{uv} = \sin\theta. \tag{6.13}$$

Equation (6.13), which may be thought of as a non-linear wave equation, is one of the basic partial differential equations of soliton theory.

6.3 The Codazzi–Mainardi equations

As we have already noted (several times!), the main thrust of the proof of the Theorema Egregium is to show that $LN - M^2$ is determined by E, F, G and the derivatives of E, F and G. In this section, we consider the natural question of whether there are any other relations between the coefficients of the two fundamental forms, other than that provided by noting that $LN - M^2$ is equal to the expression on the right hand side of the Gauss formula (6.9). This will lead us to a discussion of Bonnet's Theorem, part of which says that a surface in $\mathbb{R}^3$ is essentially completely determined up to rigid motions of $\mathbb{R}^3$ by its first and second fundamental forms. This is the analogue for surfaces in $\mathbb{R}^3$ of the Fundamental Theorem of the Local Theory of Plane Curves.

Equations (6.1), (6.2) and (6.3) imply certain consistency conditions determined by the fact that mixed partial derivatives commute. For instance, since $(\boldsymbol{x}_{uu})_v = (\boldsymbol{x}_{uv})_u$ we have that

$$(\boldsymbol{x}_{uu})_v.\boldsymbol{N} = (\boldsymbol{x}_{uv})_u.\boldsymbol{N},$$

so, using (6.1) and (6.2) we find that

$$\begin{aligned} 0 &= (\Gamma^1_{11}\boldsymbol{x}_u + \Gamma^2_{11}\boldsymbol{x}_v + L\boldsymbol{N})_v.\boldsymbol{N} - (\Gamma^1_{12}\boldsymbol{x}_u + \Gamma^2_{12}\boldsymbol{x}_v + M\boldsymbol{N})_u.\boldsymbol{N} \\ &= \Gamma^1_{11}\boldsymbol{x}_{uv}.\boldsymbol{N} + \Gamma^2_{11}\boldsymbol{x}_{vv}.\boldsymbol{N} + L_v - \Gamma^1_{12}\boldsymbol{x}_{uu}.\boldsymbol{N} - \Gamma^2_{12}\boldsymbol{x}_{uv}.\boldsymbol{N} - M_u \\ &= M\Gamma^1_{11} + N\Gamma^2_{11} + L_v - L\Gamma^1_{12} - M\Gamma^2_{12} - M_u, \end{aligned}$$

so that

$$L_v - M_u = L\Gamma^1_{12} - M\left(\Gamma^1_{11} - \Gamma^2_{12}\right) - N\Gamma^2_{11}. \tag{6.14}$$

Similarly, by considering the inner product of $(\boldsymbol{x}_{vv})_u - (\boldsymbol{x}_{uv})_v$ with $\boldsymbol{N}$, we obtain

$$M_v - N_u = L\Gamma^1_{22} - M\left(\Gamma^1_{12} - \Gamma^2_{22}\right) - N\Gamma^2_{12}. \tag{6.15}$$

Equations (6.14) and (6.15) are called the *Codazzi–Mainardi equations*.

In the case of an orthogonal local parametrisation these may be written relatively simply in terms of the coefficients of the first and second fundamental forms,

$$L_v - M_u = \frac{E_v}{2}\left(\frac{L}{E} + \frac{N}{G}\right) - \frac{M}{2}\left(\frac{E_u}{E} - \frac{G_u}{G}\right), \tag{6.16}$$

$$M_v - N_u = -\frac{G_u}{2}\left(\frac{L}{E} + \frac{N}{G}\right) - \frac{M}{2}\left(\frac{E_v}{E} - \frac{G_v}{G}\right). \tag{6.17}$$

In the case of an isothermal local parametrisation with $E = G = \lambda^2$ and $F = 0$, we saw in Lemma 2 of §5.4 that the mean curvature H is given by $2H = (L + N)/\lambda^2$, from which it follows that the Codazzi–Mainardi equations become

$$L_v - M_u = 2H\lambda\lambda_v\,, \qquad N_u - M_v = 2H\lambda\lambda_u. \tag{6.18}$$

Returning to the general situation, the above working shows that the Gauss formula (6.9) and Codazzi–Mainardi equations (6.14) and (6.15) are necessarily satisfied by a surface in $\mathbb{R}^3$. The following theorem, which, as mentioned earlier, is analogous to the Fundamental Theorem of the Local Theory of Plane Curves discussed in Chapter 1, shows that they are also sufficient.

Theorem 1 (Bonnet) *Let U be an open subset of $\mathbb{R}^2$ and let $E, F, G;\ L, M, N$ be smooth real-valued functions defined on U with $E > 0$, $G > 0$ and $EG - F^2 > 0$. Assume that $E, F, G;\ L, M, N$ satisfy the Gauss formula* (6.9) *and Codazzi–Mainardi equations* (6.14) *and* (6.15), *where $K = (LN - M^2)/(EG - F^2)$ and the $\{\Gamma^i_{jk}\}$ are defined to be the solutions to* (6.4). *Then, if $p \in U$, there is an open neighbourhood V of p in U, and a smooth map $\boldsymbol{x} : V \to \mathbb{R}^3$ such that*

(i) $\boldsymbol{x}(V)$ is a surface S in $\mathbb{R}^3$,

(ii) $\boldsymbol{x}$ *is a parametrisation of* S *such that* E, F, G *and* L, M, N *are the coefficients of the first and second fundamental forms of* S*, where* L, M, N *are determined using the unit normal* $\boldsymbol{N}$ *in the direction of* $\boldsymbol{x}_u \times \boldsymbol{x}_v$.

Moreover, if V *is connected,* $\boldsymbol{x}$ *(and hence* S*) is uniquely determined by* $E, F, G; L, M, N$ *up to rigid motions of* $\mathbb{R}^3$.

We shall not prove Bonnet's Theorem, since this would carry us into the realms of existence and uniqueness theorems for solutions of certain partial differential equations.

Example 2 (Tchebycheff parametrisations) We saw in Example 2 of §6.2 that a Tchebycheff parametrisation of a surface with constant Gaussian curvature $K = -1$ gave a solution θ of the sine-Gordon equation (6.13). Conversely, in Exercise 6.10 you are asked to use Bonnet's Theorem to show that if $\theta(u, v)$ is a solution of the sine-Gordon equation, then there exists a surface S with $K = -1$ which is covered by a Tchebycheff parametrisation $\boldsymbol{x}(u, v)$ such that

(a) $F = \cos\theta$,
(b) the coordinate curves of $\boldsymbol{x}$ are the asymptotic curves of S.

Moreover, the local parametrisation $\boldsymbol{x}$, and hence the surface S, is determined uniquely by θ up to rigid motions (and possibly a reflection) of $\mathbb{R}^3$.

6.4 Surfaces of constant Gaussian curvature †

Gaussian curvature is an important function on a surface S, so, as remarked in the introduction to this chapter, it is natural to try to find as much information as possible concerning surfaces for which this function is constant. In this optional section we give a brief review of the area, but have (rather reluctantly) restricted ourselves to the proof of just one of the results, a theorem due to Liebmann. However, a proof of Minding's Theorem (Theorem 6) is given in §7.7.

Theorem 1 (Liebmann) *Let* S *be a compact connected surface in* $\mathbb{R}^3$ *with constant Gaussian curvature* K*. Then* $K > 0$ *and* S *is a sphere.*

Proof We first recall Theorem 4 of §5.10, which says that every compact surface in $\mathbb{R}^3$ has an elliptic point. Hence, if K is constant then $K > 0$.

We next recall that if κ_1 and κ_2 are the principal curvatures, then $K = \kappa_1\kappa_2$ and the mean curvature $H = (\kappa_1 + \kappa_2)/2$. Hence $H^2 - K = (\kappa_1 - \kappa_2)^2/4$ is non-negative and is equal to zero only at umbilics. If $H^2 - K$ is everywhere zero, then every point of S is an umbilic, and, by Theorem 1 of §5.8, S is an open subset of a sphere. However, since S is compact, it is also closed, and since the sphere is connected, S is the whole of the sphere.

We now assume that $H^2 - K$ is not identically zero. Since S is compact, the continuous function $H^2 - K$ attains a maximum value at some point p of S, and we shall show that the assumption that $(H^2 - K)(p) > 0$, or, equivalently, that p is not an umbilic, implies that $K(p) \leq 0$. This contradicts the first paragraph of the proof.

In order to proceed with our proof, we use the fact (which we shall not prove) that there exists a local parametrisation $\boldsymbol{x}(u, v)$ around any non-umbilic point of a surface in $\mathbb{R}^3$ whose coordinate curves are lines of curvature. It follows immediately from Lemma 1 of §5.6 that such a parametrisation is characterised by having $F = M = 0$, and, using Exercise 6.8, the Codazzi–Mainardi equations may be written as

$$(\kappa_1 - \kappa_2)E_v = -2E(\kappa_1)_v \,, \quad (\kappa_1 - \kappa_2)G_u = 2G(\kappa_2)_u, \tag{6.19}$$

where κ_1, κ_2 are the eigenvalues corresponding to principal directions $\boldsymbol{x}_u, \boldsymbol{x}_v$ respectively. We also assume, by interchanging u and v, and reversing the direction of the unit normal N, if necessary, that $\kappa_1 > \kappa_2 > 0$ on some open neighbourhood U of p in S.

In the following, we shall be differentiating the principal curvature functions κ_1 and κ_2 several times, so to avoid confusion with suffices, we shall write λ and μ rather than κ_1 and κ_2 for the two principal curvatures. Then $\lambda > \mu > 0$ on U, and, since $(\lambda - \mu)^2$ takes its maximum value at p, so does $\lambda - \mu$.

However, since K is constant,

$$(\lambda - \mu)_u = \left(\lambda - \frac{K}{\lambda}\right)_u = \lambda_u\left(1 + \frac{K}{\lambda^2}\right),$$

so, in particular, $\lambda_u(p) = 0$.

Differentiating again we find that, at p,

$$(\lambda - \mu)_{uu}(p) = \lambda_{uu}(p)\left(1 + \frac{K}{\lambda^2}\right)(p),$$

so, since $\lambda - \mu$ takes its maximum value at p, we see that $\lambda_{uu}(p) \leq 0$. Similar methods show that $\lambda_v(p) = \mu_u(p) = \mu_v(p) = 0$, $\lambda_{vv}(p) \leq 0$, and $\mu_{uu}(p) \geq 0$.

It now follows from (6.19) that $E_v(p) = 0$ and $G_u(p) = 0$, so formula (6.10) for the Gaussian curvature in the case of an orthogonal parametrisation gives that

$$K(p) = -\frac{1}{2EG}\left(E_{vv} + G_{uu}\right)(p). \tag{6.20}$$

However, differentiating (6.19) at p, we obtain

$$(\lambda - \mu)E_{vv}(p) = -2E\lambda_{vv}(p)\,, \quad (\lambda - \mu)G_{uu}(p) = 2G\mu_{uu}(p),$$

so that $E_{vv}(p)$ and $G_{uu}(p)$ are both non-negative. It now follows from (6.20) that $K(p) \leq 0$, and we have established the contradiction needed to complete the proof of the theorem. □

We now consider non-compact surfaces with constant Gaussian curvature, and begin by considering the case $K = 0$. We have seen in Exercise 5.8 that a connected surface of revolution has $K = 0$ if and only if it is an open subset of a plane, a cone or a cylinder. As a generalisation of this, we saw in Exercise 5.16 that if p is a non-umbilic point on a surface S with $K = 0$ then there exists an open neighbourhood of p which is a ruled surface with the property that the unit normal is constant along each line of the ruling.

Such a surface is called a *developable* surface. Conversely, if S is a developable surface then 0 is an eigenvalue of dN, so that $K = 0$ for such a surface.

In Exercise 5.5, several types of example of developable surfaces are given; namely, tangent developables of regular curves in $\mathbb{R}^3$, generalised cones, and generalised cylinders. We recall, in particular, that a generalised cylinder is the image of the map into $\mathbb{R}^3$ given by

$$\boldsymbol{x}(u, v) = \boldsymbol{\alpha}(u) + v\boldsymbol{e}\,, \quad v \in \mathbb{R},$$

where $\boldsymbol{\alpha}(u)$ is a regular curve in $\mathbb{R}^3$ whose tangent vector $\boldsymbol{\alpha}'(u)$ is never parallel to the non-zero vector $\boldsymbol{e}$.

The following theorem of Massey classifies all *closed* surfaces in $\mathbb{R}^3$ with $K = 0$. We recall (from §5.10) that a closed surface is a surface that is a closed subset of its containing Euclidean space. Intuitively speaking, a closed surface has no edges to fall off. All compact surfaces are closed, as are, for instance, cylinders, catenoids and helicoids. We note that the tangent developable of a regular curve $\boldsymbol{\alpha}$ is not closed (all points of $\boldsymbol{\alpha}$ are omitted, since these are singular points), and nor is a generalised cone (the vertex is a singular point).

Theorem 2 (Massey) *Let S be a closed connected surface in $\mathbb{R}^3$ whose Gaussian curvature is identically zero. Then S is a generalised cylinder.*

We choose to omit the proof, but would like to remark that it is accessible, and uses the Codazzi–Mainardi equations.

We next consider surfaces in $\mathbb{R}^3$ with constant negative Gaussian curvature. Exercise 5.8 gave a method of finding all surfaces of revolution with constant negative Gaussian curvature, and these include the pseudosphere (see Example 7 of §5.6). However, none of these examples are closed; for instance, the pseudosphere is not closed since it does not include the unit circle $x^2 + y^2 = 1, z = 0$. In fact, we have the following.

Theorem 3 (Hilbert) *There is no closed surface in $\mathbb{R}^3$ with constant negative Gaussian curvature.*

A crucial step in the proof of Hilbert's Theorem is to show that any surface in $\mathbb{R}^3$ with $K = -1$ may be covered by a system of Tchebycheff parametrisations whose coordinate curves are asymptotic curves. In the following example we outline a method of constructing such a parametrisation of the pseudosphere, which, as we saw in Example 7 of §5.6, has constant Gaussian curvature $K = -1$.

Example 4 (Pseudosphere) Let $\boldsymbol{x}(u, v)$ be the usual parametrisation of the pseudosphere S as a surface of revolution given by

$$\boldsymbol{x}(u, v) = (\operatorname{sech} v \cos u, \operatorname{sech} v \sin u, v - \tanh v), \quad -\pi < u < \pi, \ \ v > 0.$$

Exercise 5.19 invited you to prove that the asymptotic curves are given by $u \pm v = \text{constant}$, and that the angle of intersection θ of the two asymptotic curves through $\boldsymbol{x}(u, v)$ is given by $\cos\theta = 2\operatorname{sech}^2 v - 1$. So, if we define a new parametrisation of S by taking

$$\tilde{\boldsymbol{x}}(\tilde{u}, \tilde{v}) = \boldsymbol{x}(\tilde{u} + \tilde{v}, \tilde{u} - \tilde{v})\,, \quad -\pi < \tilde{u} + \tilde{v} < \pi\,, \ \ \tilde{u} > \tilde{v},$$

(this is a *change of variables* as described in the optional §3.8), it follows that the coordinate curves of $\tilde{\boldsymbol{x}}$ are the asymptotic curves of S and the angle of intersection of the two asymptotic curves through $\tilde{\boldsymbol{x}}(\tilde{u}, \tilde{v})$ is given by

$$\cos \tilde{\theta}(\tilde{u}, \tilde{v}) = 2 \operatorname{sech}^2(\tilde{u} - \tilde{v}) - 1. \tag{6.21}$$

A routine calculation shows that the coefficients of the first fundamental form with respect to $\boldsymbol{x}(u, v)$ are given by

$$E = \operatorname{sech}^2 v\,, \quad F = 0\,, \quad G = \tanh^2 v,$$

and, since

$$\tilde{\boldsymbol{x}}_{\tilde{u}} = \boldsymbol{x}_u + \boldsymbol{x}_v\,, \quad \tilde{\boldsymbol{x}}_{\tilde{v}} = \boldsymbol{x}_u - \boldsymbol{x}_v, \tag{6.22}$$

another routine calculation shows that $\tilde{E} = \tilde{G} = 1$, so that $\tilde{\boldsymbol{x}}$ is a Tchebycheff parametrisation of the pseudosphere whose coordinate curves are the asymptotic curves of S.

The theory given in Example 2 of §6.2 now predicts that $\tilde{\theta}(\tilde{u}, \tilde{v})$ should satisfy the sine-Gordon equation

$$\tilde{\theta}_{\tilde{u}\tilde{v}} = \sin \tilde{\theta}, \tag{6.23}$$

and this may be checked from (6.21) by direct calculation.

We now consider closed connected surfaces of constant positive Gaussian curvature. It turns out (although we do not develop the tools to prove it) that such surfaces are necessarily compact, so the following theorem is an immediate consequence of Liebmann's Theorem.

Theorem 5 *Let S be a closed connected surface in $\mathbb{R}^3$ with constant positive Gaussian curvature. Then S is a sphere.*

Finally, we state Minding's Theorem. Unlike the other theorems in this section, Minding's Theorem is a local theorem; it doesn't depend on any global assumptions such as "closed" or "compact". It says that if two surfaces S, $\tilde{S}$ have the same **constant** Gaussian curvature then, locally at least, they are isometric.

Theorem 6 (Minding) *Let S and $\tilde{S}$ be surfaces in $\mathbb{R}^3$ having the same constant Gaussian curvature. If $p \in S$ and $\tilde{p} \in \tilde{S}$ then there is an isometry from an open neighbourhood of p in S onto an open neighbourhood of $\tilde{p}$ in $\tilde{S}$.*

In particular, then, a developable surface is locally isometric to a flat plane. This means that a sufficiently small piece of any developable surface may be formed from a flat piece of metal without stretching or compressing the sheet.

We need some material on geodesics (to be found in Chapter 7) before we can construct the isometry needed to prove Minding's Theorem, so we postpone the proof until §7.7.

6.5 A generalisation of Gaussian curvature †

The Theorema Egregium tells us that the Gaussian curvature K of a surface in $\mathbb{R}^3$ depends on only the intrinsic metric properties of the surface. This suggests the possibility of defining a notion of Gaussian curvature for a surface in $\mathbb{R}^n$ rather than just for a surface in $\mathbb{R}^3$, and in this section we indicate a way of achieving this. As mentioned in the introduction to this chapter, those wishing to concentrate on surfaces in $\mathbb{R}^3$ may choose to omit this section.

Let $\boldsymbol{x}(u, v)$ be a local parametrisation of a surface S in $\mathbb{R}^n$, and let $\mathcal{N}(\boldsymbol{x}_{uu})$, $\mathcal{N}(\boldsymbol{x}_{uv})$, $\mathcal{N}(\boldsymbol{x}_{vv})$ denote the components of $\boldsymbol{x}_{uu}$, $\boldsymbol{x}_{uv}$, $\boldsymbol{x}_{vv}$, respectively, orthogonal to S. It may be checked using the methods employed in §3.9 that the expression $\mathcal{N}(\boldsymbol{x}_{uu}).\mathcal{N}(\boldsymbol{x}_{vv}) - \mathcal{N}(\boldsymbol{x}_{uv}).\mathcal{N}(\boldsymbol{x}_{uv})$ involving the inner products of these vectors transforms under a change of local parametrisation in the same way as does $EG - F^2$. It follows that if we set

$$K = \frac{\mathcal{N}(\boldsymbol{x}_{uu}).\mathcal{N}(\boldsymbol{x}_{vv}) - \mathcal{N}(\boldsymbol{x}_{uv}).\mathcal{N}(\boldsymbol{x}_{uv})}{EG - F^2}, \tag{6.24}$$

then K is independent of choice of local parametrisation and hence defines a function on S. This is the *Gaussian curvature* of a surface in $\mathbb{R}^n$; when $n = 3$, it clearly reduces to our original definition of Gaussian curvature of a surface in $\mathbb{R}^3$.

If we modify the material in §6.1 and §6.2 by replacing LN, MN, and NN by $\mathcal{N}(\boldsymbol{x}_{uu})$, $\mathcal{N}(\boldsymbol{x}_{uv})$, $\mathcal{N}(\boldsymbol{x}_{vv})$, respectively, and replacing $LN - M^2$ by $\mathcal{N}(\boldsymbol{x}_{uu}).\mathcal{N}(\boldsymbol{x}_{vv}) - \mathcal{N}(\boldsymbol{x}_{uv}).\mathcal{N}(\boldsymbol{x}_{uv})$, then §6.1 and §6.2 hold in this more general situation. In particular, the Gauss formula (6.9) is true, and it then follows that:

the Theorema Egregium and its corollaries all hold if we assume that S is a surface in $\mathbb{R}^n$ and $\tilde{S}$ is a surface in $\mathbb{R}^m$.

This is a highly significant and non-trivial fact which lies at the heart of differential geometry.

Example 1 (Veronese surface) Recall from Example 2 of §4.6 that the image in $\mathbb{R}^5$ of the map $f : S^2(1) \to \mathbb{R}^5$ given by

$$f(x, y, z) = \left(yz, zx, xy, \frac{1}{2}(x^2 - y^2), \frac{1}{2\sqrt{3}}(x^2 + y^2 - 2z^2) \right), \quad x^2 + y^2 + z^2 = 1,$$

is a surface S in $\mathbb{R}^5$ called the *Veronese* surface. We showed in that example that f is a local isometry onto S and that f gives a bijective correspondence between the *real projective plane* $\mathbb{R}P^2$ of lines through the origin of $\mathbb{R}^3$ and S. The generalisation of Corollary 3 of the Theorema Egregium shows that the Veronese surface is a surface in $\mathbb{R}^5$ of constant Gaussian curvature 1.

Example 2 (Flat torus) Recall from Example 4 of §4.3 that for each pair r_1, r_2 of positive real numbers, the subset of $\mathbb{R}^4$ defined by

$$S^1(r_1) \times S^1(r_2) = \{(x_1, x_2, x_3, x_4) \in \mathbf{R}^4 : {x_1}^2 + {x_2}^2 = {r_1}^2,\ {x_3}^2 + {x_4}^2 = {r_2}^2\}$$

is a surface in $\mathbb{R}^4$. We called this a flat torus, and showed that the map $f : \mathbb{R}^2 \to S^1(r_1) \times S^1(r_2)$ defined by

$$f(u, v) = \left(r_1 \cos\frac{u}{r_1}, r_1 \sin\frac{u}{r_1}, r_2 \cos\frac{v}{r_2}, r_2 \sin\frac{v}{r_2}\right)$$

is a local isometry. Thus the Gaussian curvature of the flat torus is zero, which provides the motivation for its name.

In Exercise 4.14 we saw that there is a conformal diffeomorphism of this flat torus onto a torus of revolution in $\mathbb{R}^3$. However, the generalisation of Corollary 3 of the Theorema Egregium implies that there can be no isometry of $S^1(r_1) \times S^1(r_2)$ onto any surface in $\mathbb{R}^3$, for any such surface would be compact and thus by Theorem 4 of §5.10 would have an elliptic point. This means that, as a surface with metric, $S^1(r_1) \times S^1(r_2)$ is an object that can only be encountered once we consider surfaces in $\mathbb{R}^n$ for $n > 3$.

Examples 1 and 2 show that neither Liebmann's Theorem nor Massey's Theorem generalise to the case of surfaces in higher dimensional Euclidean spaces. However, Minding's Theorem does hold in this more general situation. Indeed, it holds in the even more general setting outlined in the next example.

Example 3 (Hyperbolic plane) In Example 5 of §3.4, we put a metric on the upper half-plane,

$$H = \{(u, v) \in \mathbb{R}^2 : v > 0\}$$

with $E = G = 1/v^2$, $F = 0$. This gives us the hyperbolic plane H, which is an abstract surface with metric with no reference to any containing $\mathbb{R}^n$. However, we may use equations (6.4) to define the Christoffel symbols $\{\Gamma_{ij}^k\}$, and then use the Gauss curvature equation (6.9) to define a notion of Gaussian curvature K for surfaces with metric. If we do this, a short calculation using (6.11) shows that the hyperbolic plane has constant Gaussian curvature $K = -1$.

Although we shall not justify any of the following, it turns out that the corollaries of the Theorema Egregium still hold in the more general situation of surfaces with metric. It now comes as no surprise that the hyperbolic plane H has constant Gaussian curvature, since it follows from Example 4 of §4.7 that the isometry group of H is transitive. That this constant is equal to -1 is also to be expected since we saw in Exercise 4.18 that there is a local isometry from part of the hyperbolic plane to the pseudosphere, which, as we saw in Example 7 in §5.6, has constant Gaussian curvature $K = -1$. Although the *Nash Embedding Theorem* implies that there is an isometry from H to a surface in some sufficiently high dimensional Euclidean space, Hilbert's Theorem says that there is no local isometry from the whole of H to a surface in $\mathbb{R}^3$.

Exercises

6.1 The Theorema Egregium says that surfaces in $\mathbb{R}^n$ which are locally isometric have the same Gaussian curvature at corresponding points. Is the same thing true for the mean curvature of surfaces in $\mathbb{R}^3$? Give a proof or a counterexample.

Note that Exercises 6.2 to 6.5 hold with exactly the same proof for surfaces in $\mathbb{R}^n$.

6.2 Let $\boldsymbol{x}(u, v)$ be a local parametrisation of a surface S in $\mathbb{R}^3$ such that $E = 1$ and $F = 0$. Show that $\Gamma^2_{12} = \Gamma^2_{21} = \dfrac{G_u}{2G}$, $\Gamma^1_{22} = -\dfrac{1}{2}G_u$, $\Gamma^2_{22} = \dfrac{G_v}{2G}$, and that all the other Christoffel symbols are zero. Hence show that the Gaussian curvature K of S is given by

$$K = -\frac{(\sqrt{G})_{uu}}{\sqrt{G}}.$$

6.3 If the coefficients of the first fundamental form of a surface S in $\mathbb{R}^3$ are given by

$$E = 2 + v^2\,, \quad F = 1\,, \quad G = 1,$$

show that the Gaussian curvature K of S is given by

$$K = -\frac{1}{(1+v^2)^2}.$$

6.4 Let $\boldsymbol{x}(u, v)$ be a Tchebycheff parametrisation of a surface S in $\mathbb{R}^3$ with $E = G = 1$ and $F = \cos uv$. Show that the Gaussian curvature K of S is given by

$$K = -\frac{1}{\sin uv}.$$

6.5 Let $\boldsymbol{x}(u, v)$ be a Tchebycheff parametrisation of a surface S in $\mathbb{R}^3$. Show that the Gaussian curvature K of S is given by

$$K = \frac{F_{uv}(1-F^2) + F F_u F_v}{(1-F^2)^2}.$$

6.6 It is natural to ask whether Corollary 3 to the Theorema Egregium has a converse. That is to say, if $f : S \to \tilde{S}$ is a smooth map with the property that, for all $p \in S$, $\tilde{K}(f(p)) = K(p)$, is it true that f is a local isometry? In fact, this is clearly not true; simply consider any smooth map from the plane to itself or from the unit sphere to itself. However, these could perhaps be rather special examples since K is constant here. We now investigate a rather more substantial counterexample.

Let S be the helicoid $x \sin z = y \cos z$, and let $\tilde{S}$ be the surface of revolution obtained by rotating the curve

$$\boldsymbol{\alpha}(v) = (v, 0, \log v)\,, \quad v > 0,$$

around the z-axis. Let

$$U = \{(u, v) \in \mathbb{R}^2 : -\pi < u < \pi,\ v > 0\},$$

and let $\boldsymbol{x} : U \to S$, $\tilde{\boldsymbol{x}} : U \to \tilde{S}$ be the local parametrisations defined by

$$\boldsymbol{x}(u, v) = (v \cos u, v \sin u, u)\,, \quad (u, v) \in U,$$
$$\tilde{\boldsymbol{x}}(u, v) = (v \cos u, v \sin u, \log v)\,, \quad (u, v) \in U.$$

Show that the correspondence $\boldsymbol{x}(u, v) \leftrightarrow \tilde{\boldsymbol{x}}(u, v)$ is not an isometry, but does have the property that, for all $(u, v) \in U$, $\tilde{K}\left(\tilde{\boldsymbol{x}}(u, v)\right) = K\left(\boldsymbol{x}(u, v)\right)$.

6.7 Let S be a connected surface in $\mathbb{R}^3$ covered by a single parametrisation $\boldsymbol{x}(u, v)$ whose coefficients E, F, G, L, M, N of the first and second fundamental forms are all constant. If $L = M = N = 0$ show that S is an open subset of a plane. Otherwise, follow the route indicated below to show that $EN - 2FM + GL \neq 0$ and S is (an open subset of) a cylinder of radius $|c|$, where

$$c = \frac{EG - F^2}{EN - 2FM + GL}.$$

(i) Show that $K = 0$.
(ii) Show that if not all of L, M, N are zero, then we may assume, by interchanging u and v if necesary, that $L \neq 0$.
From now on, we assume that $L \neq 0$.
(iii) Show that $\boldsymbol{a} = M\boldsymbol{x}_u - L\boldsymbol{x}_v$ is a non-zero constant.
(iv) Show that $|\boldsymbol{a}|^2 = L(EN - 2FM + GL)$, so, in particular, $EN - 2FM + GL \neq 0$.
(v) Show that

$$\boldsymbol{b} = \boldsymbol{x} - \frac{\boldsymbol{x}.\boldsymbol{a}}{\boldsymbol{a}.\boldsymbol{a}}\boldsymbol{a} + c\boldsymbol{N}$$

is constant, where $c = (EG - F^2)/(EN - 2FM + GL)$.
(vi) Show that $\boldsymbol{a}$ is orthogonal to $\boldsymbol{b}$.
(vii) Without loss of generality, assume that $\boldsymbol{a} = (0, 0, a_3)$ and $\boldsymbol{b} = (b_1, b_2, 0)$, and show that, writing $\boldsymbol{x} = (x_1, x_2, x_3)$, S is an open subset of the cylinder

$$(x_1 - b_1)^2 + (x_2 - b_2)^2 = c^2.$$

6.8 Let $\boldsymbol{x}(u, v)$ be a local parametrisation of a surface S in $\mathbb{R}^3$ whose image contains no umbilic points. If the coordinate curves are also lines of curvature show that the Codazzi–Mainardi equations may be written as

$$(\kappa_1 - \kappa_2)E_v = -2E(\kappa_1)_v\,,$$
$$(\kappa_1 - \kappa_2)G_u = 2G(\kappa_2)_u\,,$$

where κ_1, κ_2 are the principal curvatures corresponding to principal directions $\boldsymbol{x}_u$, $\boldsymbol{x}_v$ respectively. These equations are used in the proof of Liebmann's Theorem (Theorem 1 of §6.4).

6.9 Let S be a surface in $\mathbb{R}^3$ with constant mean curvature H, and suppose that S admits an isothermal local parametrisation whose coordinate curves are also lines of curvature. Use the equations obtained in Exercise 6.8 to show that if the image of the local parametrisation is connected then the Codazzi–Mainardi equations reduce to $L - N =$ constant. (We shall see in §9.13 that if a surface has constant mean curvature then such a local parametrisation exists on some open neighbourhood of any non-umbilic point.)

6.10 Let $\theta(u, v)$ be a solution of the sine-Gordon equation $\theta_{uv} = \sin\theta$. Show that there exists a surface S with constant Gaussian curvature $K = -1$ which is covered by a Tchebycheff parametrisation $\boldsymbol{x}(u, v)$ such that

(a) $F = \cos\theta$,

(b) the coordinate curves of $\boldsymbol{x}$ are the asymptotic curves of S.

Show also that $\boldsymbol{x}$, and hence S, is determined uniquely by θ up to rigid motions (and possibly a reflection) of $\mathbb{R}^3$.

The following two exercises use material in the optional §6.4.

6.11 Show that the sphere is the only compact connected surface in $\mathbb{R}^3$ with constant mean curvature and everywhere positive Gaussian curvature. (*Hint:* use techniques employed in the proof of Liebmann's Theorem.)

6.12 Let S be a closed connected surface in $\mathbb{R}^3$. If S has constant Gaussian and mean curvatures show that S is either a sphere, a plane, or a (round) cylinder.

7 Geodesic curvature and geodesics

In §5.7 we defined the geodesic and normal curvatures of a regular curve $\boldsymbol{\alpha}$ on a surface S in $\mathbb{R}^3$. The normal curvature κ_n is defined using the component of the acceleration of $\boldsymbol{\alpha}$ orthogonal to S, and was studied in Chapter 5. The geodesic curvature κ_g, on the other hand, is determined by the component of the acceleration of $\boldsymbol{\alpha}$ tangential to S, and we shall see that, unlike κ_n, geodesic curvature is an intrinsic property.

In fact, although we shall not justify this remark, geodesic curvature may be defined for curves on higher dimensional analogues of surfaces (which are modelled on open subsets of $\mathbb{R}^m$ rather than $\mathbb{R}^2$); we do not even need a containing Euclidean space $\mathbb{R}^n$; all we need is a metric. Spaces modelled on open subsets of $\mathbb{R}^m$ are called *manifolds*, and if they have a metric they are known as *Riemannian manifolds*. These are named in honour of **Bernhard Riemann**, a student of Gauss, whose thesis laid the foundations for the major branch of modern mathematics known as Riemannian geometry.

Curvature for curves in the plane, as discussed in Chapter 1, is a special case of geodesic curvature for curves on a surface. When suitably parametrised, curves on a surface with zero geodesic curvature are called *geodesics*; they are the analogues of straight lines in the plane and as we shall show, the analogies are quite strong. For instance, a line segment in a plane is the path in the plane of shortest length between its end points, and a (sufficiently short) geodesic on a surface is the path of shortest length on the surface between its end points.

There is also a close relationship with mechanics. In particular, Newton's second law of motion states that the acceleration of a body is directly proportional to, and in the same direction as, the net force acting on the body. In the absence of an external force there is no acceleration and the corresponding motion is in a straight line. For surfaces in $\mathbb{R}^3$, Newton's second law gives a characterisation of geodesics as the paths followed by smooth particles moving freely on the surface, for if there is no tangential force there is no tangential acceleration and conversely. Furthermore, the equations of motion in the Lagrangian formulation of mechanics are just the geodesic equations (see §7.3), and conservation of energy corresponds to the fact that every geodesic is parametrised so that the speed of travel along the curve is constant (see §7.2).

§7.1 to §7.5 contain basic material, and are needed for the final two chapters of the book. §7.6 to §7.9 are optional, and a selection of some or all of them could be made, depending on time and taste; they are independent of each other, except that §7.8 uses the material in §7.7.

7.1 Geodesic curvature

Recall that for a regular curve $\boldsymbol{\alpha}(t)$ on a surface S in $\mathbb{R}^3$ the geodesic curvature κ_g and the normal curvature κ_n are defined by

$$\frac{d\boldsymbol{t}}{ds} = \kappa_g \boldsymbol{N} \times \boldsymbol{t} + \kappa_n \boldsymbol{N} ,$$

where s is an arc length parameter along $\boldsymbol{\alpha}$, $\boldsymbol{t} = d\boldsymbol{\alpha}/ds$ is the unit tangent vector to $\boldsymbol{\alpha}$, and $\boldsymbol{N}$ is the unit normal to S (defined, as usual, up to sign). It follows that, denoting inner product in $\mathbb{R}^n$ by a dot as usual,

$$\kappa_g = \frac{d\boldsymbol{t}}{ds}.(\boldsymbol{N} \times \boldsymbol{t}) . \tag{7.1}$$

Note that replacing $\boldsymbol{N}$ by $-\boldsymbol{N}$ changes the sign of both the geodesic and nomal curvatures.

We saw in §5.7 that the normal curvature κ_n at a point of a regular curve $\boldsymbol{\alpha}$ on S is equal to the second fundamental form acting on the unit tangent vector to $\boldsymbol{\alpha}$ at that point. In particular, rather surprisingly, κ_n depends on only the tangent vector to $\boldsymbol{\alpha}$ **at the point in question** and may be interpreted as the minimum amount of bending that a regular curve must do in order to stay on S. There is no similar restriction on κ_g; the geodesic curvature may be regarded as a measure of the extra bending that $\boldsymbol{\alpha}$ does within S.

A short calculation using (7.1) and techniques discussed in Chapter 1 shows the following.

Lemma 1 *The geodesic curvature of a regular curve $\boldsymbol{\alpha}(t)$ on S is given by*

$$\kappa_g = \boldsymbol{\alpha}''.(\boldsymbol{N} \times \boldsymbol{\alpha}')/|\boldsymbol{\alpha}'|^3 , \tag{7.2}$$

where, as usual, $'$ denotes differentiation with respect to t.

Example 2 (Circles on a sphere) Each circle on a sphere in $\mathbb{R}^3$ is obtained as the intersection of the sphere with a plane. For ease, we consider the circle C of intersection of the unit sphere $S^2(1)$ with the plane $z = \sin v_0$ for some constant v_0 with $-\pi/2 < v_0 < \pi/2$. Then C may be parametrised as

$$\boldsymbol{\alpha}(t) = (\cos v_0 \cos t, \cos v_0 \sin t, \sin v_0) ,$$

in which case

$$\boldsymbol{\alpha}'(t) = (-\cos v_0 \sin t, \cos v_0 \cos t, 0) ,$$

so that $|\boldsymbol{\alpha}'(t)| = \cos v_0$. Also,

$$\boldsymbol{\alpha}''(t) = (-\cos v_0 \cos t, -\cos v_0 \sin t, 0) ,$$

while $\boldsymbol{N}(\boldsymbol{\alpha}(t)) = \boldsymbol{\alpha}(t)$.

A short calculation using (7.2) now shows that the circle C in $S^2(1)$ has constant geodesic curvature $\kappa_g = \tan v_0$. In particular, the equator has zero geodesic curvature.

More generally, a circle on $S^2(1)$ which is the intersection of $S^2(1)$ with a plane at perpendicular distance $\sin v_0$ from the origin has constant geodesic curvature $\tan v_0$.

Conversely, as you are invited to prove in Exercise 7.2, every curve of constant geodesic curvature on a sphere is (part of) a circle. In particular, the curves of zero geodesic curvature on a sphere are those curves which are the intersection of the sphere with a plane through the centre of the sphere. These are the *great circles* on the sphere.

The next result is immediate, since if two surfaces in $\mathbb{R}^3$ touch tangentially at some point, then the normals to the surfaces coincide at that point.

Proposition 3 *Let S, $\tilde{S}$ be two surfaces in $\mathbb{R}^3$ which touch tangentially along the trace of a regular curve $\boldsymbol{\alpha}$. Then the geodesic curvature of $\boldsymbol{\alpha}$ as a curve on S is equal to the geodesic curvature of $\boldsymbol{\alpha}$ as a curve on $\tilde{S}$.*

Example 4 (Torus of revolution) For positive real numbers $a > b$, let $T_{a,b}$ be the torus of revolution obtained by rotating the circle $(x - a)^2 + z^2 = b^2$ about the z-axis. A sphere with centre on the z-axis and suitable radius R will touch the torus tangentially along the trace C of a regular curve $\boldsymbol{\alpha}$. If the sphere has centre at the origin, then taking $R = a - b$ or $R = a + b$ gives such a sphere; it touches the torus along the equator of the sphere and along the innermost or outermost parallel of the torus. Hence these particular parallels have zero geodesic curvature on the torus.

For a sphere as illustrated in Figure 7.1 with centre on the z-axis but not at the origin, there are two spheres which touch $T_{a,b}$ tangentially, and the curves C are again circles. It follows from Example 2 that the geodesic curvature is again constant but this time non-zero.

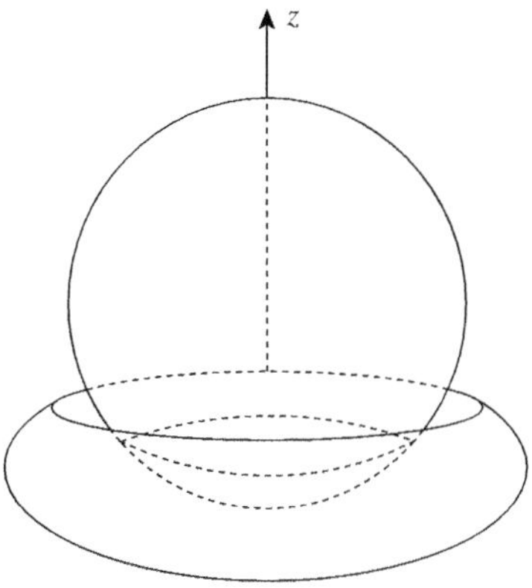

Figure 7.1 A sphere touching $T_{a,b}$ tangentially

Finally, consider the case of a sphere with radius b and centre on the core circle $x^2+y^2 = a^2$, $z = 0$. Such a sphere touches the torus along a great circle of the sphere and a meridian of the torus. Again, the geodesic curvature is zero.

Similar examples can be given for all surfaces of revolution; this is explored in Exercise 7.5.

The notion of geodesic curvature may be readily extended to regular curves on a surface S in $\mathbb{R}^n$. First, however, we have to generalise the concept of orientation, which we have defined as a smooth choice of unit normal $\boldsymbol{N}$ to a surface S in $\mathbb{R}^3$. This leads to a specific choice for the positive direction of rotation in each tangent space which varies smoothly over S; namely the rotation $\boldsymbol{X} \mapsto \boldsymbol{N} \times \boldsymbol{X}$ should be rotation through $\pi/2$ (rather than $3\pi/2$). For instance, the unit normal $(0, 0, 1)$ to the xy-plane in $\mathbb{R}^3$ leads to anticlockwise as being the positive direction of rotation for tangent vectors at each point of the plane. We generalise this to a surface S in $\mathbb{R}^n$ by defining an *orientation* of S to be a choice for the positive direction of rotation in the tangent spaces of S which varies smoothly over S.

We note that a local parametrisation of S determines an orientation of the corresponding coordinate neighbourhood; the positive direction of rotation being chosen so that rotation from $\boldsymbol{x}_u$ to $\boldsymbol{x}_v$ is less than π. For a surface in $\mathbb{R}^3$, the orientation determined by a local parametrisation $\boldsymbol{x}(u, v)$ is the same as that determined by the unit normal $\boldsymbol{N}$ in the direction of $\boldsymbol{x}_u \times \boldsymbol{x}_v$. Following the terminology we have used for surfaces in $\mathbb{R}^3$, a surface in $\mathbb{R}^n$ which admits an orientation is said to be *orientable*, and when a choice of orientation has been made then S is said to be *oriented*.

We define geodesic curvature κ_g for a regular curve $\boldsymbol{\alpha}$ on an oriented surface S in $\mathbb{R}^n$ by setting

$$\kappa_g = \frac{d^2\boldsymbol{\alpha}}{ds^2} . \boldsymbol{X} ,$$

where s is an arc length parameter along $\boldsymbol{\alpha}$, and $\boldsymbol{X}$ is the unit tangent vector obtained by rotating $d\boldsymbol{\alpha}/ds$ through $\pi/2$ in the direction determined by the orientation. We note that $\kappa_g \boldsymbol{X}$ is the component of $d^2\boldsymbol{\alpha}/ds^2$ tangential to S, and if we choose the opposite orientation on S then κ_g changes sign.

The following lemma gives a generalisation of formula (7.2), and is proved in a similar way.

Lemma 5 *Let $\boldsymbol{\alpha}(t)$ be a regular curve on an oriented surface S in $\mathbb{R}^n$, and let $\boldsymbol{X}$ be the unit tangent vector obtained by rotating $\boldsymbol{\alpha}'/|\boldsymbol{\alpha}'|$ through $\pi/2$ in the direction determined by the orientation of S. Then*

$$\kappa_g = \frac{1}{|\boldsymbol{\alpha}'|^2} \boldsymbol{\alpha}'' . \boldsymbol{X} . \tag{7.3}$$

7.2 Geodesics

In this section we consider certain curves, namely *geodesics*, on a surface in $\mathbb{R}^n$, and give some examples. We also give a characterisation in terms of geodesic curvature and the parametrisation of the curve. In subsequent sections we provide a much fuller description of geodesics and their properties.

A smooth curve $\boldsymbol{\alpha}(t)$ on a surface S in $\mathbb{R}^n$ is called a *geodesic* on S if its acceleration vector $\boldsymbol{\alpha}''$ is orthogonal to S at each point of $\boldsymbol{\alpha}$.

Example 1 (Plane) Let $\boldsymbol{\alpha}(t)$ be a regular curve in a plane P in $\mathbb{R}^n$. Then $\boldsymbol{\alpha}'$ and $\boldsymbol{\alpha}''$ are tangential to P so that $\boldsymbol{\alpha}$ is a geodesic on P if and only if $\boldsymbol{\alpha}'' = 0$, that is if and only if $\boldsymbol{\alpha}$ is a straight line parametrised so that $|\boldsymbol{\alpha}'|$ is constant.

As may be seen from the above example, in deciding whether a curve is a geodesic the actual parametrisation of the curve, as well as its trace, is important. Indeed, if $\boldsymbol{\alpha}(t)$ is a geodesic on S then $\boldsymbol{\alpha}''$ is orthogonal to S, so, in particular, $\boldsymbol{\alpha}''.\boldsymbol{\alpha}' = 0$, or, equivalently, $|\boldsymbol{\alpha}'|$ is constant.

If $s(t)$ denotes arc length along any smooth curve $\boldsymbol{\alpha}(t)$ then, quoting equation (1.2), $ds/dt = |\boldsymbol{\alpha}'|$, so it follows that $|\boldsymbol{\alpha}'|$ is constant, λ, say, if and only if $s(t) = \lambda t + c$, for some constant c. For this reason, if $|\boldsymbol{\alpha}'|$ is constant we say that $\boldsymbol{\alpha}$ is parametrised *proportional to arc length*; we have just seen that this is a necessary condition for a smooth curve $\boldsymbol{\alpha}(t)$ to be a geodesic.

Example 2 (Unit sphere) Let $\boldsymbol{N}$ be the outward unit normal to the unit sphere $S^2(1)$ in $\mathbb{R}^3$, so that if $p \in S^2(1)$ then $\boldsymbol{N}(p) = p$. A curve $\boldsymbol{\alpha}$ on $S^2(1)$ is a geodesic if and only if

$$\boldsymbol{\alpha}'' = \mu\boldsymbol{\alpha}\,,$$

for some function μ. But then

$$(\boldsymbol{\alpha} \times \boldsymbol{\alpha}')' = \boldsymbol{\alpha}' \times \boldsymbol{\alpha}' + \boldsymbol{\alpha} \times \boldsymbol{\alpha}'' = 0\,,$$

so that

$$\boldsymbol{\alpha} \times \boldsymbol{\alpha}' = \boldsymbol{c}\,,$$

where $\boldsymbol{c}$ is a constant vector which is non-zero unless $\boldsymbol{\alpha}$ is itself constant (Figure 7.2). In particular, $|\boldsymbol{\alpha}'| = |\boldsymbol{c}|$, so that $\boldsymbol{\alpha}$ is parametrised proportional to arc length, and

$$\boldsymbol{\alpha}.\boldsymbol{c} = 0\,,$$

so that $\boldsymbol{\alpha}$ is part of the great circle obtained by intersecting $S^2(1)$ with the plane through the origin in $\mathbb{R}^3$ orthogonal to $\boldsymbol{c}$.

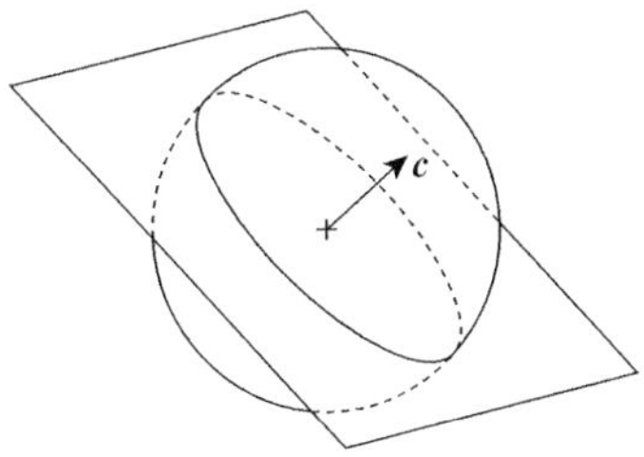

Figure 7.2 Great circles are geodesics

Conversely, a great circle on $S^2(1)$ which is parametrised proportional to arc length may be written in the form

$$\boldsymbol{\alpha}(t) = \boldsymbol{X}\cos(\lambda t) + \boldsymbol{Y}\sin(\lambda t)\,,$$

where λ is a non-zero constant and $\boldsymbol{X}$ and $\boldsymbol{Y}$ are a suitable pair of orthonormal vectors in $\mathbb{R}^3$. Then $\boldsymbol{\alpha}'' = -\lambda^2\boldsymbol{\alpha}$, which is orthogonal to $S^2(1)$ at each point of $\boldsymbol{\alpha}$, so that $\boldsymbol{\alpha}$ is a geodesic on $S^2(1)$.

This example shows that a non-constant curve on $S^2(1)$ is a geodesic if and only if it is a great circle parametrised proportional to arc length. This accords with our interpretation of a geodesic on a surface as the path followed by a smooth particle moving freely on the surface; if we attach a particle to the origin of $\mathbb{R}^3$ by a massless rod of unit length (so as to constrain the particle to travel on the unit sphere) then an impulse applied to the particle will set it going round and round a great circle at constant speed.

Example 3 (Surface of revolution) A meridian $\boldsymbol{\alpha}(t)$ of a surface S of revolution lies on a plane P containing the axis of rotation. Hence $\boldsymbol{\alpha}'$ and $\boldsymbol{\alpha}''$ also lie on P. However, P intersects S orthogonally, so that P is spanned by $\boldsymbol{\alpha}'$ and $\boldsymbol{N}$. Thus if $\boldsymbol{\alpha}$ is parametrised proportional to arc length then $\boldsymbol{\alpha}''$ is a scalar multiple of $\boldsymbol{N}$. Hence all meridians, when parametrised proportional to arc length, are geodesics.

Example 4 (Cylinder) Let S be the cylinder in $\mathbb{R}^3$ with equation $x^2 + y^2 = r^2$ and let $\boldsymbol{\alpha} : \mathbb{R} \to S$ be the curve defined by

$$\boldsymbol{\alpha}(t) = (r\cos t, r\sin t, \lambda t + \mu)\,, \quad t \in \mathbb{R}\,,$$

where λ, μ are constant. Then $\boldsymbol{\alpha}$ is a helix if $\lambda \neq 0$ and a circle if $\lambda = 0$. Also

$$\boldsymbol{\alpha}''(t) = (-r\cos t, -r\sin t, 0)\,,$$

which is normal to S at $\boldsymbol{\alpha}(t)$. Thus $\boldsymbol{\alpha}$ is a geodesic.

It is clear that any constant curve on a surface is a geodesic. The following proposition provides a characterisation of all other geodesics.

Proposition 5 *A regular curve $\boldsymbol{\alpha}(t)$ on a surface S in $\mathbb{R}^n$ is a geodesic if and only if it is parametrised proportional to arc length and has zero geodesic curvature.*

Proof For surfaces in $\mathbb{R}^3$ this follows from (7.2) since $\boldsymbol{\alpha}''$ is orthogonal to S if and only if $\boldsymbol{\alpha}''$ is orthogonal to both $\boldsymbol{\alpha}'$ and $\boldsymbol{N} \times \boldsymbol{\alpha}'$. For surfaces in $\mathbb{R}^n$ we may use (7.3) rather than (7.2). □

This result is illustrated by Example 2 of §7.1 and Example 2 in this section; in the former we saw that great circles are the curves of zero geodesic curvature on $S^2(1)$, while the latter showed that the geodesics on $S^2(1)$ are the great circles parametrised proportional to arc length.

7.3 Differential equations for geodesics

In this section we will obtain two equivalent sets of differential equations for geodesics in a coordinate neighourhood on a surface S in $\mathbb{R}^n$. We will then use these to obtain existence and uniqueness results for geodesics.

Proposition 1 *Let $\boldsymbol{x}(u, v)$ be a local parametrisation of S and let $\boldsymbol{\alpha}(t) = \boldsymbol{x}(u(t), v(t))$ be a smooth curve. Then $\boldsymbol{\alpha}$ is a geodesic on S if and only if both the following equations are satisfied:*

$$u''E + \frac{1}{2}u'^2 E_u + u'v'E_v + v''F + v'^2(F_v - \frac{1}{2}G_u) = 0\,, \tag{7.4}$$

$$v''G + \frac{1}{2}v'^2 G_v + u'v'G_u + u''F + u'^2(F_u - \frac{1}{2}E_v) = 0\,. \tag{7.5}$$

Proof We have that

$$\boldsymbol{\alpha}' = u'\boldsymbol{x}_u + v'\boldsymbol{x}_v\,,$$

so that

$$\boldsymbol{\alpha}'' = u''\boldsymbol{x}_u + u'(u'\boldsymbol{x}_{uu} + v'\boldsymbol{x}_{uv}) + v''\boldsymbol{x}_v + v'(u'\boldsymbol{x}_{uv} + v'\boldsymbol{x}_{vv})\,. \tag{7.6}$$

Thus, using the expressions discussed in §6.1 for $\boldsymbol{x}_{uu}.\boldsymbol{x}_u$ and other similar quantities in terms of the coefficients of the first fundamental form and their derivatives, we find that

$$\boldsymbol{\alpha}''.\boldsymbol{x}_u = u''E + \frac{1}{2}u'^2 E_u + u'v'E_v + v''F + v'^2(F_v - \frac{1}{2}G_u)\,. \tag{7.7}$$

A similar calculation shows that $\boldsymbol{\alpha}''.\boldsymbol{x}_v$ is equal to the left hand side of (7.5) and, since $\boldsymbol{\alpha}$ is a geodesic if and only if $\boldsymbol{\alpha}''.\boldsymbol{x}_u = \boldsymbol{\alpha}''.\boldsymbol{x}_v = 0$, it follows that $\boldsymbol{\alpha}$ is a geodesic if and only if equations (7.4) and (7.5) are both satisfied. □

The two equations given in the statement of Proposition 1 form a system of second order ordinary differential equations for $u(t), v(t)$. They are not linear and cannot usually be explicitly solved. However, the existence and uniqueness theorem for solutions of systems of this type enables us to prove the following result.

Theorem 2 *Let $p \in S$ and let $\boldsymbol{X} \in T_pS$.*

(i) There is a unique geodesic $\boldsymbol{\alpha} : (a, b) \to S$ with initial point $\boldsymbol{\alpha}(0) = p$ and initial vector $\boldsymbol{\alpha}'(0) = \boldsymbol{X}$ which is maximal in the sense that the domain of the geodesic cannot be further extended. Here, a is either $-\infty$ or a negative real number, while b is either ∞ or a positive real number.

(ii) Any geodesic $\boldsymbol{\beta}$ with initial point p and initial vector $\boldsymbol{X}$ is the restriction of $\boldsymbol{\alpha}$ to some subinterval of (a, b).

This accords with our physical interpretation of geodesics as the paths followed by smooth particles moving freely on a surface; in the absence of external forces, a trajectory is determined by the initial position and velocity. It might not be possible to follow the trajectory for all time, however; for instance, if S is the open unit disc then a freely moving particle will soon fall off the edge!

Example 3 (Paraboloid of revolution) Let S be the paraboloid of revolution with equation $z = x^2 + y^2$. We saw in Example 3 of §7.2 that, when parametrised proportional to arc length, all meridians are geodesics. Clearly, at $p = (0, 0, 0)$ there is exactly one such meridian with a given non-zero tangent vector as initial vector. The uniqueness part of Theorem 2 shows that all geodesics through $(0,0,0)$ are of this form.

We now discuss some consequences of Theorem 2. It is clear from the definition (and also follows from (7.4) and (7.5)) that if $\boldsymbol{\alpha}(t)$ is a geodesic then, for any real constant λ, so is the curve $\tilde{\boldsymbol{\alpha}}(t) = \boldsymbol{\alpha}(\lambda t)$. The following lemma is now immediate from the uniqueness part of Theorem 2.

Lemma 4 *If X is a tangent vector at a point p of a surface S in $\mathbb{R}^n$, let $\boldsymbol{\alpha}_X(t)$ be the unique maximal geodesic with initial point p and initial vector X. Then, for any real number λ,*

$$\boldsymbol{\alpha}_{\lambda X}(t) = \boldsymbol{\alpha}_X(\lambda t)\,.$$

Again, this agrees with intuition. If a particle leaves a particular point in a given direction with a given initial speed, then a particle setting off from the same point in the same direction but with, say, twice the initial speed travels along the same path but travels along it twice as fast.

Proposition 1 also shows that, although we defined geodesics extrinsically, that is to say we used the containing Euclidean space, geodesics are actually determined by intrinsic quantities. This is rather surprising, and, as in the case of Gaussian curvature, points to the fundamental importance of the concept of geodesic in the study of general Riemannian manifolds. It also leads to the following result, which is similar to Corollary 3 of the Theorema Egregium.

Proposition 5 *Let $f : S \to \tilde{S}$ be a local isometry of surfaces. Then f maps geodesics on S to geodesics on $\tilde{S}$, that is to say if $\boldsymbol{\alpha}(t)$ is a geodesic on S then $f\boldsymbol{\alpha}(t)$ is a geodesic on $\tilde{S}$.*

Example 6 (Cylinder) Let S be the cylinder in $\mathbb{R}^3$ with equation $x^2 + y^2 = r^2$, and let $f : \mathbb{R}^2 \to S$ be the local isometry defined by

$$f(x, y) = \left(r\cos\frac{x}{r}, r\sin\frac{x}{r}, y\right)\,.$$

Then the geodesics of S, being the images under f of the geodesics in $\mathbb{R}^2$, are of the following type:

(a) the meridians of S (which are the images of the lines $x =$ constant in $\mathbb{R}^2$);
(b) the parallels of S (which are the images of the lines $y =$ constant in $\mathbb{R}^2$);
(c) helices (which are the images of lines in $\mathbb{R}^2$ of the form $ax + by = c$, where a, b, c are constant with $a, b \neq 0$). These can be parametrised proportional to arc length as:

$$\boldsymbol{\beta}(t) = (r\cos t, r\sin t, \lambda t + \mu)\,,$$

where λ, μ are constants with $\lambda \neq 0$.

Geodesics of types (b) and (c) have already been discussed in Example 4 of §7.2.

We note that, in contrast with the case of a plane, there are infinitely many geodesics joining any two points of a cylinder (Figure 7.3). This is an example of the way in which

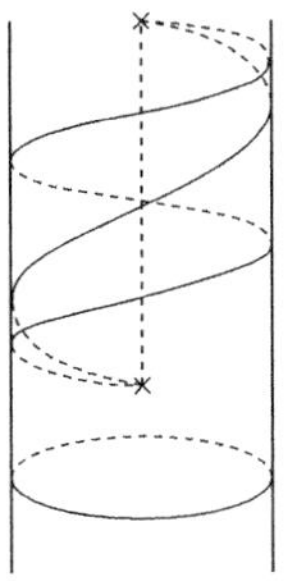

Figure 7.3 Three geodesics with the same start point and the same end point

the topology of a surface is reflected in the behaviour of its geodesics. Indeed, consideration of geodesic behaviour is often very useful in the investigation of the topology of Riemannian manifolds.

A surface for which every maximal geodesic has domain the whole of $\mathbb{R}$ is said to be *complete*. The cylinder of Example 6 is complete, but the surface obtained by considering that part of the cylinder lying between the planes $z = 1$ and $z = -1$ is not complete. It is not hard to prove that all compact surfaces are complete; in fact, all surfaces which are closed subsets of $\mathbb{R}^n$ are complete.

Example 7 (Cone) An acetate sheet can be bent, but not stretched or compressed, and the action of rolling up such a sheet to make a cylinder (which is modelled mathematically in Example 6) gives a local isometry from the plane to the cylinder. Proposition 5 shows that lines drawn on the sheet become geodesics on the cylinder, which enables us to "see" the geodesics on the cylinder described in Example 6. In a similar way, by rolling up our acetate sheet to make a cone, we may "see" the geodesics on a cone, which typically look as illustrated in Figure 7.4.

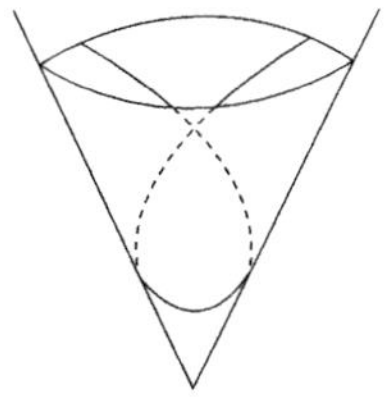

Figure 7.4 A geodesic on a cone

It is interesting to investigate how the self-intersection properties of the geodesics on a cone depend on the angle at the base of the cone. This and other properties of geodesics on a cone are explored in Exercises 7.15 and 7.20.

Example 8[†] (Veronese surface and real projective plane) This example uses material in the optional §4.6, and may be omitted if desired. Let $f : S^2(1) \to S$ be the local isometry of $S^2(1)$ onto the Veronese surface S in $\mathbb{R}^5$ defined in Example 2 of §4.6. We recall that the real projective plane $\mathbb{R}P^2$ is the set of lines through the origin of $\mathbb{R}^3$, and that f enables us to identify the real projective plane with the Veronese surface. Each plane P through the origin of $\mathbb{R}^3$ intersects $S^2(1)$ in a great circle, and the image under f of this great circle is a geodesic on S, covered twice. This geodesic corresponds in $\mathbb{R}P^2$ to the set of lines through the origin in the plane P. It is clear that any two geodesics on S meet in a unique point (since any two planes through the origin intersect in a unique line), and, through any two points there is a unique geodesic. The resulting geometry, called *projective geometry*, is of great historical importance.

The geodesic equations (7.4) and (7.5) of Proposition 1 may be written in an alternative form using the Christoffel symbols $\{\Gamma^i_{jk}\}$ introduced in §6.1 (where, for surfaces in $\mathbb{R}^n$, the quantities LN, MN, NN are replaced by the components of $\boldsymbol{x}_{uu}, \boldsymbol{x}_{uv}, \boldsymbol{x}_{vv}$, respectively, orthogonal to S).

Proposition 9 *Let $\boldsymbol{x}(u, v)$ be a local parametrisation of S and let $\boldsymbol{\alpha}(t) = \boldsymbol{x}(u(t), v(t))$ be a smooth curve. Then $\boldsymbol{\alpha}$ is a geodesic on S if and only if both of the following equations are satisfied:*

$$u'' + \Gamma^1_{11}u'^2 + 2\Gamma^1_{12}u'v' + \Gamma^1_{22}v'^2 = 0\,, \tag{7.8}$$

$$v'' + \Gamma^2_{11}u'^2 + 2\Gamma^2_{12}u'v' + \Gamma^2_{22}v'^2 = 0\,. \tag{7.9}$$

Proof Writing u_1, u_2 in place of u, v for the local parametrisation, we let

$$\boldsymbol{\alpha}(t) = \boldsymbol{x}(u_1(t), u_2(t))\ .$$

Then, writing $\boldsymbol{x}_i, \boldsymbol{x}_{ij}$ for the first and second partial derivatives of $\boldsymbol{x}$, equation (7.6) becomes

$$\boldsymbol{\alpha}'' = \sum_{k=1}^{2} u_k'' \boldsymbol{x}_k + \sum_{i,j=1}^{2} u_i' u_j' \boldsymbol{x}_{ij}\,,$$

so that the component $(\boldsymbol{\alpha}'')^{\tan}$ of $\boldsymbol{\alpha}''$ tangential to S is given in terms of the Christoffel symbols $\{\Gamma^k_{ij}\}$ by

$$(\boldsymbol{\alpha}'')^{\tan} = \sum_{k=1}^{2}\left(u_k'' + \sum_{i,j=1}^{2}\Gamma^k_{ij}u_i'u_j'\right)\boldsymbol{x}_k\,.$$

Thus $(\boldsymbol{\alpha}'')^{\tan} = 0$ if and only if

$$u_k'' + \sum_{i,j=1}^{2}\Gamma^k_{ij}u_i'u_j' = 0\,, \quad k = 1, 2\,, \tag{7.10}$$

which gives an alternative form of the geodesic equations. If we write these in the more familiar $\boldsymbol{x}(u, v)$ notation, we obtain equations (7.8) and (7.9). □

Example 10† (Hyperbolic plane) This example uses material in the optional Example 5 of §3.4, and may be omitted if desired. Let H be the upper half-plane model of the hyperbolic plane as first discussed in Example 5 of §3.4. The coefficients of the first fundamental form are given by

$$E = G = \frac{1}{v^2}, \quad F = 0,$$

and the non-zero Christoffel symbols are given by

$$\Gamma^1_{12} = \Gamma^1_{21} = -\Gamma^2_{11} = \Gamma^2_{22} = -\frac{1}{v}.$$

The geodesic equations (7.8) and (7.9) become

$$u'' - 2\frac{u'v'}{v} = 0, \tag{7.11}$$

$$v'' + \frac{u'^2 - v'^2}{v} = 0, \tag{7.12}$$

as may also be obtained directly from (7.4) and (7.5).

We now show that, when parametrised proportional to arc length, the lines $u =$ constant are geodesics on H. In fact, for such curves $v' = \pm kv$ for some constant k so that $v'' = k^2 v$ and equations (7.11) and (7.12) are satisfied.

For any real number u_0, the semicircles illustrated in Figure 7.5 centred on $(u_0, 0)$ traced out by

$$(u(t), v(t)) = (u_0 + r\cos\theta(t), r\sin\theta(t)), \quad 0 < \theta(t) < \pi, \ r > 0,$$

are also geodesics when parametrised proportional to arc length. Indeed, in this case,

$$\theta'^2 = k^2 \sin^2\theta,$$

for some constant k. Hence $\theta' = \pm k \sin\theta$ and

$$\theta'' = k^2 \sin\theta \cos\theta.$$

But then an easy calculation shows that equations (7.11) and (7.12) are both satisfied so that all these curves are geodesics. Moreover, through any given point of H in any given direction there is one of these curves, so that by the existence and uniqueness theorem (Theorem 2), we have now found every geodesic on H.

The above assumes, of course, that we already knew the traces of the geodesics on H. In Exercise 7.16 you are invited to find the geodesics without having this prior knowledge.

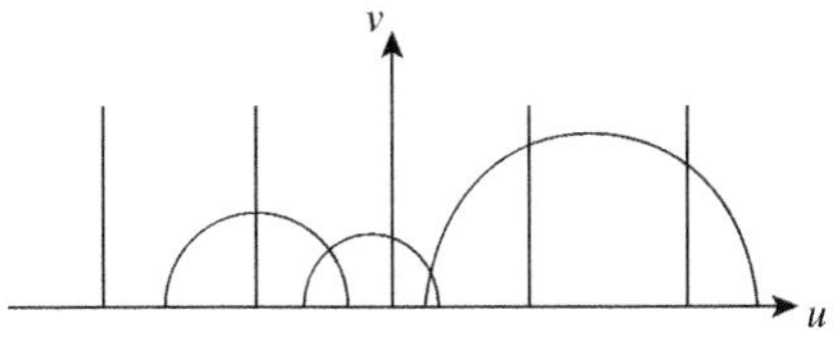

Figure 7.5 Some geodesics on the hyperbolic plane

7.4 Geodesics as curves of stationary length

Let $\boldsymbol{\alpha} : [a,b] \to S$ be a regular curve which minimises the length of all regular curves on S from $p_0 = \boldsymbol{\alpha}(a)$ to $p_1 = \boldsymbol{\alpha}(b)$ (hereafter called a *length minimising curve on S*). We shall show that, when parametrised proportional to arc length, a length-minimising curve $\boldsymbol{\alpha}$ is a geodesic on S. In fact, we shall characterise geodesics on S as curves parametrised proportional to arc length which, in a sense made precise below, are **stationary** points of the length functional applied to the space of all regular curves on S with given end points.

It turns out that any sufficiently short piece of geodesic is a length **minimising** curve on S (see Proposition 1 of the optional §7.8); however, the examples of great circles on $S^2(1)$ and helices on the cylinder show that long geodesics do not necessarily have this length minimising property.

We begin by considering how the length of a smooth curve $\boldsymbol{\alpha}$ on S changes as we deform the curve while staying on S. We do this by considering suitable 1-parameter families of curves on S with $\boldsymbol{\alpha}$ as initial curve. Specifically, a 1-parameter family $\{\boldsymbol{\alpha}_r\}_{-\epsilon<r<\epsilon}$ of smooth curves from a closed interval $[a,b]$ to S is called a *smooth variation of* $\boldsymbol{\alpha}$ *through curves on S* if $\boldsymbol{\alpha}_0 = \boldsymbol{\alpha}$ and the map $\boldsymbol{H}(r,t)$ defined by

$$\boldsymbol{H}(r,t) = \boldsymbol{\alpha}_r(t)\,, \quad -\epsilon < r < \epsilon\,, \quad a \le t \le b\,,$$

is a smooth map whose image lies on S. If $L(r)$ is the length of $\boldsymbol{\alpha}_r$, then L is a smooth function of r, and we now obtain an expression for the derivative $L'(0)$ in terms of the *variation vector field* $\boldsymbol{X}$ (Figure 7.6), which is defined by setting $\boldsymbol{X}(t) = \partial \boldsymbol{H}/\partial r|_{(0,t)}$. In the course of the proof it will be useful to note that, when $r = 0$,

$$\partial \boldsymbol{H}/\partial r = \boldsymbol{X}\,, \quad \partial \boldsymbol{H}/\partial t = \boldsymbol{\alpha}'\,, \quad \partial^2 \boldsymbol{H}/\partial t^2 = \boldsymbol{\alpha}''\,. \tag{7.13}$$

Lemma 1 *Let* $\boldsymbol{\alpha} : [a,b] \to S$ *be a regular curve parametrised proportional to arc length, and let* $|\boldsymbol{\alpha}'(t)| = c$. *If* $\boldsymbol{X}$ *is the variation vector field of a smooth variation* $\{\boldsymbol{\alpha}_r\}$ *of* $\boldsymbol{\alpha}$ *through curves on S, then*

$$L'(0) = \frac{1}{c}\left\{\boldsymbol{X}(b).\boldsymbol{\alpha}'(b) - \boldsymbol{X}(a).\boldsymbol{\alpha}'(a) - \int_a^b \boldsymbol{X}.\boldsymbol{\alpha}''\,dt\right\}\,.$$

Proof We shall use the result (often called "differentiating under the integral sign") which says that if $f(r,t)$ is smooth then

$$\frac{d}{dr}\left(\int_a^b f\,dt\right) = \int_a^b \frac{\partial f}{\partial r}\,dt\,.$$

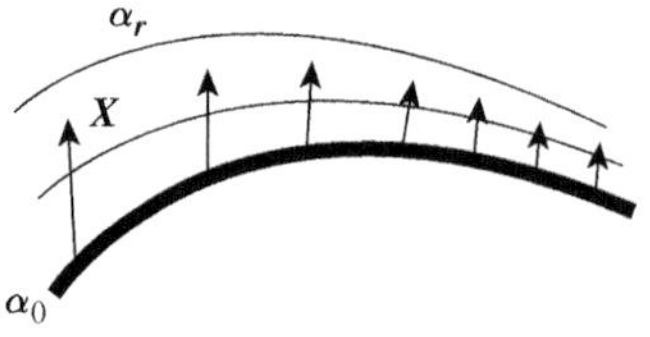

Figure 7.6 Variation vector field of a smooth variation

In particular, since

$$L(r) = \int_a^b \left(\frac{\partial \boldsymbol{H}}{\partial t} . \frac{\partial \boldsymbol{H}}{\partial t} \right)^{1/2} dt ,$$

we have that

$$L'(0) = \frac{1}{c} \int_a^b \frac{\partial}{\partial r} \left(\frac{\partial \boldsymbol{H}}{\partial t} \right) . \frac{\partial \boldsymbol{H}}{\partial t} dt ,$$

where here, and in what follows, the integrand on the right hand side of the equation is evaluated at $r = 0$.

Interchanging the order of differentiation, we now find that

$$\begin{aligned} L'(0) &= \frac{1}{c} \int_a^b \frac{\partial}{\partial t} \left(\frac{\partial \boldsymbol{H}}{\partial r} \right) . \frac{\partial \boldsymbol{H}}{\partial t} dt \\ &= \frac{1}{c} \left\{ \int_a^b \frac{\partial}{\partial t} \left(\frac{\partial \boldsymbol{H}}{\partial r} . \frac{\partial \boldsymbol{H}}{\partial t} \right) dt - \int_a^b \frac{\partial \boldsymbol{H}}{\partial r} . \frac{\partial^2 \boldsymbol{H}}{\partial t^2} dt \right\} , \end{aligned}$$

and the result now follows from (7.13). □

Since the image of $\boldsymbol{H}$ lies on S, $\boldsymbol{X}(t) \in T_{\boldsymbol{\alpha}(t)}S$ for each $t \in [a,b]$. Conversely, we have the following.

Lemma 2 *Let $\boldsymbol{X} : [a,b] \to \mathbb{R}^n$ be a smooth map with $\boldsymbol{X}(t) \in T_{\boldsymbol{\alpha}(t)}S$ for all $t \in [a,b]$. Then there is a smooth variation $\{\boldsymbol{\alpha}_r\}$ of $\boldsymbol{\alpha}$ through curves on S which has $\boldsymbol{X}$ as its variation vector field.*

Proof We save ourselves some technical difficulties by assuming that $\boldsymbol{\alpha}$ lies in the image of a local parametrisation $\boldsymbol{x}(u,v)$. In this case, there are smooth functions $u(t)$, $v(t)$, $X_1(t)$, $X_2(t)$, such that $\boldsymbol{\alpha}(t) = \boldsymbol{x}\,(u(t), v(t))$ and $\boldsymbol{X} = X_1\boldsymbol{x}_u + X_2\boldsymbol{x}_v$. Then

$$\boldsymbol{H}(r,t) = \boldsymbol{x}\,(u(t) + rX_1(t), v(t) + rX_2(t))$$

provides the required smooth variation of $\boldsymbol{\alpha}$. □

A smooth variation $\{\boldsymbol{\alpha}_r\}$ of $\boldsymbol{\alpha}$ is said to have *fixed end points* if, as illustrated in Figure 7.7, $\boldsymbol{\alpha}_r(a) = \boldsymbol{\alpha}(a)$ and $\boldsymbol{\alpha}_r(b) = \boldsymbol{\alpha}(b)$ for all r. We now state and prove the main result of this section. In this theorem, when we say that $\boldsymbol{\alpha}$ is a stationary point of the length functional applied to the space of all regular curves on S joining $\boldsymbol{\alpha}(a)$ to $\boldsymbol{\alpha}(b)$, we mean that $L'(0) = 0$ for every smooth variation of $\boldsymbol{\alpha}$ with fixed end points through curves on S.

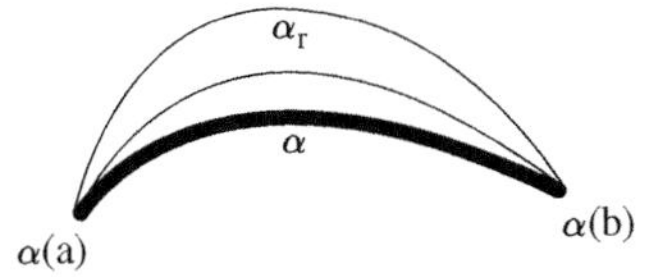

Figure 7.7 Variation with fixed end points

Theorem 3 *Let* $\boldsymbol{\alpha} : [a, b] \to S$ *be a regular curve on* S *parametrised proportional to arc length. Then* $\boldsymbol{\alpha}$ *is a geodesic if and only if* $\boldsymbol{\alpha}$ *is a stationary point of the length functional applied to the space of all regular curves on* S *joining* $\boldsymbol{\alpha}(a)$ *to* $\boldsymbol{\alpha}(b)$.

Proof Assume $\boldsymbol{\alpha}$ is a geodesic, and let $\boldsymbol{X}$ be the variation vector field of a smooth variation of $\boldsymbol{\alpha}$ with fixed end points through curves on S. Then $\boldsymbol{X}(a) = \boldsymbol{X}(b) = 0$, and $\boldsymbol{\alpha}''$ is orthogonal to S, so it follows from Lemma 1 that $L'(0) = 0$.

Conversely, assume that $\boldsymbol{\alpha}$ is not a geodesic and let $\boldsymbol{Y}(t)$ be the orthogonal projection of $\boldsymbol{\alpha}''(t)$ onto the tangent space of S at $\boldsymbol{\alpha}(t)$. Then $\boldsymbol{Y}.\boldsymbol{\alpha}''$ is a non-negative function which is not identically zero. Now let $\lambda = \pi/(b - a)$, and let $\boldsymbol{X}(t) = \sin(\lambda(t - a))\,\boldsymbol{Y}(t)$. Then $\boldsymbol{X}(a) = \boldsymbol{X}(b) = 0$ and $\boldsymbol{X}.\boldsymbol{\alpha}''$ is also a non-negative function which is not identically zero. The smooth variation given in Lemma 2 with variation vector field $\boldsymbol{X}$ has fixed end points, and it follows from Lemma 1 that $L'(0)$ is non-zero (in fact it is negative). □

Remark 4 Similar techniques show that a curve $\boldsymbol{\alpha}$, parametrised proportional to arc length, is a geodesic if and only if $L'(0) = 0$ for all smooth variations of $\boldsymbol{\alpha}$ through curves on S (not necessarily with fixed end-points) whose variation vector field $\boldsymbol{X}$ is orthogonal to $\boldsymbol{\alpha}$.

As mentioned at the beginning of this section, it may be shown that any sufficiently short piece of a geodesic on a surface S is a length minimising curve on S, whereas long geodesics do not necessarily have this property. An important and useful theorem of Hopf and Rinow says that any two points of a complete connected surface S may be joined by a length minimising curve. The example of the annulus $\{(x, y, 0) : 1 < x^2 + y^2 < 3\}$ shows that this is not always the case without the assumption of completeness – the idea is that, for example, no matter how close to the central hole you take a curve in the annulus joining $(-\sqrt{2}, 0, 0)$ to $\sqrt{2}, 0, 0)$, you can always take a curve that goes a little closer (Figure 7.8).

7.5 Geodesic curvature is intrinsic

We noted in §7.3 that geodesics are determined by intrinsic quantities, and in this section we generalise this by showing that geodesic curvature of a regular curve is intrinsic.

Let $\boldsymbol{x}(u, v)$ be a local parametrisation of a surface S in $\mathbb{R}^n$ and let $\boldsymbol{\alpha}(t) = \boldsymbol{x}(u(t), v(t))$ be a regular curve. Equation (7.7), and a similar one for $\boldsymbol{\alpha}''.\boldsymbol{x}_v$, show that the component $(\boldsymbol{\alpha}'')^{\tan}$ of $\boldsymbol{\alpha}''$ tangential to S depends on only the coefficients of the first fundamental form and their derivatives. The intrinsic nature of κ_g now follows from (7.2) (or (7.3) for surfaces in $\mathbb{R}^n$). This leads to the following generalisation of Proposition 5 of §7.3.

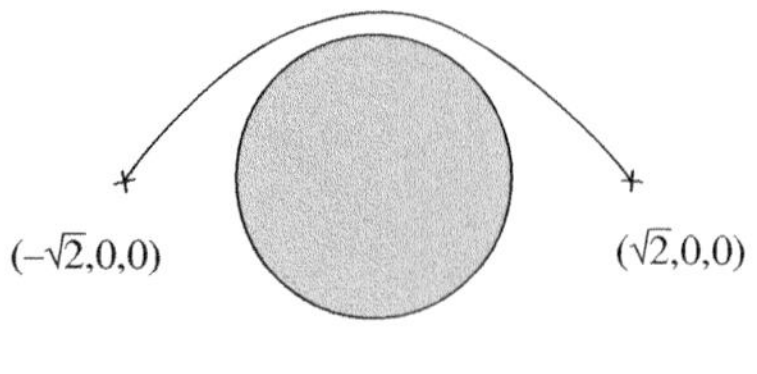

Figure 7.8 There are always shorter curves

Proposition 1 *Let $f : S \to \tilde{S}$ be a local isometry of surfaces, and let $\boldsymbol{\alpha}(t)$ be a regular curve on S. Then (up to sign) the geodesic curvature of $f\boldsymbol{\alpha}$ as a curve on $\tilde{S}$ is equal to the geodesic curvature of $\boldsymbol{\alpha}$.*

The following proposition gives an alternative proof of the intrinsic nature of geodesic curvature, in that it produces a formula for κ_g which depends only on angles and the coefficients of the first fundamental form. This proposition is used in the the proof of the Gauss–Bonnet Theorem for a triangle in Chapter 8.

Proposition 2 *Let $\boldsymbol{x}(u, v)$ be an orthogonal local parametrisation of a surface S in $\mathbb{R}^n$, and let $\boldsymbol{\alpha}(t) = \boldsymbol{x}\,(u(t), v(t))$ be a curve lying in the image of $\boldsymbol{x}$. Then the geodesic curvature κ_g of $\boldsymbol{\alpha}$ (using the orientation determined by $\boldsymbol{x}$) is given by*

$$\kappa_g = \frac{1}{2\sqrt{EG}}\left(G_u\frac{dv}{ds} - E_v\frac{du}{ds}\right) + \frac{d\phi}{ds}\,, \tag{7.14}$$

where d/ds denotes differentiation with respect to an arc length parameter s along $\boldsymbol{\alpha}$, and ϕ is the angle from $\boldsymbol{x}_u$ to $\boldsymbol{\alpha}'$ measured in the direction determined by the local parametrisation.

Proof We give the proof for surfaces in $\mathbb{R}^3$; the generalisation to surfaces in $\mathbb{R}^n$ is straightforward.

Let $\boldsymbol{e}_1 = \boldsymbol{x}_u/\sqrt{E}$, and $\boldsymbol{e}_2 = \boldsymbol{x}_v/\sqrt{G}$. Then $\{\boldsymbol{e}_1, \boldsymbol{e}_2\}$ is an orthonormal basis of the tangent space of S, and, if $\boldsymbol{t} = d\boldsymbol{\alpha}/ds$ is the unit tangent vector to $\boldsymbol{\alpha}$ then $\boldsymbol{t} = \cos\phi\;\boldsymbol{e}_1 + \sin\phi\;\boldsymbol{e}_2$. Hence

$$\frac{d\boldsymbol{t}}{ds} = (-\sin\phi\;\boldsymbol{e}_1 + \cos\phi\;\boldsymbol{e}_2)\frac{d\phi}{ds} + \cos\phi\,\frac{d\boldsymbol{e}_1}{ds} + \sin\phi\,\frac{d\boldsymbol{e}_2}{ds}\,.$$

Taking $\boldsymbol{N} = \boldsymbol{x}_u \times \boldsymbol{x}_v/|\boldsymbol{x}_u \times \boldsymbol{x}_v|$, we have $\boldsymbol{N} \times \boldsymbol{t} = -\sin\phi\;\boldsymbol{e}_1 + \cos\phi\;\boldsymbol{e}_2$, and a short calculation using (7.1) shows that

$$\kappa_g = \frac{d\phi}{ds} + \frac{d\boldsymbol{e}_1}{ds}.\boldsymbol{e}_2\,. \tag{7.15}$$

However,

$$\begin{aligned}
\frac{d\boldsymbol{e}_1}{ds}.\boldsymbol{e}_2 &= \left(\frac{\partial\boldsymbol{e}_1}{\partial u}\frac{du}{ds} + \frac{\partial\boldsymbol{e}_1}{\partial v}\frac{dv}{ds}\right).\boldsymbol{e}_2\\
&= \left(\frac{\partial}{\partial u}\left(\frac{\boldsymbol{x}_u}{\sqrt{E}}\right)\frac{du}{ds} + \frac{\partial}{\partial v}\left(\frac{\boldsymbol{x}_u}{\sqrt{E}}\right)\frac{dv}{ds}\right).\frac{\boldsymbol{x}_v}{\sqrt{G}}\\
&= \frac{1}{\sqrt{EG}}\left((\boldsymbol{x}_{uu}.\boldsymbol{x}_v)\frac{du}{ds} + (\boldsymbol{x}_{uv}.\boldsymbol{x}_v)\frac{dv}{ds}\right).
\end{aligned}$$

We now recall that $\boldsymbol{x}_{uu}.\boldsymbol{x}_v = F_u - E_v/2$ and $\boldsymbol{x}_{uv}.\boldsymbol{x}_v = G_u/2$, so, since $F = 0$, we have that

$$\frac{d\boldsymbol{e}_1}{ds}.\boldsymbol{e}_2 = \frac{1}{2\sqrt{EG}}\left(G_u\frac{dv}{ds} - E_v\frac{du}{ds}\right).$$

The result now follows from (7.15). □

7.6 Geodesics on surfaces of revolution †

The symmetry properties of surfaces of revolution make the study of their geodesics more tractible than is the case for more general surfaces in $\mathbb{R}^3$. Although it is usually still not possible to solve the geodesic equations explicitly, some interesting information may be obtained.

The material in this section is interesting in its own right, and should also help you to get a feel for the behaviour of geodesics on surfaces. However, it is not necessary for an understanding of the rest of the book, so may be omitted if necessary.

Recall that if we parametrise a surface of revolution in $\mathbb{R}^3$ using a local parametrisation in the standard form, namely

$$\boldsymbol{x}(u, v) = (f(v)\cos u, f(v)\sin u, g(v)) , \quad -\pi < u < \pi , \quad f(v) > 0 \; \forall v ,$$

then the coefficients of the first fundamental form are given by

$$E = f^2 , \quad F = 0 , \quad G = {f_v}^2 + {g_v}^2 .$$

Note that we have used the suffix v to denote differentiation with respect to v, since we reserve $'$ for the derivative with respect to t along a curve $\boldsymbol{\alpha}(t) = \boldsymbol{x}(u(t), v(t))$. Since E and G are functions of v only, the geodesic equations (7.4) and (7.5) become

$$u''E + u'v'E_v = 0$$

and

$$v''G + \frac{1}{2}{v'}^2 G_v - \frac{1}{2}{u'}^2 E_v = 0 . \tag{7.16}$$

The first equation integrates up to give

$$u'E = \text{constant}. \tag{7.17}$$

We have already seen in Example 3 of §7.2 that all meridians are geodesics when parametrised proportional to arc length, and another proof of this may be given by noting that, along such a curve, both u and ${v'}^2 G$ are constant so that (7.16) and (7.17) both hold.

We now characterise those parallels which are also geodesics (Figure 7.9).

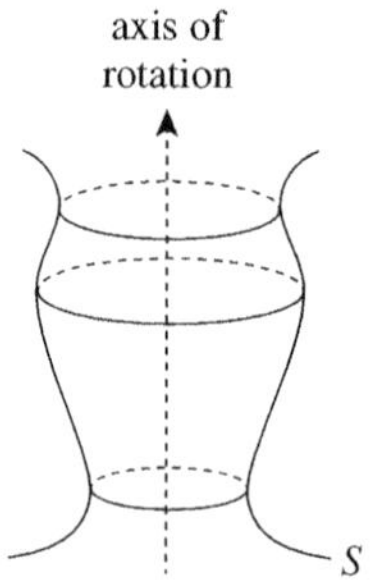

Figure 7.9 These parallels are geodesics

Lemma 1 *The parallel through $p_0 = (f(v_0), 0, g(v_0))$, when parametrised proportional to arc length, is a geodesic if and only if $f_v(v_0) = 0$, that is to say if and only if the tangent vector $(f_v(v_0), 0, g_v(v_0))$ to the generating curve at p_0 is parallel to the axis of rotation of the surface.*

Proof The parallel $\boldsymbol{\alpha}(t) = \boldsymbol{x}(u(t), v_0)$ is parametrised proportional to arc length if and only if $u'^2 E$ is constant. Since E is constant along $\boldsymbol{\alpha}$, (7.16) and (7.17) show that $\boldsymbol{\alpha}$ is a geodesic if and only if $E_v(v_0) = 0$. However, $E = f^2$, so this holds if and only if $f_v(v_0) = 0$. □

If we think of S as being made from a smooth material and the parallels of S obtained by stretching an elastic band around S, then it is intuitively clear (Figure 7.9) that the equilibrium positions of the band are exactly the parallels discussed in the above lemma. This is in agreement with the interpretation given in Remark 4 of §7.4; geodesics are curves $\boldsymbol{\alpha}$ of stationary length for all smooth variations of $\boldsymbol{\alpha}$ through curves on S whose variation vector field is orthogonal to $\boldsymbol{\alpha}$.

As can be seen from Figure 7.9, some geodesic parallels are in *stable* equilibrium, while some are *unstable*. The parallel corresponding to $v = v_0$ is an unstable geodesic if f has a non-degenerate local maximum at v_0 (simply push the parallel up or down slightly to obtain shorter curves), and (a little more difficult to prove) is a stable geodesic if f has a non-degenerate local minimum at v_0. The global minimising properties of geodesics on Riemannian manifolds form part of an active area of current research.

We now illustrate a method of obtaining information concerning those geodesics on a surface of revolution which are neither parallels nor meridians. We make use of *Clairaut's relation* which, for a curve parametrised proportional to arc length, is equivalent to one of the geodesic equations, namely (7.17).

Proposition 2 (Clairaut's relation) *Let $\boldsymbol{\alpha}(t)$ be a geodesic on the surface of revolution S obtained by rotating $(f(v), 0, g(v))$, $f(v) > 0\ \forall v$, about the z-axis, and let $\theta(t)$ be the angle at which $\boldsymbol{\alpha}$ intersects the parallel through $\boldsymbol{\alpha}(t)$. Then, along $\boldsymbol{\alpha}$,*

$$f \cos\theta = c\,,$$

for some constant c.

Proof We write $\boldsymbol{\alpha}(t) = \boldsymbol{x}(u(t), v(t))$, where $\boldsymbol{x}(u, v)$ is the standard local parametrisation of S as a surface of revolution. Then

$$\cos\theta = \frac{\boldsymbol{x}_u.(u'\boldsymbol{x}_u + v'\boldsymbol{x}_v)}{|\boldsymbol{x}_u||\boldsymbol{\alpha}'|} = \frac{u'\sqrt{E}}{|\boldsymbol{\alpha}'|}\,,$$

so that

$$f\cos\theta = \frac{u'E}{|\boldsymbol{\alpha}'|}\,,$$

which is constant by (7.17) and Proposition 5 of §7.2. □

Note that if $\boldsymbol{\alpha}(t) = \boldsymbol{x}(u(t), v(t))$, then $f(v(t))$ is the distance of $\boldsymbol{\alpha}(t)$ from the axis of rotation of S. It follows from Clairaut's relation that as a geodesic $\boldsymbol{\alpha}$ gets closer to the axis

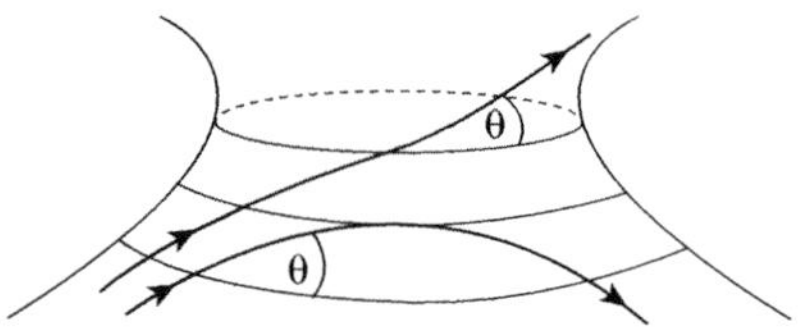

Figure 7.10 Clairaut's relation

of rotation of S then $\boldsymbol{\alpha}'$ becomes closer to, and may indeed become, horizontal (Figure 7.10).

Corollary 3 *With notation as in the statement of Clairaut's relation, if t_0 is such that $\boldsymbol{\alpha}$ is tangential to the parallel at $\boldsymbol{\alpha}(t_0)$ then $\boldsymbol{\alpha}$ attains its minimum distance from the axis of rotation at $\boldsymbol{\alpha}(t_0)$.*

Proof The hypothesis implies that $\cos\theta(t_0) = \pm 1$. Hence, if $\boldsymbol{\alpha}(t) = \boldsymbol{x}(u(t), v(t))$, Clairaut's relation shows that

$$f(v(t))\cos\theta = \pm f(v(t_0)) .$$

Since $f(v) > 0$ for all v, the result follows by taking the modulus of both sides of the above equation. □

We now give some examples of the use of Clairaut's relation.

Example 4 (Torus of revolution) We investigate the behaviour of geodesics on the torus $T_{a,b}$, which is obtained as usual by rotating the curve $(f(v), 0, g(v))$ about the z-axis, where

$$f(v) = a + b\cos v\,, \quad g(v) = b\sin v\,, \quad a > b > 0\,.$$

We first consider a geodesic on $T_{a,b}$ which intersects the outermost parallel (the parallel through the point $(a+b, 0, 0)$) at angle θ_0. Clairaut's relation gives that, along the geodesic,

$$f\cos\theta = (a+b)\cos\theta_0\,.$$

If $\theta_0 = 0$, then the geodesic is simply the outermost parallel. More generally, if θ_0 is sufficiently small, specifically, if

$$\cos\theta_0 \geq \frac{a}{a+b}\,,$$

then $f\cos\theta \geq a$ so that, along the geodesic,

$$f \geq \frac{a}{\cos\theta} \geq a\,.$$

Hence, as illustrated in Figure 7.11, the geodesic lies entirely in the region in which $f \geq a$ or, equivalently, the region $-\frac{\pi}{2} \leq v \leq \frac{\pi}{2}$. These geodesics don't cut through either the top or bottom parallels (those through $(a, 0, \pm b)$).

A similar argument (or use Corollary 3) shows that a geodesic which is initially tangential to the top (or bottom) parallel stays in the region with $-\frac{\pi}{2} \leq v \leq \frac{\pi}{2}$. Notice,

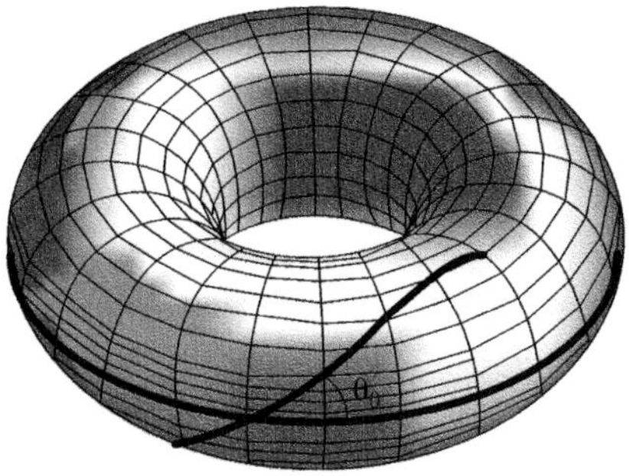

Figure 7.11 Geodesics on a torus of revolution

however, that without some further argument we cannot conclude that this geodesic cuts the outermost parallel, nor that a geodesic starting on the outermost parallel for which θ_0 satisfies

$$\cos\theta_0 = \frac{a}{a+b}$$

meets the top and bottom parallels tangentially, although both of these statements are true. Further properties of geodesics on $T_{a,b}$ are explored in Exercise 7.25.

For the next example, we recall that a *closed* curve in $\mathbb{R}^n$ is a regular curve $\boldsymbol{\alpha} : [a,b] \to \mathbb{R}^n$ such that $\boldsymbol{\alpha}$ and all its derivatives agree at the end points of the interval; that is,

$$\boldsymbol{\alpha}(a) = \boldsymbol{\alpha}(b)\,, \quad \boldsymbol{\alpha}'(a) = \boldsymbol{\alpha}'(b)\,, \quad \boldsymbol{\alpha}''(a) = \boldsymbol{\alpha}''(b), \dots .$$

Example 5 We show that there are no closed geodesics on the surface of revolution S with equation $z^2 = 1/(x^2+y^2)$, $z > 0$. This is the surface illustrated in Figure 7.12 obtained by rotating the curve $z = 1/x$, $x > 0$, $y = 0$ about the z-axis.

We parametrise the generating curve by $v \mapsto (f(v), 0, g(v))$, with $f(v) = v$, $g(v) = 1/v$, $v > 0$, and let $\boldsymbol{x}(u,v)$ be the corresponding parametrisation of S as a surface of revolution. Suppose, then, that there is a closed geodesic $\boldsymbol{\alpha} : [a,b] \to S$. Since $[a,b]$ is compact, Weierstrass's Extremal Value Theorem shows there is some point $t_0 \in [a,b]$

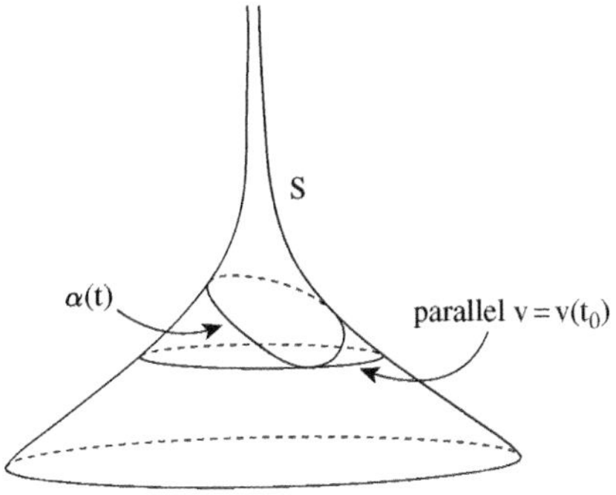

Figure 7.12 There are no closed geodesics on S

at which $\boldsymbol{\alpha}$ attains its maximum distance from the z-axis. This implies that, if we write $\boldsymbol{\alpha}(t) = \boldsymbol{x}(u(t), v(t))$, then $f(v(t))$ attains its maximum value at $t = t_0$. However, in this example $f(v) = v$, so $v(t)$ attains its maximum value at $t = t_0$.

It follows that $v'(t_0) = 0$, so that $\boldsymbol{\alpha}$ is tangential to the parallel at $\boldsymbol{\alpha}(t_0)$. Hence, from Corollary 3, $v(t)$ also attains its minimum at t_0, so that $v(t)$ is constant. Thus $\boldsymbol{\alpha}$ is a parallel, which is impossible by Lemma 1, since the tangent vector to the generating curve of S is never parallel to the z-axis.

In Exercises 7.21 and 7.22 you are asked to find all the closed geodesics on the catenoid and on the hyperboloid of revolution of one sheet.

7.7 Geodesic coordinates †

In this optional section we use geodesics to define two very useful local parametrisations centred on a point p of a surface S. For ease of discussion, we assume that S is a surface in $\mathbb{R}^n$, although the constructions we describe, and the results we obtain, are valid more generally. We are then able to prove Minding's Theorem.

We first describe *geodesic cartesian coordinates*. Let $\{\boldsymbol{e}_1, \boldsymbol{e}_2\}$ be an orthonormal basis of T_pS, and let

$$\tilde{\boldsymbol{x}}(u, v) = \boldsymbol{\alpha}_{u\boldsymbol{e}_1+v\boldsymbol{e}_2}(1)\,, \tag{7.18}$$

where, as usual, $\boldsymbol{\alpha}_X$ is the maximal geodesic on S with initial point p and initial vector $X \in T_pS$. Then $\tilde{\boldsymbol{x}}(u, v)$ is defined whenever 1 is in the domain of definition of $\boldsymbol{\alpha}_{u\boldsymbol{e}_1+v\boldsymbol{e}_2}$, and it may be proved using (a slight extension of) Theorem 2 of §7.3 that $\tilde{\boldsymbol{x}}$ is smooth on its domain of definition U, and there exists $r_0 > 0$ such that U contains the open disc $D_{r_0} = \{(u, v) \in \mathbb{R}^2 : u^2 + v^2 < r_0{}^2\}$.

We may work out the derivative of $\tilde{\boldsymbol{x}}$ at $(0, 0)$ by using Lemma 4 of §7.3 to note that $\tilde{\boldsymbol{x}}(u, 0) = \boldsymbol{\alpha}_{\boldsymbol{e}_1}(u)$ and $\tilde{\boldsymbol{x}}(0, v) = \boldsymbol{\alpha}_{\boldsymbol{e}_2}(v)$. Hence

$$\tilde{\boldsymbol{x}}_u(0, 0) = \boldsymbol{e}_1\,, \quad \tilde{\boldsymbol{x}}_v(0, 0) = \boldsymbol{e}_2\,. \tag{7.19}$$

It now follows from Theorem 3 of §2.5 that, taking r_0 smaller if necessary, $\tilde{\boldsymbol{x}} : D_{r_0} \to S$ is a local parametrisation of an open neighbourhood, $\mathcal{D}_{r_0}$, of p in S. We call $\mathcal{D}_{r_0}$ the *normal neighbourhood of radius r_0 centred on p*, and note that, letting $\tilde{E}$, $\tilde{F}$ and $\tilde{G}$ denote the coefficients of the first fundamental form of $\tilde{\boldsymbol{x}}$, we have

$$\tilde{E}(0, 0) = 1\,, \quad \tilde{F}(0, 0) = 0\,, \quad \tilde{G}(0, 0) = 1\,. \tag{7.20}$$

It follows from Lemma 4 of §7.3 that

$$\tilde{\boldsymbol{x}}(r\cos\theta, r\sin\theta) = \boldsymbol{\alpha}_{\cos\theta\,\boldsymbol{e}_1+\sin\theta\,\boldsymbol{e}_2}(r)\,, \quad 0 < r < r_0\,, \quad \theta \in \mathbb{R}\,, \tag{7.21}$$

so the images under $\tilde{\boldsymbol{x}}$ of lines through the origin give the geodesics on S radiating from p. The images under $\tilde{\boldsymbol{x}}$ of circles in D_{r_0} with centre at the origin are the *geodesic circles* centred on p; the rate at which their circumference increases with r measures the rate of divergence of the geodesics radiating from p. In the next section we obtain a characterisation of the Gaussian curvature of S in terms of this rate of divergence;

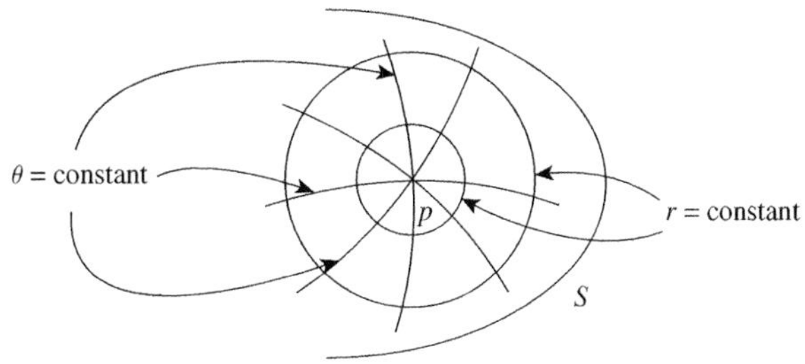

Figure 7.13 Coordinate curves for geodesic polar coordinates

this illustrates once again the intrinsic nature and geometrical significance of Gaussian curvature.

Polar coordinates (r, θ) are often useful in the plane, and, for similar reasons, we shall consider *geodesic polar coordinates*, $\boldsymbol{x}(r, \theta)$, defined on a surface S. They are constructed from geodesic cartesian coordinates $\tilde{\boldsymbol{x}}(u, v)$ by taking

$$\boldsymbol{x}(r, \theta) = \tilde{\boldsymbol{x}}(r\cos\theta, r\sin\theta)\,, \quad 0 < r < r_0\,, \quad -\pi < \theta < \pi\,. \tag{7.22}$$

Then $\boldsymbol{x}$ is also a local parametrisation of $\mathcal{D}_{r_0}$ (except that, as usual with polar coordinates, $\tilde{\boldsymbol{x}}(u, 0)$, $u \leq 0$, is omitted). As illustrated in Figure 7.13, the two sets of coordinate curves of $\boldsymbol{x}(r, \theta)$ give the geodesics radiating from p and the geodesic circles centred on p.

Example 1 (Surface of revolution) Let S be a surface of revolution with a pole p which is not a singular point (see Example 6 of §4.2). The sphere $S^2(r)$ and the paraboloid of revolution $z = x^2 + y^2$ give two examples of this. The geodesics radiating from p give the meridians, and the geodesic circles centred on p are the parallels. In fact (see Exercise 7.26), geodesic polar coordinates (centred on the pole) coincide with the standard parametrisation of S as a surface of revolution when the generating curve is parametrised by arc length starting from the pole. For the paraboloid of revolution, a normal neighbourhood centred on the pole may be chosen to have any positive radius, while for the sphere $S^2(r)$, the maximum radius is πr.

For us, the main advantage of geodesic cartesian coordinates is that the coefficients $\tilde{E}$, $\tilde{F}$, $\tilde{G}$ of the first fundamental form are smooth functions describing the metric on the whole of $\mathcal{D}_{r_0}$. In contrast, the coefficients E, F, G of the first fundamental form of $\boldsymbol{x}(r, \theta)$ give us only the metric on $\mathcal{D}_{r_0} \setminus \{\tilde{\boldsymbol{x}}(u, 0) : u \leq 0\}$. However, as we show in the next proposition, E, F and G have some very nice properties which more than make up for this.

In order to state the proposition, it is convenient to introduce a function $g(r, \theta)$ defined using the coefficients of the first fundamental form of $\tilde{\boldsymbol{x}}$, namely

$$g(r, \theta) = r\sqrt{\tilde{E}\tilde{G} - \tilde{F}^2}\,, \quad -r_0 < r < r_0\,, \quad \theta \in \mathbb{R}\,, \tag{7.23}$$

where the square root on the right hand side is evaluated at $(r\cos\theta, r\sin\theta)$. Since $\tilde{E}\tilde{G} - \tilde{F}^2 > 0$ on D_{r_0}, the function g is smooth. The first equality below is immediate, and the second follows from (7.20).

$$g(0, \theta) = 0\,, \quad g_r(0, \theta) = 1\,. \tag{7.24}$$

Proposition 2 *Let $\mathcal{D}_{r_0}$ be a normal neighbourhood centred on a point $p \in S$ and let E, F, G be the coefficients of the first fundamental form of the corresponding system $\boldsymbol{x}(r,\theta)$ of geodesic polar coordinates. Then, for $r > 0$,*

$$E = 1\,, \quad F = 0\,, \quad \sqrt{G} = g\,. \tag{7.25}$$

Proof It follows from (7.22) and (7.21) that, for $r > 0$,

$$\boldsymbol{x}_r = \boldsymbol{\alpha}'_{\cos\theta \boldsymbol{e}_1 + \sin\theta \boldsymbol{e}_2}(r)\,, \tag{7.26}$$

and

$$\boldsymbol{x}_{rr} = \boldsymbol{\alpha}''_{\cos\theta \boldsymbol{e}_1 + \sin\theta \boldsymbol{e}_2}(r)\,, \tag{7.27}$$

while, by the chain rule,

$$\boldsymbol{x}_\theta = r(-\sin\theta\,\tilde{\boldsymbol{x}}_u + \cos\theta\,\tilde{\boldsymbol{x}}_v)\,. \tag{7.28}$$

We see from (7.26) that $\boldsymbol{x}_r$ is the tangent vector of a geodesic parametrised by arc length, so we have that $E = 1$. Also, the type of working we have seen several times shows that

$$F_r = \frac{\partial}{\partial r}(\boldsymbol{x}_r.\boldsymbol{x}_\theta) = \boldsymbol{x}_{rr}.\boldsymbol{x}_\theta + \frac{1}{2}E_\theta\,.$$

However, we have seen that E is constant, while (7.27) shows that $\boldsymbol{x}_{rr}$ is the acceleration vector of a geodesic, and hence is orthogonal to all tangent vectors of S. It follows that $F_r = 0$, so that F is independent of r. On the other hand, (7.28) implies that $\lim_{r\to 0^+} \boldsymbol{x}_\theta = 0$, so that $\lim_{r\to 0^+} F = 0$. This shows that $F = 0$, or, in other words, the family of geodesic circles centred on p is orthogonal to the family of geodesics radiating from p. This result is known as the *Gauss Lemma*.

We now consider G, and deal first with the case that S is a surface in $\mathbb{R}^3$, which enables us to use the vector product to simplify our calculations. In fact, it follows from (7.22) that, for $r > 0$,

$$\boldsymbol{x}_r = \cos\theta\,\tilde{\boldsymbol{x}}_u + \sin\theta\,\tilde{\boldsymbol{x}}_v\,,$$

so that, using (3.22) and (7.28),

$$\begin{aligned} EG - F^2 &= |\boldsymbol{x}_r \times \boldsymbol{x}_\theta|^2 \\ &= |(\cos\theta\,\tilde{\boldsymbol{x}}_u + \sin\theta\,\tilde{\boldsymbol{x}}_v) \times (-r\sin\theta\,\tilde{\boldsymbol{x}}_u + r\cos\theta\,\tilde{\boldsymbol{x}}_v)|^2 \\ &= r^2|\tilde{\boldsymbol{x}}_u \times \tilde{\boldsymbol{x}}_v|^2 \\ &= r^2(\tilde{E}\tilde{G} - \tilde{F}^2)\,. \end{aligned} \tag{7.29}$$

It now follows that $G = g^2$, and, since $g(r,\theta) > 0$ for $r > 0$ we see that $\sqrt{G} = g$ as claimed.

We may obtain (7.29) for surfaces in $\mathbb{R}^n$ by using equation (3.33) (from the optional §3.9), which tells us how $EG - F^2$ changes under a change of variables. □

Corollary 3 *With notation as in the statement of the previous proposition, we have that for $r > 0$,*

$$g_{rr} + Kg = 0\,, \tag{7.30}$$

where, as usual, K denotes the Gaussian curvature.

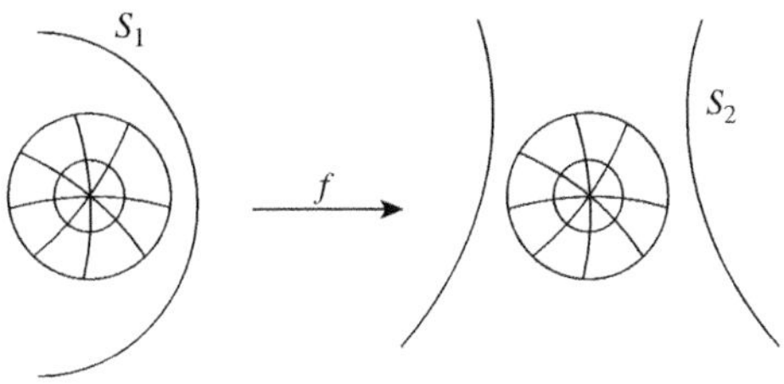

Figure 7.14 The correspondence f

Proof Since $E = 1$ and $F = 0$, Exercise 6.2 (or use equation (6.10)) gives that $K = -(\sqrt{G})_{rr}/\sqrt{G}$. Equation (7.30) now follows from Proposition 2. □

We are now going to prove Minding's Theorem, which we stated in §6.4, and we restate below (with a slight change of notation) for surfaces in Euclidean spaces of arbitrary dimension.

To help us prove the theorem, we first show how to use geodesic cartesian coordinates to define a diffeomorphism between normal neighbourhoods of surfaces. Let p be a point on a surface S_1 in $\mathbb{R}^m$ and let $\tilde{\boldsymbol{x}}(u, v)$ be a system of geodesic cartesian coordinates for a normal neighbourhood $\mathcal{D}_1$ of p. Similarly, let q be a point on a surface S_2 in $\mathbb{R}^n$ and let $\tilde{\boldsymbol{y}}(u, v)$ be a system of geodesic cartesian coordinates for a normal neighbourhood $\mathcal{D}_2$ of q. By taking smaller normal neighbourhoods if necessary, we may assume that $\mathcal{D}_1$ and $\mathcal{D}_2$ both have the same radius. Then the correspondence $f : \mathcal{D}_1 \to \mathcal{D}_2$ (first discussed in §4.5) given by

$$f\left(\tilde{\boldsymbol{x}}(u, v)\right) = \tilde{\boldsymbol{y}}(u, v) \tag{7.31}$$

provides a diffeomorphism (illustrated in Figure 7.14) from $\mathcal{D}_1$ to $\mathcal{D}_2$.

It is clear from the definition of geodesic cartesian coordinates that f maps geodesics radiating from p to geodesics radiating from q, and also maps each geodesic circle centred on p to the geodesic circle of the same radius centred on q. It also follows from the definition (7.22) of geodesic polar coordinates that if $\boldsymbol{x}(r, \theta)$, $\boldsymbol{y}(r, \theta)$ are the corresponding systems of geodesic polar coordinates centred on p and q then

$$f\left(\boldsymbol{x}(r, \theta)\right) = \boldsymbol{y}(r, \theta)\,. \tag{7.32}$$

We now use the above to prove Minding's Theorem.

Theorem 4 (Minding) *Let S_1 be a surface in $\mathbb{R}^m$ and S_2 a surface in $\mathbb{R}^n$, and assume that S_1 and S_2 have the same constant Gaussian curvature. If $p \in S_1$ and $q \in S_2$ then there is an isometry from an open neighbourhood of p in S_1 onto an open neighbourhood of q in S_2.*

Proof We shall show that if K is constant then the coefficients of the first fundamental form of a geodesic polar coordinate system are uniquely determined by K. Hence, if S_1 and S_2 are surfaces with the same constant Gaussian curvature K and if $\boldsymbol{x}(r, \theta)$ and $\boldsymbol{y}(r, \theta)$ are geodesic polar coordinates centred on $p \in S_1$ and $q \in S_2$, then $\boldsymbol{x}$ and $\boldsymbol{y}$ will have the same coefficients of the first fundamental form. Hence, using (7.32) and Proposition 1 of §4.5, the map f defined in (7.31) will be an isometry.

So, let $\boldsymbol{x}(r,\theta)$ be a geodesic polar coordinate system centred on p. We have seen that $E = 1$ and $F = 0$, and we shall show that, if K is constant, then conditions (7.24) on g determine a unique solution of (7.30). The result will then follow from Proposition 2.

Case 1: $K = 0$. The general solution of $g_{rr} = 0$ is $g(r,\theta) = rh_1(\theta) + h_2(\theta)$. Conditions (7.24) then imply that $g(r,\theta) = r$, so that $G = r^2$.

Case 2: $K = k^2 > 0$. The general solution of $g_{rr} + k^2 g = 0$ is $g(r,\theta) = h_1(\theta)\sin kr + h_2(\theta)\cos kr$. Conditions (7.24) then imply that $g(r,\theta) = (1/k)\sin kr$, so that $G = (1/K)\sin^2\sqrt{K}r$.

Case 3: $K = -k^2 < 0$. The general solution of $g_{rr} - k^2 g = 0$ is $g(r,\theta) = h_1(\theta)\sinh kr + h_2(\theta)\cosh kr$. Conditions (7.24) then imply that $g(r,\theta) = (1/k)\sinh kr$, so that $G = -(1/K)\sinh^2\sqrt{-K}r$.

This shows that if K is constant, then E, F and G are uniquely determined by K, and so Minding's Theorem now follows. □

Remark 5 In fact, we have shown rather more than claimed. The proof of Minding's Theorem shows the following. Let S_1 and S_2 be surfaces having the same constant Gaussian curvature. Let $\{\boldsymbol{v}_1, \boldsymbol{v}_2\}$ be an orthonormal basis of the tangent space of S_1 at a point $p \in S_1$, and let $\{\boldsymbol{w}_1, \boldsymbol{w}_2\}$ be an orthonormal basis of the tangent space of S_2 at a point $q \in S_2$. Then (7.19) shows that the derivative df_p of f at p maps $\boldsymbol{v}_i$ to $\boldsymbol{w}_i$, $i = 1, 2$.

In particular, if we take $S_1 = S_2$ and $p = q$, we see that there is an isometry from an open neighbourhood of p to itself whose derivative is any desired rotation of the tangent space at p. This isometry will permute the radial geodesics, and map each geodesic circle centred on p to itself (setwise). It follows from Proposition 1 of §7.5 that geodesic circles on a surface of constant Gaussian curvature have constant geodesic curvature (and in Exercise 7.29 you are asked to find an explicit expression for this).

In the next section we give some more examples of the use of geodesic polar coordinates.

7.8 Metric behaviour of geodesics †

In this optional section we use geodesic polar coordinates to further investigate the metric properties of geodesics. We first prove a basic property of geodesics, namely that a sufficiently short geodesic segment on a surface S minimises the length of all smooth curves on S between the end points of the geodesic segment. We then discuss how the Gaussian curvature gives information about the rate of divergence of geodesics radiating from a given point.

Proposition 1 *Let $\boldsymbol{\alpha}$ be a geodesic segment joining two points p and q of a surface S in $\mathbb{R}^n$, and assume that $\boldsymbol{\alpha}$ is contained in a normal neighbourhood centred on p. If $\boldsymbol{\gamma}$ is any smooth curve on S joining p to q then the length L_{γ} of $\boldsymbol{\gamma}$ is greater than or equal to the length L_{α} of $\boldsymbol{\alpha}$. Moreover, if $L_{\gamma} = L_{\alpha}$ then the trace of $\boldsymbol{\gamma}$ coincides with that of $\boldsymbol{\alpha}$.*

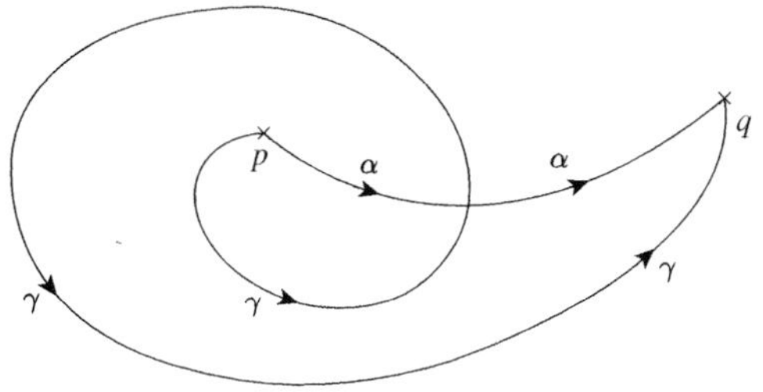

Figure 7.15 Range of values of $\theta(t)$

Proof Assume that $\boldsymbol{\alpha}$ is contained in the normal neighbourhood $\mathcal{D}_{r_0}$ of radius r_0 centred on p, and let $\boldsymbol{x}(r,\theta)$ be a corresponding system of geodesic polar coordinates. If $\boldsymbol{\alpha}$ is a segment of the geodesic $\theta = \theta_0$, and if we parametrise $\boldsymbol{\alpha}$ by arc length starting at $\boldsymbol{\alpha}(0) = p$, then $\boldsymbol{\alpha}(t) = \boldsymbol{x}(t,\theta_0)$, $0 < t \le t_0$, where $t_0 = L_\alpha < r_0$.

Let $\boldsymbol{\gamma} : [0,b] \to S$ be a smooth curve on S with $\boldsymbol{\gamma}(0) = p$ and $\boldsymbol{\gamma}(b) = \boldsymbol{\alpha}(t_0) = q$. We may assume that $\boldsymbol{\gamma}$ doesn't return to p, since, if it does, we simply discard the first part of $\boldsymbol{\gamma}$. If we also assume that the trace of $\boldsymbol{\gamma}$ is contained in $\mathcal{D}_{r_0}$, then we may find L_γ by using the usual limiting argument required when dealing with polar coordinates.

Specifically, for each $\epsilon > 0$, we set $\boldsymbol{\gamma}(t) = \boldsymbol{x}\,(r(t),\theta(t))$ for $\epsilon \le t \le b$. Here, as is usual with polar coordinates, we define the function $\theta(t)$ smoothly on $[\epsilon, b]$ without restricting the range of values taken by $\theta(t)$ to be less than 2π (technically: we can do this because the map $\theta \mapsto e^{i\theta}$ is a *covering map* from $\mathbb{R}$ to the unit circle). In Figure 7.15, the range of values taken by $\theta(t)$ will approach 3π as $\epsilon \to 0$.

Since the end point of $\boldsymbol{\gamma}$ is equal to that of $\boldsymbol{\alpha}$ we have that

$$r(b) = t_0\,, \quad \theta(b) - \theta_0 = 2\pi n \text{ for some } n \in \mathbb{Z}\,.$$

Hence,

$$\begin{aligned} L_{\boldsymbol{\gamma}|[\epsilon,b]} &= \int_\epsilon^b \sqrt{r'^2 + \theta'^2 G}\,dt \\ &\ge \int_\epsilon^b r'(t)\,dt = r(b) - r(\epsilon) = t_0 - r(\epsilon) = L_\alpha - r(\epsilon)\,. \end{aligned}$$

Since $\boldsymbol{\gamma}(0) = p$, we have that $\lim_{\epsilon\to 0} r(\epsilon) = 0$, and so, if we let $\epsilon \to 0$, we find that $L_\gamma \ge L_\alpha$ and equality occurs if and only if $\theta(t)$ is constant and hence is equal to $\theta_0 + 2\pi n$ for some $n \in \mathbb{Z}$.

If the trace of $\boldsymbol{\gamma}$ is not contained in $\mathcal{D}_{r_0}$ then an argument similar to the one above shows that the arc length along $\boldsymbol{\gamma}$ from p to the point where $\boldsymbol{\gamma}$ first leaves $\mathcal{D}_{r_0}$ is greater than or equal to r_0, and hence the length of $\boldsymbol{\gamma}$ is greater than the length of $\boldsymbol{\alpha}$. □

We now discuss the relation between Gaussian curvature and the rate of divergence of geodesics radiating from a given point p of S. We consider a normal neighbourhood $\mathcal{D}_{r_0}$ centred on p, and let $\boldsymbol{x}(r,\theta)$ be a corresponding system of geodesic polar coordinates. If $g(r,\theta)$ is the function defined in (7.23) and if $\theta_0 < \theta_1$, then the length of the arc of the geodesic circle of radius $r > 0$ between geodesics $\theta = \theta_0$ and $\theta = \theta_1$ is given by

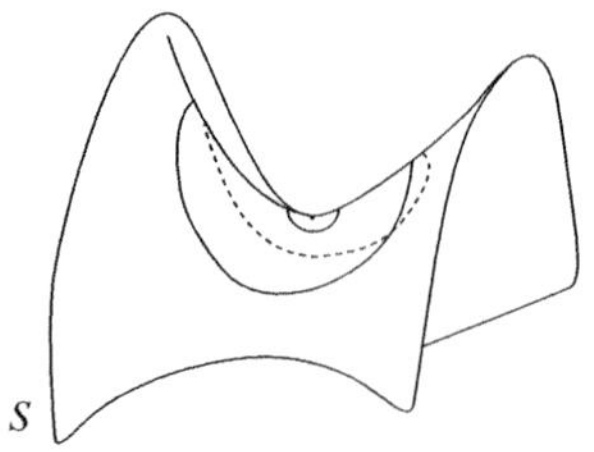

Figure 7.16 Geodesic circles when $K < 0$

$$L(r) = \int_{\theta_0}^{\theta_1} g(r,\theta)\, d\theta \,, \tag{7.33}$$

since, for $r > 0$, $\sqrt{G} = g$. Differentiating under the integral sign, we find that

$$L'(r) = \int_{\theta_0}^{\theta_1} g_r\, d\theta \,, \quad L''(r) = \int_{\theta_0}^{\theta_1} g_{rr}\, d\theta \,, \tag{7.34}$$

and it follows from (7.24) and the first equation of (7.34) that $\lim_{r\to 0^+} L'(r) = \theta_1 - \theta_0$.

If we assume that $K < 0$ (but not necessarily constant) on $\mathcal{D}_{r_0}$, then (7.30) shows that g_{rr} is positive for $0 < r < r_0$ so that $L''(r) > 0$ for all $0 < r < r_0$. Hence geodesics radiating from p diverge faster than they do in Euclidean space (where $L(r) = r(\theta_1 - \theta_0)$). In particular (Figure 7.16), the circumference of geodesic circles centred on p grows faster with r than that of the corresponding circles in the plane.

If we assume that $K > 0$ on $\mathcal{D}_{r_0}$ then $L''(r) < 0$ for all $0 < r < r_0$, so that geodesics radiating from p diverge slower than they do in Euclidean space. Indeed, after some time, these geodesics could start to converge again. For example, the geodesics radiating from the pole of a paraboloid of revolution are the meridians, and these continue to diverge but more slowly than do straight lines radiating from a point in the plane. On the other hand, the geodesics of the sphere radiating from the south pole are great circles, which start to converge again after they get past the equator.

We now obtain another intrinsic characterisation of Gaussian curvature K.

Theorem 2 *Let $L(r)$ be the circumference of the geodesic circle with centre p and radius r. Then*

$$\frac{\pi}{3} K(p) = \lim_{r\to 0^+} \frac{2\pi r - L(r)}{r^3} \,. \tag{7.35}$$

Proof In terms of geodesic polar coordinates centred on p, we have that

$$L(r) = \int_{-\pi}^{\pi} \sqrt{G}\, d\theta = \int_{-\pi}^{\pi} g(r,\theta)\, d\theta \,,$$

so we begin by writing down the first few terms in the Taylor series in r for $g(r,\theta)$ about $r = 0$, keeping θ fixed. We already know that $g(0,\theta) = 0$ and $g_r(0,\theta) = 1$. Also, using (7.30), $g_{rr}(0,\theta) = 0$ and $g_{rrr}(0,\theta) = -K(p)$.

So, for each fixed θ, the Taylor series for $g(r,\theta)$ about $r = 0$ is now seen to be

$$g(r,\theta) = r - \frac{r^3}{3!} K(p) + R_0(r,\theta) \,,$$

where $R_0(r,\theta)$ is the remainder term. If we choose a real number M such that $|g_{rrrr}(\xi,\theta)| \le M$ for all $0 \le \xi \le r$, $0 \le \theta \le 2\pi$, then the Lagrange form of the remainder for Taylor series shows that $|R_0(r,\theta)| \le Mr^4$ for all $r > 0$ and all θ.

Hence

$$L(r) = \int_{-\pi}^{\pi} g(r,\theta)\,d\theta = 2\pi\left(r - \frac{r^3}{3!}K(p)\right) + R_1(r)\,,$$

where $|R_1(r)| \le 2\pi M r^4$. It now follows that

$$\frac{L(r) - 2\pi r}{r^3} = -\frac{\pi}{3}K(p) + R_2(r)$$

where $\lim_{r\to 0^+} R_2(r) = 0$, and the result follows. □

The *geodesic disc* $\mathcal{D}_r$ with centre p and radius r is the image under $\tilde{x}$ of the disc D_r in the plane. By integrating (7.35), we may show that if $A(r)$ is the area of $\mathcal{D}_r$ then

$$\frac{\pi}{12}K(p) = \lim_{r\to 0^+} \frac{\pi r^2 - A(r)}{r^4}\,.$$

7.9 Rolling without slipping or twisting †

In this optional section, we consider the notion of rolling one surface on another without slipping. Although we present the ideas here only for the case of surfaces in $\mathbb{R}^3$, much of the discussion may be considerably generalised.

Suppose that $\boldsymbol{\alpha}(s)$ and $\tilde{\boldsymbol{\alpha}}(s)$ are curves parametrised by arc length on surfaces S and $\tilde{S}$ in $\mathbb{R}^3$. Denoting d/ds by $'$, and setting $\boldsymbol{t} = \boldsymbol{\alpha}'$, $\tilde{\boldsymbol{t}} = \tilde{\boldsymbol{\alpha}}'$, we note that for each s there is a unique rotation $A(s)$ such that

$$\tilde{\boldsymbol{t}} = A\boldsymbol{t} \quad \text{and} \quad \tilde{\boldsymbol{N}} = A\boldsymbol{N}\,,$$

where $\boldsymbol{N}(s)$ and $\tilde{\boldsymbol{N}}(s)$ are unit normals to S at $\boldsymbol{\alpha}(s)$ and $\tilde{S}$ at $\tilde{\boldsymbol{\alpha}}(s)$ respectively.

If, for each s, we define $\boldsymbol{v}(s) \in \mathbb{R}^3$ by setting

$$\boldsymbol{v} = \tilde{\boldsymbol{\alpha}} - A\boldsymbol{\alpha}\,,$$

then the family $B(s)$ of rigid motions of $\mathbb{R}^3$ given by

$$B(s)(p) = A(s)(p) + \boldsymbol{v}(s)\,, \quad p \in \mathbb{R}^3\,,$$

is uniquely determined by S, $\tilde{S}$, $\boldsymbol{\alpha}$ and $\tilde{\boldsymbol{\alpha}}$. Then $B\boldsymbol{\alpha} = \tilde{\boldsymbol{\alpha}}$, and the family $B(s)$ of rigid motions is said to roll S *without slipping* on $\tilde{S}$, the curves of contact being $\boldsymbol{\alpha}$ and $\tilde{\boldsymbol{\alpha}}$.

The rate at which S is twisting relative to $\tilde{S}$ as it rolls along $\tilde{S}$ is given by

$$A\left((\boldsymbol{N} \times \boldsymbol{t})'\right) - (\tilde{\boldsymbol{N}} \times \tilde{\boldsymbol{t}})'\,,$$

and if we define the derivative A' of A as the matrix obtained by taking the derivative of each entry of A, then the product rule for differentiation shows that this rate of twisting is given by

$$A\left((\boldsymbol{N} \times \boldsymbol{t})'\right) - (A(\boldsymbol{N} \times \boldsymbol{t}))' = -A'(\boldsymbol{N} \times \boldsymbol{t})\,.$$

We say that the rolling of S along $\tilde{S}$ takes place *without twisting* if and only if

$$A'(\boldsymbol{N} \times \boldsymbol{t}) = 0 \ \forall s \, . \tag{7.36}$$

Proposition 1 *Let S and $\tilde{S}$ be surfaces in $\mathbb{R}^3$ and assume that S is rolled without slipping along $\tilde{S}$ with curves of contact $\boldsymbol{\alpha}(s)$ on S and $\tilde{\boldsymbol{\alpha}}(s)$ on $\tilde{S}$, both parametrised by arc length. If the rolling takes place without twisting then $\boldsymbol{\alpha}$ and $\tilde{\boldsymbol{\alpha}}$ have the same geodesic curvature, that is to say,*

$$\kappa_g(s) = \tilde{\kappa}_g(s) \ \forall s \, ,$$

where κ_g is the geodesic curvature of $\boldsymbol{\alpha}$ and $\tilde{\kappa}_g$ is the geodesic curvature of $\tilde{\boldsymbol{\alpha}}$. In particular, $\boldsymbol{\alpha}$ is a geodesic on S if and only if $\tilde{\boldsymbol{\alpha}}$ is a geodesic on $\tilde{S}$.

Proof Let

$$B(s) = A(s) + \boldsymbol{v}(s)$$

be the family of rigid motions performing the rolling without slipping. We first note that, for any pair of fixed vectors $\boldsymbol{u}$ and $\boldsymbol{w}$, we have $A\boldsymbol{u}.A\boldsymbol{w} = \boldsymbol{u}.\boldsymbol{w}$, and hence, by differentiating,

$$A'\boldsymbol{u}.A\boldsymbol{w} + A\boldsymbol{u}.A'\boldsymbol{w} = 0 \, . \tag{7.37}$$

We now recall from (7.1) that the geodesic curvature $\kappa_g(s)$ of $\boldsymbol{\alpha}$ is given by $\kappa_g = \boldsymbol{t}'.(\boldsymbol{N} \times \boldsymbol{t})$. Thus, using (7.37) for the fifth equality,

$$\begin{aligned}
\tilde{\kappa}_g &= \tilde{\boldsymbol{t}}'.(\tilde{\boldsymbol{N}} \times \tilde{\boldsymbol{t}}) \\
&= (A\boldsymbol{t})'.(A\boldsymbol{N} \times A\boldsymbol{t}) \\
&= (A'\boldsymbol{t} + A\boldsymbol{t}').A(\boldsymbol{N} \times \boldsymbol{t}) \\
&= A'\boldsymbol{t}.A(\boldsymbol{N} \times \boldsymbol{t}) + \boldsymbol{t}'.(\boldsymbol{N} \times \boldsymbol{t}) \\
&= -A\boldsymbol{t}.A'(\boldsymbol{N} \times \boldsymbol{t}) + \kappa_g \\
&= \kappa_g \, ,
\end{aligned}$$

where we have used (7.36), the condition that the rolling should take place without twisting, for the final equality. □

The condition $\tilde{\kappa}_g(s) = \kappa_g(s)$ is not sufficient to ensure that when S is rolled without slipping on $\tilde{S}$ with curves of contact $\boldsymbol{\alpha}(s)$ on S and $\tilde{\boldsymbol{\alpha}}(s)$ on $\tilde{S}$, then the rolling takes place without twisting. For example, in Exercise 7.31 you are invited to prove that if a surface S is rolled without slipping along a straight line on a plane then, if there is no twisting, the curve of contact $\boldsymbol{\alpha}$ on S is not only a geodesic but also a line of curvature. It then follows (from Exercise 7.11) that $\boldsymbol{\alpha}$ is the curve of intersection of S with a plane which is perpendicular to S at each point of intersection.

Exercises

7.1 Prove equation (7.2), which says that the geodesic curvature of a regular curve $\boldsymbol{\alpha}(t)$ on a surface S in $\mathbb{R}^3$ is given by

$$\kappa_g = \frac{1}{|\boldsymbol{\alpha}'|^3}\boldsymbol{\alpha}''.(\boldsymbol{N} \times \boldsymbol{\alpha}') \, .$$

7.2 Show that a curve of constant geodesic curvature c on the unit sphere $S^2(1)$ in $\mathbb{R}^3$ is the intersection of $S^2(1)$ with a plane whose perpendicular distance from the origin is $|c|/\sqrt{1+c^2}$. (*Hint:* if $\boldsymbol{\alpha}$ is a curve of constant geodesic curvature c show that the vector $\mathbf{v} = \boldsymbol{\alpha} \times \boldsymbol{\alpha}' + c\boldsymbol{\alpha}$ is constant, where $'$ denotes differentiation with respect to arc length.)

7.3 Let C be the circle on the sphere $S^2(r)$ obtained as the intersection of $S^2(r)$ with the plane $z = r \sin v_0$ for some constant v_0 with $-\pi/2 < v_0 < \pi/2$. Show that, when C is parametrised as $\boldsymbol{\alpha}(t) = r(\cos v_0 \cos t, \cos v_0 \sin t, \sin v_0)$ and $S^2(r)$ is oriented using the outward unit normal, then C has constant geodesic curvature $\kappa_g = (1/r)\tan v_0$.

7.4 Let $(f(v), 0, g(v))$, $f(v) > 0\ \forall v$, be the generating curve of a surface of revolution S. Show that, for any parameter value v, the geodesic curvature of the parallel C_v through $(f(v), 0, g(v))$ is constant and given by

$$\kappa_g = -\frac{f'}{f\sqrt{f'^2 + g'^2}}\,,$$

when the orientation on S is that induced by the standard local parametrisation $\boldsymbol{x}(u, v)$ of a surface of revolution, and we travel around C_v in the direction of increasing u.

7.5 This exercise is designed to illustrate Proposition 3 of §7.1.

Let $(f(v), 0, g(v))$, $f(v) > 0\ \forall v$, be the generating curve of a surface of revolution S. At a point where $g'(v) \neq 0$, let $(0, 0, h(v))$ be the point at which the line normal to S at $(f(v), 0, g(v))$ meets the z-axis.

(a) Show that $h - g = ff'/g'$.
Define a function $r(v) > 0$ by setting $r(v)^2 = f(v)^2 + (g(v) - h(v))^2$.

(b) Show that the sphere $\tilde{S}$ centre $(0, 0, h(v))$, radius $r(v)$, is tangential to S along the parallel C_v through $(f(v), 0, g(v))$.

(c) Use Exercise 7.3, suitably adjusted for a sphere centre $(0, 0, h(v))$, to show that, when C_v is parametrised as $\boldsymbol{\alpha}(t) = (f(v)\cos t, f(v)\sin t, g(v))$ and $\tilde{S}$ is oriented using the outward unit normal, the geodesic curvature of C_v as a curve on $\tilde{S}$ is given by

$$\tilde{\kappa}_g = -\epsilon \frac{f'(v)}{f(v)\sqrt{f'(v)^2 + g'(v)^2}}\,,$$

where $\epsilon = 1$ if $g'(v) > 0$ and $\epsilon = -1$ if $g'(v) < 0$.

(d) Use Exercise 7.4 to show that, when the same unit normal to S, $\tilde{S}$ is chosen along C_v, the geodesic curvature of C_v is the same when it is considered as a curve on S or as a curve on $\tilde{S}$ (in accordance with Proposition 3 of §7.1).

7.6 Show that the curve

$$\boldsymbol{\alpha}(t) = e^t(\cos t, \sin t, 1)\,, \quad t \in \mathbb{R}\,,$$

lies on the cone $x^2 + y^2 = z^2$, and show that the geodesic curvature of $\boldsymbol{\alpha}$ is inversely proportional to e^t.

7.7 Let S be the surface in $\mathbb{R}^4$ covered by the single (isothermal) parametrisation

$$\boldsymbol{x}(u,v) = (u, v, u^2 - v^2, 2uv)\,, \quad u, v \in \mathbb{R}\,.$$

Find the geodesic curvature of the coordinate curves $v =$ constant.

7.8 Let $\boldsymbol{x}(u,v)$ be an orthogonal local parametrisation of a surface S in $\mathbb{R}^3$ (or, more generally, in $\mathbb{R}^n$).

(a) Show that the geodesic curvature of the coordinate curve $v =$ constant is given by

$$\kappa_g = -\frac{1}{2\sqrt{G}}(\log E)_v\,.$$

Here, we travel along the coordinate curve $v =$ constant in the direction of increasing u, and use the orientation of the surface determined by the local parametrisation.

(b) Obtain the formula of Exercise 7.4 as a special case.

(c) If, in addition to assuming $F = 0$, we also assume that $G_u = 0$, show that, when parametrised proportional to arc length, each of the coordinate curves $u =$ constant is a geodesic.

(d) Obtain the result given in Example 3 of §7.2 (namely, that, when parametrised proportional to arc length, every meridian of a surface of revolution is a geodesic) as a special case of (c).

7.9 Let $\boldsymbol{\alpha} : I \to \boldsymbol{R}^3$ be a curve parametrised by arc length with everywhere non-zero curvature, and let $\boldsymbol{b}$ be the binormal of $\boldsymbol{\alpha}$. Assume that for some $\epsilon > 0$ the map

$$\boldsymbol{x}(s,v) = \boldsymbol{\alpha}(s) + v\boldsymbol{b}(s)\,, \quad s \in I\,,\ v \in (-\epsilon, \epsilon)\,,$$

is a parametrisation of a surface S (so that S is the ruled surface swept out by the binormals of $\boldsymbol{\alpha}$). Prove that $\boldsymbol{\alpha}$ is a geodesic on S.

7.10 (a) Let S be a surface in $\mathbb{R}^3$ and suppose that P is a plane which intersects S orthogonally along the trace of a regular curve $\boldsymbol{\alpha}(t)$. If $\boldsymbol{\alpha}$ is parametrised proportional to arc length, show that $\boldsymbol{\alpha}$ is a geodesic on S.

(b) Show that, when parametrised proportional to arc length, the curves of intersection of the coordinate planes in $\mathbb{R}^3$ with the surface S defined by the equation $x^4 + y^6 + z^8 = 1$ are geodesics.

7.11 Let S be a surface in $\mathbb{R}^3$.

(a) Let $\boldsymbol{\alpha}$ be a geodesic on S which is also a line of curvature on S. Show that $\boldsymbol{\alpha}$ lies on the intersection of S and a plane P which intersects S orthogonally at all points on the trace of $\boldsymbol{\alpha}$. (*Hint:* show that $\boldsymbol{N} \times \boldsymbol{\alpha}'$ is constant along $\boldsymbol{\alpha}$.)

(b) Let $\boldsymbol{\alpha}$ be a geodesic on S with nowhere vanishing curvature κ which also lies on a plane P. Show that $\boldsymbol{\alpha}$ is a line of curvature on S (so that, by (a), P intersects S orthogonally at all points on the trace of $\boldsymbol{\alpha}$.)

7.12 Assume that

$$\boldsymbol{x}(u,v) = \frac{1}{2}\,(u - \sin u \cosh v\,,\ 1 - \cos u \cosh v\,,\ 4\sin(u/2)\sinh(v/2))$$

is a parametrisation of a surface S in $\mathbb{R}^3$.

(i) Show that the coordinate curve $u = \pi$ is a parabola.
(ii) Show that the coordinate curve $v = 0$ is a cycloid.
(iii) Show that both the above curves, when parametrised proportional to arc length, are geodesics on S. (*Hint:* it may help to prove that $\boldsymbol{x}_u$ is orthogonal to $\boldsymbol{x}_v$.)

This is **Catalan's surface** (pictured in Figure 9.1). It is a minimal surface with self-intersections, and with singular points at $(2n\pi, 0)$, $n \in \mathbb{Z}$.

7.13 Show that if all the geodesics on a connected surface S in $\mathbb{R}^3$ are plane curves, then the surface is an open subset of a plane or a sphere. (*Hint:* use Exercise 7.11 to show that every point of S is an umbilic, and then use Theorem 1 of §5.8.)

7.14 Use the existence and uniqueness theorem for geodesics (Theorem 2 of §7.3) to show that there are no closed geodesics on a helicoid in $\mathbb{R}^3$.

7.15 Let S be the cone obtained by rotating the line $z = \beta x$, $z > 0$, about the z-axis, where β is a positive constant. Use the ideas discussed in Example 7 of §7.3 to show that:

(a) if S is obtained from a sector (with vertex at the origin) of the plane using the process described in Example 7 of §7.3 in such a way that the edges of the sector join up to form a meridian of the cone, then (Figure 7.17)

$$\frac{\sqrt{1+\beta^2}}{2} = \frac{\pi}{\phi},$$

where ϕ is the angle between the bounding lines of the sector;

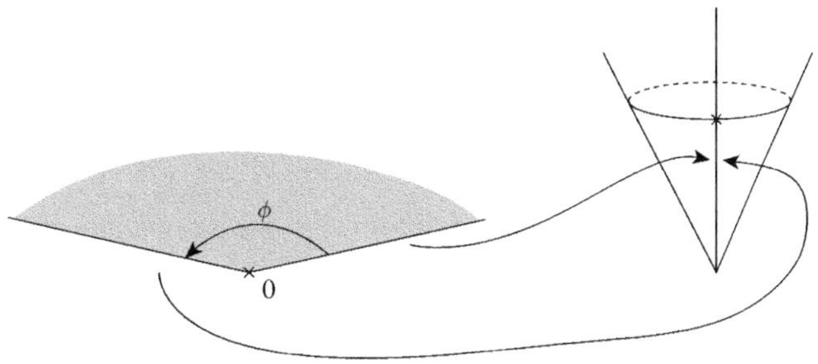

Figure 7.17 Cone

(b) if $\sqrt{1+\beta^2} \leq 2$, then no geodesics on S have self-intersections;
(c) if $2 < \sqrt{1+\beta^2} \leq 4$, then every geodesic which is not a meridian has exactly one point of self-intersection;
(d) in general, if n is the positive integer such that

$$2n < \sqrt{1+\beta^2} \leq 2(n+1),$$

then every geodesic which is not a meridian has exactly n self-intersections.

In particular, we note that, for a given cone, every geodesic which is not a meridian has exactly the same number of self-intersections.

7.16 **(Geodesics on the hyperbolic plane)** *(This exercise uses material in the optional Example 10 of §7.3.)* In this exercise, we use equations (7.11) and (7.12) to show directly that the geodesics on the hyperbolic plane H are as found in Example 10 of §7.3.

(a) Show that (7.11) integrates up to give $u' = cv^2$ for some constant c.

(b) Show that, in this case, (7.12) implies that

$$u'^2 + v'^2 = kv^2 \quad \text{for some constant } k ,$$

(or, equivalently, $(u(t), v(t))$ is parametrised proportional to hyperbolic arc length).

(c) Deduce that, when parametrised proportional to hyperbolic arc length, the curve $(u(t), v(t))$ is a geodesic on H if and only if either u = constant or $(u - u_0)^2 + v^2 = r^2$, where u_0 and r are constants with $r > 0$.

7.17 *(This exercise uses material in the optional Example 10 of §7.3.)* Determine the curves on the hyperbolic plane H of constant geodesic curvature.

7.18 *(This exercise uses material in the optional Example 4 of §4.3.)* Find all the geodesics on the flat torus $S^1(1) \times S^1(1)$ obtained by taking $r_1 = r_2 = 1$ in Example 4 of §4.3. Prove that, through the point $(1, 0, 1, 0) \in S^1(1) \times S^1(1)$ there are an infinite number of geodesics which are not closed and an infinite number which are closed.

7.19 **(Liouville's formula)** Let $\boldsymbol{x}(u, v)$ be an orthogonal local parametrisation of a surface S in $\mathbb{R}^3$ (or, more generally, in $\mathbb{R}^n$), and let $\boldsymbol{\alpha}(s) = \boldsymbol{x}\,(u(s), v(s))$ be a curve, parametrised by arc length, lying in the image of $\boldsymbol{x}$. Let μ_g and ν_g be the geodesic curvatures of the coordinate curves u = constant and v = constant respectively. Prove that the geodesic curvature κ_g of $\boldsymbol{\alpha}(s)$ (using the orientation determined by the local parametrisation) is given by

$$\kappa_g = \nu_g \cos\phi + \mu_g \sin\phi + \frac{d\phi}{ds} ,$$

where ϕ is the angle from $\boldsymbol{x}_u$ to $\boldsymbol{\alpha}'$ measured in the direction determined by the local parametrisation.

Exercises 7.20 to 7.25 use material in the optional §7.6.

7.20 Let S be the cone obtained by rotating the line $z = \beta x, z > 0$, about the z-axis, where β is a positive constant. If $\boldsymbol{\alpha}(t) = (x(t), y(t), z(t))$ is a geodesic on S which cuts the parallel $z = 1$ at an angle θ_0, $0 \le \theta_0 \le \pi/2$, show that $\boldsymbol{\alpha}$ stays entirely in the region where $z \ge \cos\theta_0$ (note: this is independent of the slope β of the generating line).

7.21 Find all the closed geodesics on the catenoid $x^2 + y^2 = \cosh^2 z$.

7.22 Find all the closed geodesics on the hyperboloid of revolution of one sheet with equation $x^2 + y^2 - z^2 = 1$.

7.23 Find all the closed geodesics on the surface of revolution in $\mathbb{R}^3$ obtained by rotating the curve $z = 1/x^2$, $x > 0$, $y = 0$, about the z-axis.

7.24 Let $\boldsymbol{\alpha}(t)$ be a geodesic on a surface of revolution S, and assume there is a value t_0 of the parameter at which $\boldsymbol{\alpha}(t)$ attains its maximum distance from the axis of rotation of S. Show that, when parametrised proportional to arc length, the parallel through $\boldsymbol{\alpha}(t_0)$ is itself a geodesic.

Does this result hold if "maximum" is replaced by "minimum"? Give either a proof or a counterexample.

7.25 Let $T_{a,b}$ denote the torus of revolution discussed in Example 4 of §7.6, and let $\boldsymbol{\alpha}(t)$ be a geodesic on $T_{a,b}$ which intersects the outermost parallel $v = 0$ at an angle θ_0 (where $0 < \theta_0 < \pi/2$). If $\boldsymbol{\alpha}(t)$ also intersects the innermost parallel $v = \pi$, show that

$$\cos\theta_0 < \frac{a-b}{a+b} \,. \tag{7.38}$$

Conversely *(a little more difficult)*, show that if (7.38) holds then a geodesic on $T_{a,b}$ which intersects the outermost parallel at an angle θ_0 also intersects the innermost parallel.

Exercises 7.26 to 7.30 use material in the optional §7.7 and §7.8.

7.26 Let S be a surface of revolution generated by rotating a curve in the xz-plane about the z-axis. Justify a statement made in Example 1 of §7.7 by showing that if p is a pole of S which is not a singular point of S then geodesic polar coordinates defined by taking $\boldsymbol{e}_1 = (1,0,0)$ centred on p coincide with the standard parametrisation of S as a surface of revolution when the generating curve is parametrised by arc length starting from p.

7.27 **(Fermi coordinates)** Let p be a point on a surface S in $\mathbb{R}^n$ and let $\boldsymbol{\alpha}(v)$ be a geodesic on S parametrised by arc length with $\boldsymbol{\alpha}(0) = p$. For $|v|$ sufficiently small, let $\boldsymbol{\beta}_v(u)$ be the geodesic on S parametrised by arc length with $\boldsymbol{\beta}_v(0) = \boldsymbol{\alpha}(v)$ and $\boldsymbol{\beta}_v{}'(0)$ orthogonal to $\boldsymbol{\alpha}'(v)$ (Figure 7.18). Let $\boldsymbol{x}(u,v) = \boldsymbol{\beta}_v(u)$, and assume that $\boldsymbol{x}$ is smooth on its domain of definition.

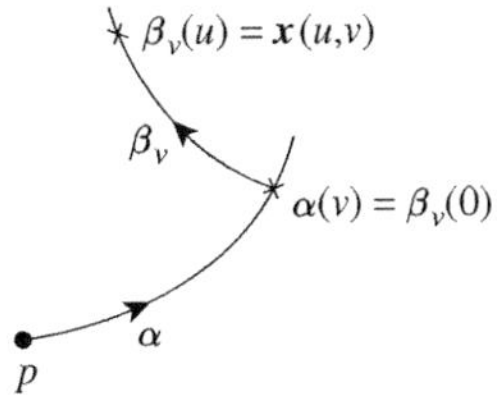

Figure 7.18 Fermi coordinates

(a) Show that there is an open neighbourhood U of $(0,0)$ in the plane such that the restriction of $\boldsymbol{x}$ to U is a local parametrisation of an open neighbourhood of p in S (giving a system of *Fermi* coordinates). Prove also that the coefficients of the first fundamental form of $\boldsymbol{x}$ satisfy

$$E = 1\,, \quad F = 0\,, \quad G(0,v) = 1\,, \quad G_u(0,v) = 0\,.$$

(b) Find the system of Fermi coordinates on the unit sphere $S^2(1)$ determined by taking $p = (1,0,0)$ and $\boldsymbol{\alpha}(v)$ the equator. Find explicitly the coefficients of the first fundamental form in this case.

(c) Show that if the Gaussian curvature K of a surface S is constant, then K determines E, F and G uniquely, thus giving an alternative proof of Minding's Theorem (Theorem 4 of §7.7) using Fermi coordinates.

7.28 Let S_1 be a surface in $\mathbb{R}^n$ and let S_2 be a surface in $\mathbb{R}^m$. If $f : S_1 \to S_2$ is a local isometry and if $p \in S_1$, show that f maps geodesics on S_1 radiating from p to geodesics on S_2 radiating from $f(p)$. Show also that f maps geodesic circles of S_1 centred on p to geodesics circles of S_2 centred on $f(p)$.

7.29 Let S be a surface in $\mathbb{R}^n$ with constant Gaussian curvature K.

(i) Find an expression in terms of K for the length $\ell(r)$ of a geodesic circle of radius r.

(ii) Find an expression in terms of K for the geodesic curvature $\kappa_g(r)$ at each point of a geodesic circle of radius r (*Hint:* you could use, for instance, the result of Exercise 7.8 to show that, when using geodesic polar coordinates, $2\kappa_g(r) = \pm(\log G)_r$).

(iii) Find an expression in terms of K for the area $A(r)$ of the geodesic disc of radius r.

Show that, when $\kappa_g(r)$ is given an appropriate sign, $\ell(r)\kappa_g(r) + K A(r) = 2\pi$.

Remark 1 You might be surprised that $\ell\kappa_g + KA$ is independent of r (and even more surprised that it is equal to 2π). However, the reason will become clear after you have looked at the Gauss–Bonnet Theorem in the next chapter.

7.30 Verify the intrinsic characterisation of Gaussian curvature given in Theorem 2 of §7.8 when p is a point of the unit sphere $S^2(1)$.

7.31 *(This exercise uses material in the optional §7.9.)* Let S and $\tilde{S}$ be surfaces in $\mathbb{R}^3$ with unit normals $\boldsymbol{N}$, $\tilde{\boldsymbol{N}}$, respectively, and suppose that S is rolled without slipping on $\tilde{S}$ with curves of contact $\boldsymbol{\alpha}(s)$ on S and $\tilde{\boldsymbol{\alpha}}(s)$ on $\tilde{S}$, both parametrised by arc length. Show that the rolling takes place without twisting if and only if $\boldsymbol{\alpha}$ and $\tilde{\boldsymbol{\alpha}}$ have the same geodesic curvature and

$$\tilde{\boldsymbol{N}}'.(\tilde{\boldsymbol{N}} \times \tilde{\boldsymbol{t}}) = \boldsymbol{N}'.(\boldsymbol{N} \times \boldsymbol{t})\,.$$

In particular, show that if a surface S is rolled without slipping along a straight line on a plane, then the rolling takes place without twisting if and only if the curve of contact on S is both a geodesic and a line of curvature. Deduce that, in this case, the curve of contact $\boldsymbol{\alpha}$ on S is the intersection of S with a plane which is orthogonal to S at all points on the trace of $\boldsymbol{\alpha}$ (see Exercise 7.11).

8 The Gauss–Bonnet Theorem

The Gauss–Bonnet Theorem is a deep theorem which lies at the heart of differential geometry and its many related areas. Here are three of the many reasons for its importance: firstly, it shows how curvature affects the geometry of surfaces in Euclidian space, providing further insight into the meaning of Gaussian curvature; secondly, it exhibits the interrelationship between curvature and topology which is at the heart of global differential geometry, providing, for instance, an example of the use of geometric quantities to calculate topological invariants; and thirdly, it may be regarded as a special case in the important modern mathematical theory of characteristic classes.

The theorem describes the relation between the topology of a surface S in $\mathbb{R}^n$ and the geometrical quantities of Gaussian curvature K of S and geodesic curvature κ_g of curves on S. If you have been reading the sections on surfaces in higher dimensional Euclidean spaces, and, in particular, have looked at §6.5, you will be aware that Gaussian curvature K may be defined for surfaces in $\mathbb{R}^n$ for all $n \geq 3$. In this chapter we shall consider surfaces in Euclidean spaces of any dimension, since the methods used and the results proved are valid in this situation; all we need is that formulae (6.10) for Gaussian curvature, and (7.14) for geodesic curvature, are valid for an orthogonal parametrisation of a surface in a Euclidean space of any dimension. If you have been concentrating on surfaces in $\mathbb{R}^3$, then simply take $n = 3$ throughout this chapter.

We begin by indicating how Gaussian curvature is related to the sum of the interior angles of triangles with geodesic sides on a surface S, and we then prove the Gauss–Bonnet Theorem for a (not necessarily geodesic) triangle on S. In order to extend this to the general case and to understand the relationship with topology, we gather together some topological preliminaries in §8.4 and then, in §8.5, use the Gauss–Bonnet Theorem for a triangle to prove the general Gauss–Bonnet Theorem. In some ways the approach is very similar to the proof of Cauchy's Theorem in complex analysis, which may be achieved by deducing the general theorem from the theorem for a triangle. Indeed, there are other similarities between the ideas used to prove the two theorems.

Finally, in §8.6, we indicate some of the many consequences and applications of the Gauss–Bonnet Theorem.

8.1 Preliminary examples

In this section we discuss the sum of the interior angles α, β and γ of a triangle with geodesic sides (which we call *geodesic triangles*) on the cylinder, the sphere, the hyperbolic plane and the pseudosphere.

Example 1 (Plane and cylinder) For a geodesic triangle in the plane we have that

$$\alpha + \beta + \gamma = \pi \ .$$

The same result holds if we take a geodesic triangle on the cylinder (Figure 8.1), for this will be the image of a geodesic triangle in the plane under the local isometry which wraps the plane round the cylinder an infinite number of times.

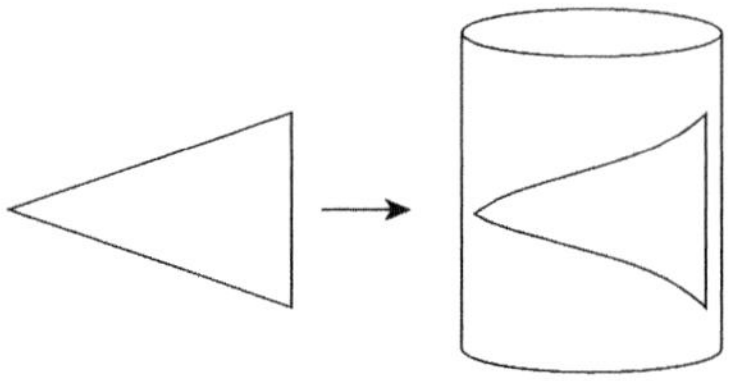

Figure 8.1 Geodesic triangles on the plane and cylinder

Example 2 (Sphere) The sides of a geodesic triangle on the sphere are segments of great circles (Figure 8.2), and some spherical geometry may be used to show that

$$\alpha + \beta + \gamma > \pi \ .$$

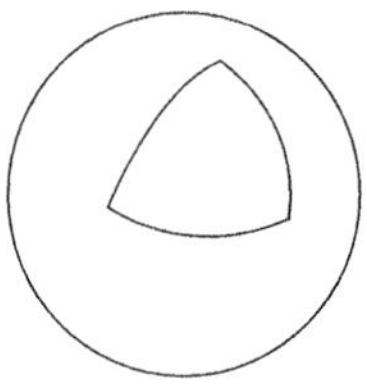

Figure 8.2 A geodesic triangle on the sphere

Example 3† (Hyperbolic plane and pseudosphere) *(This example should be omitted if the optional Examples 5 of §3.4 and 10 of §7.3 were omitted.)* As we saw in Example 10 of §7.3, the sides of a geodesic triangle in the upper half-plane model H of the hyperbolic plane are segments of lines or arcs of semicircles, where the lines and semicircles intersect the x-axis orthogonally (Figure 7.5). Since angles in the upper half-plane model of the hyperbolic plane are the same as the angles in the standard flat plane, it may be shown that for the hyperbolic plane we have

$$\alpha + \beta + \gamma < \pi \ .$$

The same holds for the pseudosphere since, as we saw in Example 5 of §3.4, the pseudosphere (minus a meridian) may be covered by a parametrisation having the same E, F and G as the open subset $\{(u, v) : -\pi < u < \pi \ , \ \ v > 1\}$ of H. Indeed (see Exercise 4.18) there is a local isometry which wraps the subset $\{(u, v) : v > 1\}$ of H round the pseudosphere an infinite number of times.

Since $K = 0$ in Example 1, $K = 1$ in Example 2 and $K = -1$ in Example 3, it would appear that, for a geodesic triangle, there is a relationship between $\alpha + \beta + \gamma - \pi$ and the Gaussian curvature K on the interior of the triangle.

8.2 Regular regions, interior angles

The Gauss–Bonnet Theorem entails the integral of the Gaussian curvature K over a particular type of region, a *regular region*, of a surface. In this section, we describe what we mean by a regular region, but first we describe the type of boundary such a region should have.

A *piecewise regular* curve is (the trace of) a continuous map $\boldsymbol{\alpha} : [a, b] \to \mathbb{R}^n$ for which there is a partition $a = t_1 < \cdots < t_n = b$ such that, for each $i = 1, \ldots, n - 1$, $\boldsymbol{\alpha}|[t_i, t_{i+1}]$ is regular; such a curve is *simple and closed* if it has no self-intersections other than that its start point $\boldsymbol{\alpha}(t_1)$ is equal to its end point $\boldsymbol{\alpha}(t_n)$. The points $\boldsymbol{\alpha}(t_1), \ldots, \boldsymbol{\alpha}(t_{n-1})$ are called the *vertices* of $\boldsymbol{\alpha}$. Figure 8.3 illustrates some piecewise regular curves.

We also need some topological definitions. If R is a subset of a surface S, then a point $p \in S$ is on the *boundary* ∂R of R if every open neighbourhood of p in S contains at least one point in R and one point not in R. The *closure* of R is the union of R and its boundary.

We may now describe those subsets of a surface S which we shall be considering in this chapter. A compact connected subset R of a surface S will be called a *regular region* if R is the closure of a bounded open subset $\breve{R}$ of S ($\breve{R}$ is called the *interior* of R), and if the boundary ∂R is a disjoint union of a finite number of piecewise regular simple closed curves. This covers a very wide class of compact subsets of surfaces; in particular, a compact connected surface is itself a regular region (with empty boundary).

If R is a regular region, the *edges* of R are the images of those intervals on which the boundary curves are regular, while the set of *vertices* of R is the union of the sets of vertices of the boundary curves. Figure 8.4 illustrates some regions which are not regular, while Figure 8.5 shows a regular region whose boundary has ten vertices and edges.

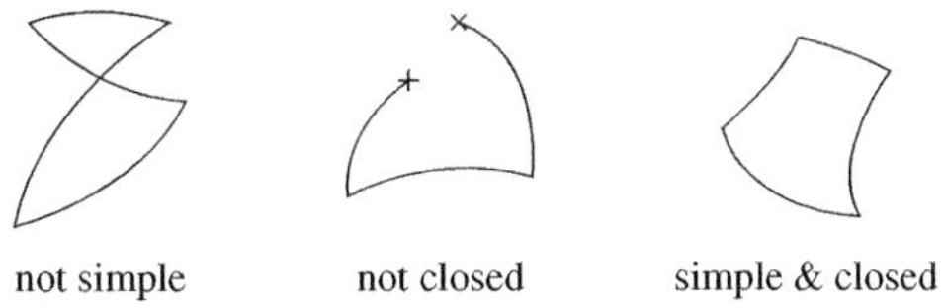

Figure 8.3 Piecewise regular curves

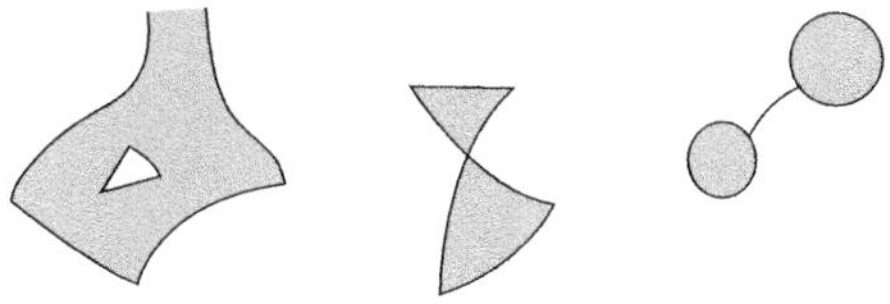

Figure 8.4 These are not regular regions

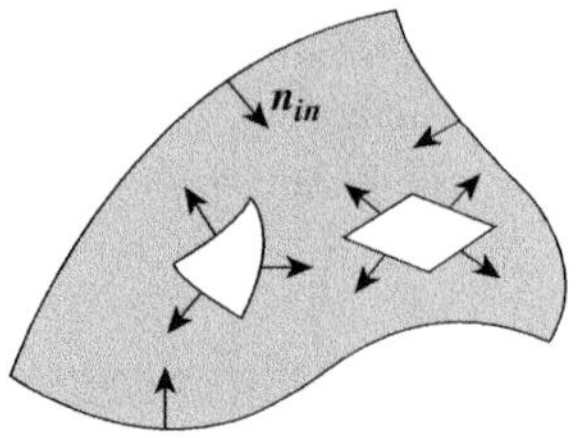

Figure 8.5 A regular region whose boundary has ten vertices and edges

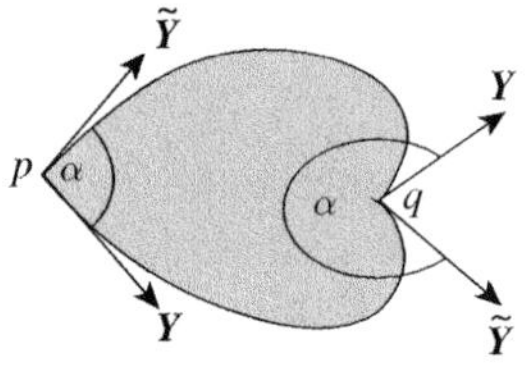

Figure 8.6 Interior angles

If $p \in \partial R$ is not a vertex, then ∂R has a well-defined tangent line at p. We let $\boldsymbol{n}_{\text{in}}(p) \in T_pS$ be the unit vector which is orthogonal to ∂R and points **into** R. Here, if $p \in \partial R$ then a vector $\boldsymbol{X} \in T_pS$ points *into* R if $\boldsymbol{X} = \boldsymbol{\beta}'(0)$ for some curve $\boldsymbol{\beta}(t)$ on S with $\boldsymbol{\beta}(0) = p$ and $\boldsymbol{\beta}(t)$ being in R for sufficiently small positive values of t.

The unit tangent vector to ∂R varies smoothly along each edge, but has discontinuities at the vertices. We now describe how to measure these discontinuities. The boundary ∂R near a vertex p is made up of two edges starting at p, and we let $\boldsymbol{Y}, \tilde{\boldsymbol{Y}} \in T_pS$ be the initial unit vectors of these edges. The two half-lines in T_pS consisting of positive scalar multiples of $\boldsymbol{Y}$ and $\tilde{\boldsymbol{Y}}$ bound two sectors of T_pS, one of which consists of vectors pointing into R.

The angle α subtended at the origin by the sector of vectors pointing into R is the *interior* angle of R at p (Figure 8.6). We assume that none of our vertices are *cusp points*; that is to say, we assume that $\boldsymbol{Y} \neq \tilde{\boldsymbol{Y}}$, so the value of α is uniquely specified by taking $0 < \alpha < 2\pi$.

By analogy with notation used in elementary geometry, we define the *exterior* angle θ at p to be given by

$$\theta = \pi - \alpha ,$$

so, in particular, $-\pi < \theta < \pi$.

We now consider the geometrical information carried by the exterior angle at a vertex p. We recall from §7.1 that an *orientation* of a surface S is a specific choice for the positive direction of rotation in each tangent space of S which varies smoothly over S; for a surface in $\mathbb{R}^3$, this is equivalent to making a smooth choice of unit normal to the surface. The *(directed) angle* from a non-zero vector $\boldsymbol{X}$ to a non-zero vector $\boldsymbol{Y}$ is then defined to be the angle of rotation, measured in the positive direction, from $\boldsymbol{X}$ to $\boldsymbol{Y}$ (so that, as usual, angle is defined up to addition of integer multiples of 2π).

An orientation defined on an open neighbourhood of a vertex p on ∂R allows us to specify a direction of travel along the two edges through p. We travel in the direction given by the unit tangent vectors to the edges which, when rotated through the angle $\pi/2$

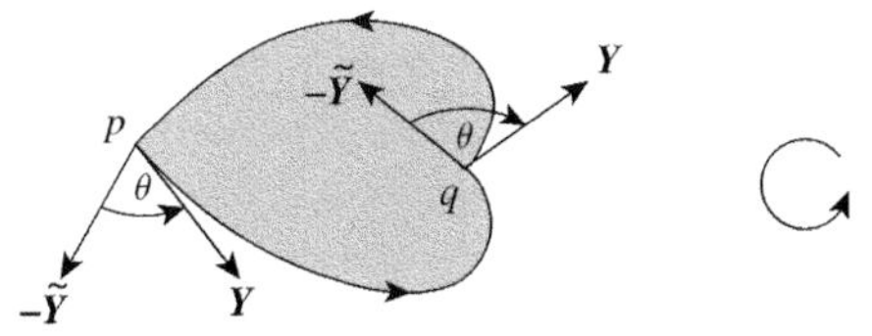

Figure 8.7 Exterior angles

(measured in the positive direction of rotation) give the inward pointing vector $\boldsymbol{n}_{\text{in}}$ (Figure 8.7). By relabelling if necessary, we may assume that $\boldsymbol{Y}$ is the outgoing unit vector at p, so that $-\tilde{\boldsymbol{Y}}$ is the incoming unit vector.

If we rotate $-\tilde{\boldsymbol{Y}}$ through angle π (measured in the positive direction of rotation) we obtain $\tilde{\boldsymbol{Y}}$, and if we rotate $\boldsymbol{Y}$ through the interior angle α we also obtain $\tilde{\boldsymbol{Y}}$. Since the exterior angle $\theta = \pi - \alpha$, we have the following lemma, which gives a geometrical interpretation of the exterior angle at a vertex.

Lemma 1 *If we choose an orientation on an open neighbourhood of a vertex p then the exterior angle θ at p is the angle of rotation (measured in the positive direction of rotation) from the incoming to the outgoing tangent vectors to the boundary at p.*

In Figure 8.7 we have redrawn Figure 8.6, marking on it the exterior angles at the two vertices for the indicated choice of orientation. The exterior angle at p is positive, but at q it is negative (the rotation is in the opposite direction to that determined by the orientation).

8.3 Gauss–Bonnet Theorem for a triangle

A *triangle* T on a surface S in $\mathbb{R}^n$ is a regular region of S whose boundary ∂T is made up of three edges, and which is homeomorphic to a closed disc (that is to say, there is a continuous bijection, whose inverse is also continuous, from T to a closed disc). Then T is connected, and, intuitively speaking, the interior of T has no holes and no points omitted (Figure 8.8).

The Gauss–Bonnet Theorem for a triangle gives a relationship between the sum of the interior angles of a triangle on a surface S in $\mathbb{R}^n$, the geodesic curvature of the sides of the triangle, and the Gaussian curvature on the interior of the triangle. In order to prove the theorem, we need to discuss the Theorem of Turning Tangents for a triangle, and to do this, we must first consider questions of orientation.

It may be shown that any triangle on S is the image under a local parametrisation $\boldsymbol{x} : D \to S$ of a triangle contained in the open unit disc $D \subset \mathbb{R}^2$, so we may use $\boldsymbol{x}(u, v)$ to give T an orientation; the positive direction of rotation being chosen so that rotation from $\boldsymbol{x}_u$ to $\boldsymbol{x}_v$ is less than π (or, for a surface in $\mathbb{R}^3$, take $\boldsymbol{N} = (\boldsymbol{x}_u \times \boldsymbol{x}_v)/|\boldsymbol{x}_u \times \boldsymbol{x}_v|$). As described in §8.2, this enables us to specify a *positive* direction of travel around ∂T; we travel in the direction given by the unit tangent vectors to the edges which, when rotated through the angle $\pi/2$ (measured in the positive direction) give the inward pointing vector $\boldsymbol{n}_{\text{in}}$ (so, for

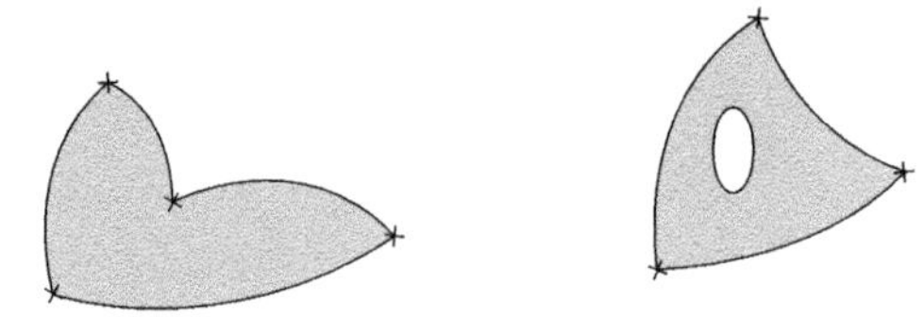

Figure 8.8 These regions are not triangles

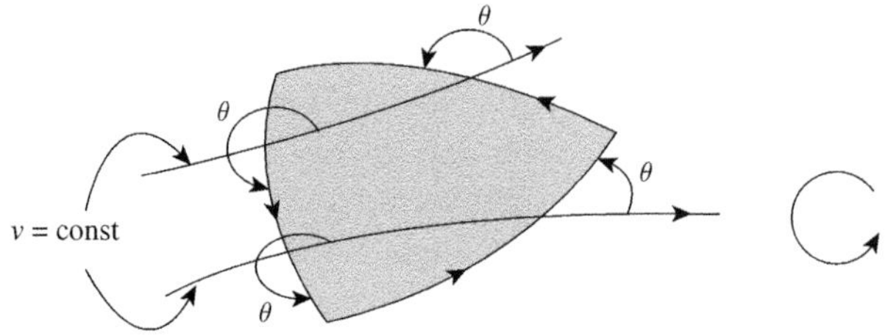

Figure 8.9 The angle θ

instance, for a triangle in the plane $z = 0$, taking $\boldsymbol{N} = (0, 0, 1)$ gives anticlockwise as the positive direction of travel).

Let T be a triangle on S and assume that $\boldsymbol{x}(u, v)$ is a local parametrisation as described above. Let θ be a function, differentiable on each of the three edges, which gives the angle from the coordinate vector field $\boldsymbol{x}_u$ to the unit tangent vector to the boundary ∂T when the boundary is traversed in the positive direction (Figure 8.9).

Then $d\theta/ds$ measures the rate of rotation of the unit tangent vector to the edge relative to $\boldsymbol{x}_u$ as we travel along each edge at unit speed, and, as noted in Lemma 1 of §8.2, the exterior angles $\theta_1, \theta_2, \theta_3$ at the vertices measure the angle at the vertex from the incoming to the outgoing tangent vectors to the boundary. It would seem reasonable that the total rotation of the tangent vector as we traverse the whole of ∂T should be through one complete revolution, namely 2π (try following the tangent vector with your pen as you travel once round the boundary of the triangle), which would mean that

$$\int_{\partial T} \frac{d\theta}{ds} ds + \theta_1 + \theta_2 + \theta_3 = 2\pi \ , \tag{8.1}$$

where $\int_{\partial T} \frac{d\theta}{ds} ds$ denotes the sum of these integrals along each of the regular curves making up ∂T. We shall use a similar convention concerning integrals round the boundary of a regular region in several of the following results.

It turns out that (8.1) is true. This is the Theorem of Turning Tangents, which we state in terms of interior angles.

Theorem 1 (Turning Tangents) *Let T be a triangle contained in the image of a local parametrisation $\boldsymbol{x} : D \to S$ of a surface S in $\mathbb{R}^n$, and let θ be the angle from the coordinate vector field $\boldsymbol{x}_u$ to the unit tangent vector to the boundary ∂T when the boundary is traversed in the positive direction as determined by the local parametrisation. If s is an arc length parameter in the positive direction for each of the edges, and if α, β, γ are the interior angles of the triangle, then*

$$\int_{\partial T} \frac{d\theta}{ds} ds = \alpha + \beta + \gamma - \pi .$$

The Theorem of Turning Tangents is due to Heinz Hopf. It is essentially a theorem in topology and, unfortunately, the proof is outside the scope of this book.

We now return to our discussion of the Gauss–Bonnet Theorem for a triangle. This theorem involves the integral round ∂T of the geodesic curvature κ_g of the regular curves making up the boundary ∂T of a triangle T. We recall that κ_g is defined only up to sign, but we can decide on the sign of κ_g by using the unit vector $\boldsymbol{n}_{\text{in}}$ pointing into the triangle described in §8.2. We define κ_g so that, if $\boldsymbol{\alpha}$ is a parametrisation of an edge of T, and if s is an arc length parameter along $\boldsymbol{\alpha}$, then the component of the acceleration vector $d^2\boldsymbol{\alpha}/ds^2$ tangential to S is equal to $\kappa_g \boldsymbol{n}_{\text{in}}$, or, in symbols,

$$\kappa_g = \frac{d^2\boldsymbol{\alpha}}{ds^2}.\boldsymbol{n}_{\text{in}} . \tag{8.2}$$

This uniquely defines the sign of κ_g since the acceleration vector at a point of an edge is independent of the direction in which we traverse the edge. If $\boldsymbol{\alpha}(t)$ parametrises an edge of T not necessarily by arc length then

$$\kappa_g = \frac{1}{|\boldsymbol{\alpha}'|^2}\boldsymbol{\alpha}''.\boldsymbol{n}_{\text{in}} . \tag{8.3}$$

We now state and prove the main result of this section.

Theorem 2 (Gauss–Bonnet Theorem for a triangle) *Let T be a triangle on a surface S in $\mathbb{R}^n$, and let α, β, γ be the interior angles of T. Then*

$$\int_{\partial T} \kappa_g ds + \iint_T K dA = \alpha + \beta + \gamma - \pi , \tag{8.4}$$

where the sign of κ_g along ∂R is determined as described above by the unit vectors $\boldsymbol{n}_{\text{in}}$ pointing into the triangle.

Proof In order to simplify the calculations in the proof, we shall assume that T is the image under an **orthogonal** local parametrisation $\boldsymbol{x} : D \to S$ of a triangle Γ contained in the open unit disc $D \subset \mathbb{R}^2$. It may be shown that such a local parametrisation exists for any triangle. We use the orientation determined by $\boldsymbol{x}$ and travel round ∂T in the direction determined by this orientation (so that, if $\boldsymbol{X}$ is the unit tangent vector in the direction of travel, then rotating $\boldsymbol{X}$ through $\pi/2$ gives $\boldsymbol{n}_{\text{in}}$) and let θ be the angle from the coordinate vector field $\boldsymbol{x}_u$ to $\boldsymbol{X}$.

Then, with the choice of sign of κ_g determined using $\boldsymbol{n}_{\text{in}}$ as described above, Proposition 2 of §7.5 states that the geodesic curvature of each edge is given by

$$\kappa_g = \frac{d\theta}{ds} + \frac{1}{2\sqrt{EG}}\left(G_u \frac{dv}{ds} - E_v \frac{du}{ds}\right) .$$

Thus, integrating round ∂T,

$$\int_{\partial T} \kappa_g ds = \int_{\partial T} \frac{d\theta}{ds} ds + \int_{\partial T} \frac{1}{2\sqrt{EG}}\left(G_u \frac{dv}{ds} - E_v \frac{du}{ds}\right) ds .$$

We now apply Green's Theorem in the plane, which states that if $P(u, v)$, $Q(u, v)$ are smooth functions defined on D then

$$\int_{\partial\Gamma}\left(P\frac{du}{ds}+Q\frac{dv}{ds}\right)ds=\iint_{\Gamma}\left(\frac{\partial Q}{\partial u}-\frac{\partial P}{\partial v}\right)du\,dv\,.$$

Hence, using the expression for K in orthogonal coordinates obtained in Example 1 of §6.2 for the second equality, we obtain

$$\begin{aligned}\int_{\partial T}&\frac{1}{2\sqrt{EG}}\left(G_u\frac{dv}{ds}-E_v\frac{du}{ds}\right)ds\\&=\iint_{\Gamma}\frac{1}{2\sqrt{EG}}\left\{\frac{\partial}{\partial u}\left(\frac{G_u}{\sqrt{EG}}\right)+\frac{\partial}{\partial v}\left(\frac{E_v}{\sqrt{EG}}\right)\right\}\sqrt{EG}\,dudv\\&=-\iint_T K\,dA\,,\end{aligned}$$

and the result follows from the Theorem of Turning Tangents. □

The following three corollaries are pertinent to the examples in §8.1.

Corollary 3 *If T is a geodesic triangle with interior angles α, β and γ, then*

$$\iint_T K\,dA=\alpha+\beta+\gamma-\pi\,.$$

The difference $\alpha+\beta+\gamma-\pi$ between the sum of the interior angles and π is called the *angular excess* of the geodesic triangle T (or, if negative, the *angular defect*).

As a second corollary, we have the following characterisation of surfaces with zero Gaussian curvature.

Corollary 4 *The Gaussian curvature of a surface is identically zero if and only if the sum of the interior angles of every geodesic triangle is equal to π.*

Proof Suppose for some point $p \in S$ that $K(p) > 0$. Then there is an open neighbourhood U of p in S such that $K > 0$ on U. But then for a geodesic triangle T contained in U with interior angles α, β, γ we have $\alpha+\beta+\gamma > \pi$ which is a contradiction. A similar argument holds if there is a point $p \in S$ such that $K(p) < 0$. □

Corollary 5 *If T is a geodesic triangle with interior angles α, β and γ on a surface S of constant curvature $K = K_0 \neq 0$, then*

$$\text{Area}(T)=\frac{1}{K_0}(\alpha+\beta+\gamma-\pi)\,.$$

In particular, if $K = \pm 1$ then $\text{Area}(T) = \pm(\alpha+\beta+\gamma-\pi)$.

The assumption that a triangle should be a region on the surface which is homeomorphic to a closed disc is very important; without it the Gauss–Bonnet Theorem for a triangle would be false. The following example (which may be omitted if preferred) illustrates some of the issues involved.

Example 6† (Cone) Although a cone has zero Gaussian curvature, if the interior of a geodesic triangle on the cone contains the vertex of the cone then $\alpha+\beta+\gamma > \pi$. Indeed, if the cone is formed as described in Example 7 of §7.3 and Exercise 7.15 by identifying the edges of a sector with angle $\phi = 2\pi - \theta_0$ then (Figure 8.10)

$$\alpha + \beta + \gamma = \pi + \theta_0 \,.$$

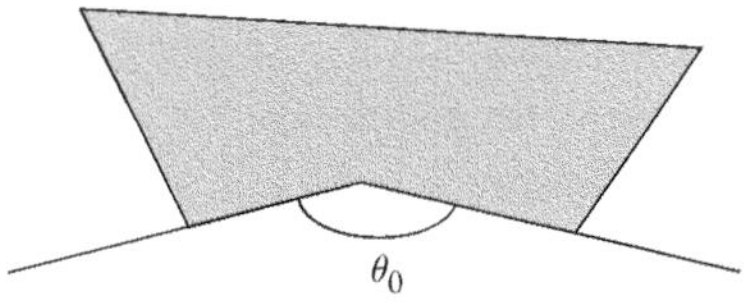

Figure 8.10 Maps to a geodesic triangle containing the vertex of a cone

This does not contradict the Gauss–Bonnet Theorem for a triangle because the interior of the geodesic triangle contains the vertex of the cone, and, as we saw in Example 4 of §3.1, if the vertex is included then the cone fails to be a surface. If we remove the vertex then T has a point omitted and so isn't homeomorphic to a disc, but we could get over this problem by modifying the cone to make it into a surface by removing a portion near the vertex and putting on a smooth cap as illustrated in Figure 8.11. Then T would once again be a geodesic triangle on the resulting surface, but we would pick up some positive Gaussian curvature from the cap. Indeed, the Gauss–Bonnet Theorem for a triangle shows that the total curvature of the cap (i.e. the integral of K over the cap) must be equal to $\alpha + \beta + \gamma - \pi = \theta_0$.

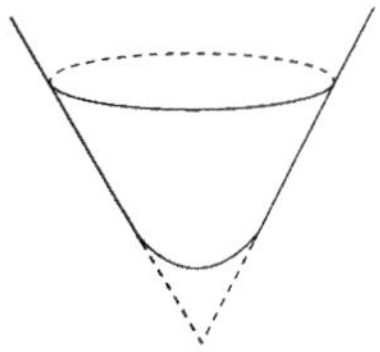

Figure 8.11 Smoothing the vertex with a smooth cap

The above example illustrates how important it is that none of the points in the triangle should be singular points of the surface, and that the triangle must be homeomorphic to a closed disc.

8.4 Classification of surfaces

In this section we describe how a regular region of a surface may be partitioned into triangles, and then use this to define an integer, the *Euler characteristic* of the region, which turns out to be independent of the way the region is partitioned. This will enable us in the next section to use the Gauss–Bonnet Theorem for a triangle to prove the Gauss–Bonnet

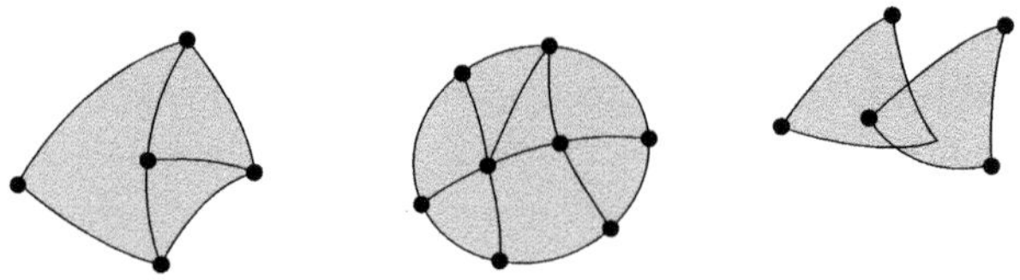

Figure 8.12 These are not triangulations

Theorem for a regular region of a surface. The treatment of the topology in this chapter is, of necessity, not as detailed as in a specialist topology textbook. Rather, we seek to give an intuitive idea of the concepts involved.

So, let R be a subset of a surface S which can be *triangulated*; that is to say, R can be decomposed as a union of a finite number of triangles any two of which are disjoint, or meet in a common edge, or meet in a common vertex. Such a decomposition is called a *triangulation*; Figure 8.12 illustrates some decompositions which are not triangulations.

Example 1 (Closed disc) This may be triangulated as in Figure 8.13.

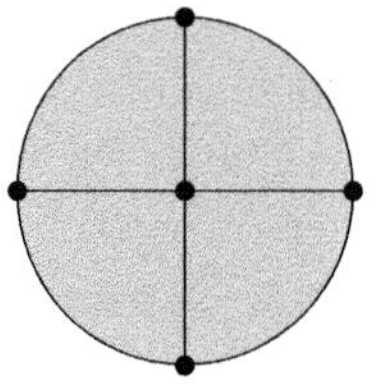

Figure 8.13 Triangulation of a closed disc

Example 2 (Sphere) This may be triangulated as in Figure 8.14.

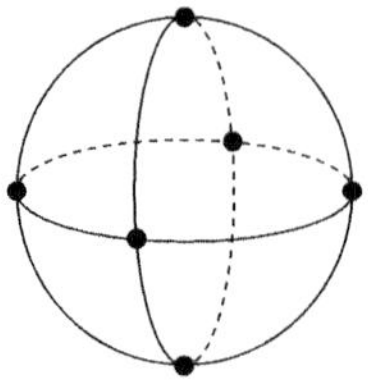

Figure 8.14 Triangulation of a sphere

Example 3 (Annulus and cylinder of finite length) The annulus may be triangulated as in Figure 8.15. Also, by putting this example into perspective, we obtain a triangulation of that part

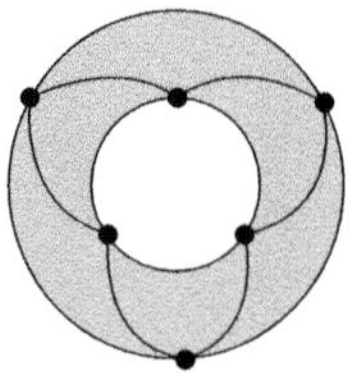

Figure 8.15 Triangulation of an annulus

of the cylinder $x^2 + y^2 = r^2$ between the planes $z = a$ and $z = b$. Here, a, b and $r > 0$ are real constants.

We quote the following topological theorem.

Theorem 4 *Every regular region of a surface can be triangulated, although, of course, in many different ways. However, if we denote by F the number of triangles (often called faces), by E the number of edges, and by V the number of vertices, then $V - E + F$ is independent of the triangulation.*

The integer $V - E + F$ is called the *Euler characteristic* $\chi(R)$ of the regular region R. The above theorem assures us that this is independent of the way we triangulate R.

We see from the triangulations given earlier that

$$\chi(\text{disc}) = 5 - 8 + 4 = 1\,,$$
$$\chi(\text{sphere}) = 6 - 12 + 8 = 2\,,$$
$$\chi(\text{annulus}) = \chi(\text{finite cylinder}) = 6 - 12 + 6 = 0\,.$$

We have now covered the topology required to understand the Gauss–Bonnet Theorem and its proof. We need a little more, however, to appreciate the consequences of the Gauss–Bonnet Theorem discussed in §8.6.

We first recall from §4.3 that if S and $\tilde{S}$ are surfaces, then a smooth bijective map $f : S \to \tilde{S}$ is called a *diffeomorphism* if its inverse map is also smooth. If such a map f exists, then S and $\tilde{S}$ are said to be *diffeomorphic*. As far as properties concerning differentiability are concerned, diffeomorphic surfaces are essentially indistinguishable.

It is a remarkable fact that, up to diffeomorphism, a compact connected surface in $\mathbb{R}^n$ is determined by just two pieces of information, namely its Euler characteristic and whether or not it is orientable. This is known as the Classification Theorem for compact surfaces, and we state it without proof.

Theorem 5 (Classification Theorem for compact surfaces)

(i) *A compact connected surface S may be triangulated, and $\chi(S) \leq 2$.*
(ii) *Two compact connected surfaces are diffeomorphic if and only if they have the same Euler characteristic and are either both orientable or both non-orientable.*
(iii) *A compact connected orientable surface has Euler characteristic 2, 0 or a negative even integer.*

We can say rather more for surfaces in $\mathbb{R}^3$ by using a deep result (quoted as Theorem 3 of §5.12) on compact connected surfaces in $\mathbb{R}^3$ which have no self-intersections; namely that such a surface S divides $\mathbb{R}^3$ into two connected components, an unbounded piece (the *outside*) and a bounded piece (the *inside*). It follows that S is orientable, since a unit normal may be assigned smoothly over the whole of the surface (either the outward unit normal or the inward unit normal). Thus, using the Classification Theorem for compact surfaces, we have the following.

Theorem 6 *A compact connected surface S (without self-intersections) in $\mathbb{R}^3$ is orientable. In particular:*

(i) its Euler characteristic is 2, 0 or a negative even integer;
(ii) two such surfaces are diffeomorphic if and only if they have the same Euler characteristic.

In the final part of this section, we attempt to give some insight into the structure of compact connected orientable surfaces by describing how to construct such a surface for every possible Euler characteristic $\chi(S) = 2, 0, -2, \ldots$. This material is optional, since it is not necessary for an understanding of the Gauss–Bonnet Theorem nor for the applications we cover.

We first describe a process for constructing a new surface from a given one by "adding a handle". To do this, we cut two discs from a surface S and attach a cylinder of finite length as illustrated in Figure 8.16. For instance, if we add a handle to a sphere then we obtain a torus, and adding a further handle gives a double torus. Adding a third handle gives us a pretzel (Figure 8.17).

We now describe how the process of adding a handle to a compact surface S changes the Euler characteristic. If we triangulate a cylinder of finite length as in Example 3, and triangulate S so that each of the removed discs is the face of a triangle (with all interior angles equal to π), then the number of faces of the resulting triangulation is two fewer than the number of the faces of the triangulation of S added to the number of the faces of the triangulation of the cylinder of finite length, while the number of edges (resp. vertices) is six fewer than the number of edges (resp. vertices) of the triangulation of S added to the number of the edges (resp. vertices) of the triangulation of the cylinder of finite length.

We obtain the following lemma by recalling that the Euler characteristic of a cylinder of finite length is zero.

Lemma 7 *Let $\tilde{S}$ be the surface obtained by adding a handle to a compact surface S. Then*

$$\chi(\tilde{S}) = \chi(S) - 2 .$$

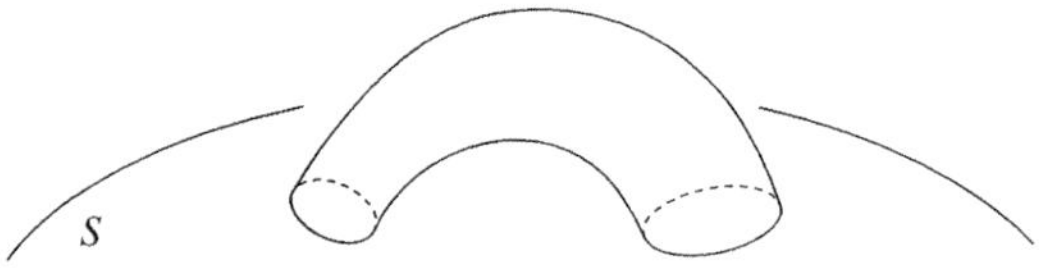

Figure 8.16 Adding a handle

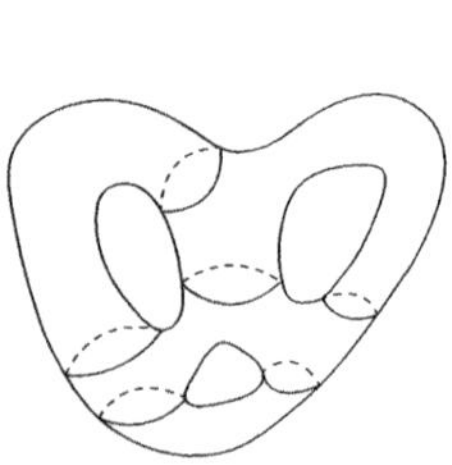

Figure 8.17 A pretzel

We have seen that the Euler characteristic of the sphere is equal to two, so, for any positive integer g, we may construct a compact surface of Euler characteristic $2-2g$ by adding g handles to the sphere. We now see, for instance, that the torus has Euler characteristic 0, the double torus has Euler characteristic -2, while the pretzel has Euler characteristic -4. These results may, with some difficulty, be checked by producing suitable triangulations of the various surfaces.

It follows from the Classification Theorem for compact surfaces that every compact connected orientable surface S (and so, in particular, every compact connected surface in $\mathbb{R}^3$) is diffeomorphic to a sphere with a certain number of handles attached. This number is called the *genus* of the surface. So, if S has genus g then

$$\chi(S) = 2 - 2g\ .$$

For example, a torus has genus one, and a pretzel has genus three.

8.5 The Gauss–Bonnet Theorem

In this section we use the Gauss–Bonnet Theorem for a triangle to prove the Gauss–Bonnet Theorem for a regular region. We give some consequences of the theorem in the next section.

Let R be a regular region of a surface S in $\mathbb{R}^n$ whose boundary ∂R is made up of m edges, meeting at vertices $\{v_1, \ldots, v_m\}$, say, on ∂R. For each vertex v_j, $j = 1, \ldots, m$, we may define the exterior angle θ_j at v_j using the process described in §8.2. As before, we assume that ∂R has no cusps, so that $-\pi < \theta_j < \pi$.

Theorem 1 (Gauss–Bonnet Theorem) *Let R be a regular region of a surface S in $\mathbb{R}^n$. If $\theta_1, \ldots, \theta_m$ are the exterior angles at the vertices of ∂R then*

$$\int_{\partial R} \kappa_g ds + \iint_R K dA + \sum_{j=1}^{m} \theta_j = 2\pi\chi(R)\,,$$

where the sign of κ_g along ∂R is determined by the unit vectors $\mathbf{n}_{\text{in}}$ pointing into the region.

In particular, if S is a compact surface then

$$\iint_S K dA = 2\pi\chi(S)\,.$$

Before we prove the theorem, we would like to point out what an amazing theorem it is. For instance, consider the final statement. The Gaussian curvature K is a geometric quantity, very sensitive to deformations of the surface, while the Euler characteristic is a topological property so doesn't change as the surface is deformed. So, for example, however we may pull or stretch a sphere (thought of as being made of a rubber membrane) to form a surface S then the total Gaussian curvature of the deformed sphere (ie $\iint_S K dA$) will stay fixed at 4π.

Proof Let V, E and F be the numbers of vertices, edges and triangles respectively of a triangulation of R, and label the triangles as $\Delta_1, \ldots, \Delta_F$. For $i = 1, \ldots, F$, let α_i, β_i and γ_i denote the interior angles of the triangle Δ_i.

Those edges and vertices of the triangulation on the boundary of R will be called *external edges* and *external vertices*, the other edges and vertices will be called *internal*. Every vertex of ∂R is an external vertex of the triangulation, and we shall regard every other external vertex of the triangulation as a vertex of ∂R with exterior angle equal to 0. The sum of the exterior angles at the external vertices of the triangulation is then equal to $\sum_{j=1}^{m} \theta_j$, the sum of the exterior angles at the vertices of ∂R.

We apply the Gauss–Bonnet Theorem for a triangle to each of the triangles Δ_i in the triangulation, and add the results. We note that each internal edge is traversed twice, but the two integrals of κ_g along each internal edge cancel each other since the inward pointing normals are in opposite directions (Figure 8.18), so the corresponding geodesic curvatures have opposite signs.

Hence, we find that

$$\int_{\partial R} \kappa_g ds + \iint_R K dA = \sum_{i=1}^{F} (\alpha_i + \beta_i + \gamma_i) - F\pi \, . \tag{8.5}$$

Suppose that the triangulation has M external edges and hence M external vertices. If v_i is an external vertex of the triangulation, the sum of the interior angles at v_i of those triangles with a vertex at v_i gives the interior angle of the boundary at v_i. Hence, this sum, when added to the exterior angle of the boundary at v_i, gives π.

Since the interior angles at each internal vertex add up to 2π, we have

$$\sum_{i=1}^{F} (\alpha_i + \beta_i + \gamma_i) + \sum_{j=1}^{m} \theta_j = 2\pi(V - M) + \pi M \, ,$$

so that

$$2\pi V = \sum_{i=1}^{F} (\alpha_i + \beta_i + \gamma_i) + \sum_{j=1}^{m} \theta_j + \pi M \, .$$

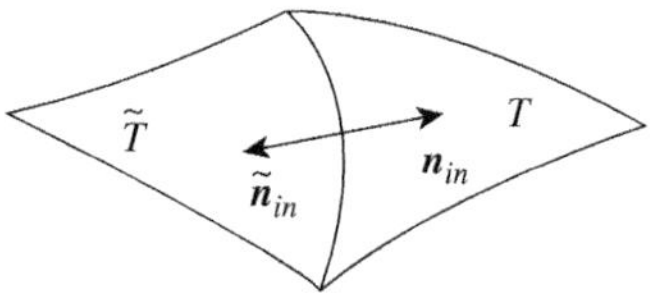

Figure 8.18 Integrals along internal edges cancel

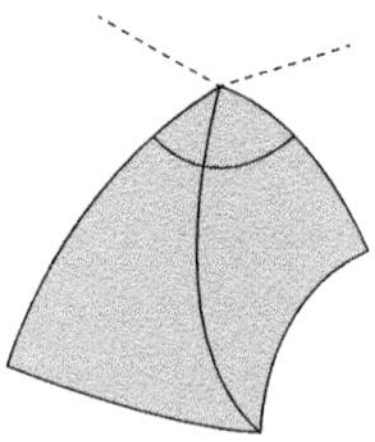

Figure 8.19 Interior angles at external vertex

Thus, using (8.5),

$$\int_{\partial R} \kappa_g ds + \iint_R K dA + \sum_{j=1}^{m} \theta_j = 2\pi V - \pi(F + M) . \tag{8.6}$$

We next note that each triangle has three edges, each internal edge belongs to exactly two triangles, and each external edge belongs to exactly one triangle. Thus

$$3F = 2(E - M) + M = 2E - M ,$$

so that

$$F + M = 2(E - F) ,$$

and the result follows from (8.6) and the definition of Euler characteristic. □

8.6 Consequences of Gauss–Bonnet

We now give some examples of the way in which the Gauss–Bonnet Theorem may be used to establish other results in differential geometry. Most of these results indicate how the Gaussian curvature of a surface affects the topology and the behaviour of the geodesics. A notable exception is the remarkable theorem of Jacobi (Theorem 7) on closed curves in $\mathbb{R}^3$.

The first result is specifically for surfaces in $\mathbb{R}^3$; the flat torus in $\mathbb{R}^4$ shows that the result doesn't hold for general compact connected orientable surfaces. The other results hold more generally for surfaces in $\mathbb{R}^n$.

Theorem 1 *Let S be a compact connected surface in $\mathbb{R}^3$. If $K \geq 0$ everywhere on S, then S is diffeomorphic to the sphere $S^2(1)$.*

Proof Since S is compact, Theorem 4 of §5.10 tells us that S must have an elliptic point (i.e. a point where $K > 0$), so, since K is continuous, it follows that if $K \geq 0$ everywhere on S then $\iint_S K dA > 0$. Hence, by the Gauss–Bonnet Theorem, $\chi(S) > 0$. It now follows from Theorem 6 of §8.4 that S is diffeomorphic to $S^2(1)$. □

The next theorem is easily proved using some of the ideas used in the proof of the previous theorem. Just use Theorem 5 (rather than Theorem 6) of §8.4.

Theorem 2 *Let S be a compact connected orientable surface in $\mathbb{R}^n$.*

(i) If $K > 0$ everywhere on S then S is diffeomorphic to $S^2(1)$.
(ii) If $K = 0$ everywhere on S then S is diffeomorphic to a torus.
(iii) If $K < 0$ everywhere on S then the Euler characteristic $\chi(S)$ of S is a negative even integer.

We note that surfaces satisfying either (ii) or (iii) above cannot be surfaces in $\mathbb{R}^3$.

Theorem 3 *Let S be a surface in $\mathbb{R}^n$ with $K \leq 0$ everywhere. Then any two geodesics γ_1, γ_2 through a point $p \in S$ cannot meet at a point $q \neq p$ in such a way that the traces of γ_1, γ_2 between p and q form the boundary of a regular region of S homeomorphic to a closed disc. In particular there does not exist a closed geodesic on S which bounds a regular region of S homeomorphic to a closed disc.*

Proof We assume that the traces of γ_1, γ_2 bound a regular region R of S homeomorphic to a closed disc. The uniqueness theorem for geodesics (Theorem 2 of §7.3) shows that neither of the two vertices of R are cusps, so that the exterior angles θ_1 and θ_2 of R at p and q satisfy $-\pi < \theta_j < \pi$, $j = 1, 2$. However, if R is homeomorphic to a closed disc, then the Gauss–Bonnet Theorem would imply

$$\iint_R K\,dA + \theta_1 + \theta_2 = 2\pi\chi(R) = 2\pi,$$

so that $\iint_R K\,dA > 0$. But this contradicts $K \leq 0$, and establishes the theorem. □

In order to prove our next three theorems we shall need the *Jordan Curve Theorem*, the statement of which is illustrated in Figure 8.20.

Theorem 4 (Jordan Curve Theorem) *Let Γ be the trace of a piecewise regular simple closed curve in $\mathbb{R}^2$. Then $\mathbb{R}^2 \setminus \Gamma$ is the disjoint union of two connected sets, one of which, the inside $\check{R}$, is bounded. Moreover, Γ is the boundary of $\check{R}$, and $R = \check{R} \cup \Gamma$ is a regular region homeomorphic to a closed disc.*

Theorem 5 *Let S be a compact connected orientable surface in $\mathbb{R}^n$ with $K > 0$ everywhere. Then any two simple closed geodesics on S intersect.*

Proof By Theorem 2, S is diffeomorphic to $S^2(1)$. Suppose the traces of the geodesics are Γ_1 and Γ_2. If they do not intersect, then the Jordan Curve Theorem shows that Γ_1 and Γ_2 form the boundary of a regular region R which is homeomorphic to an annulus. But then

$$\iint_R K\,dA = 2\pi\chi(R) = 0,$$

which contradicts $K > 0$. □

Theorem 6 *Let S be a surface in $\mathbb{R}^n$ with $K < 0$ everywhere which is diffeomorphic to a cylinder. Then S has at most one simple closed geodesic.*

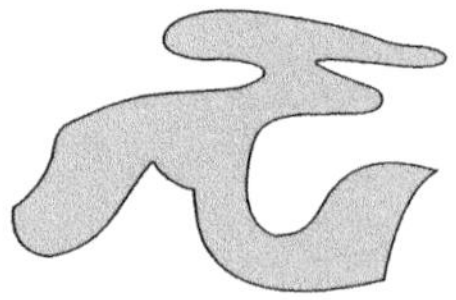

Figure 8.20 The inside of a simple closed curve

Proof Suppose that there exists a simple closed geodesic with trace Γ. Since S is diffeomorphic to a cylinder there exists a diffeomorphism $f : S \to \mathbb{R}^2 \setminus \{(0,0)\}$ and, by the Jordan Curve Theorem, $f(\Gamma)$ is the boundary of a regular region Q of $\mathbb{R}^2$ homeomorphic to a closed disc. We note that, by Theorem 3, Q must contain $(0,0)$.

If there is a second simple closed geodesic with trace Γ', say, then $f(\Gamma')$ is also the boundary of a regular region Q' of $\mathbb{R}^2$ containing $(0,0)$ which is homeomorphic to a disc. If Γ and Γ' intersect then they must do so in at least two points since Γ and Γ' cannot meet tangentially.

Thus there exist points $p_1, p_2 \in \Gamma \cap \Gamma'$ with the property that traversing $f(\Gamma)$ in an anticlockwise direction from $f(p_1)$, the next point in $f(\Gamma) \cap f(\Gamma')$ is $f(p_2)$. But then (Figure 8.21) suitable arcs of $f(\Gamma)$, $f(\Gamma')$ joining $f(p_1)$ to $f(p_2)$ form the trace of a piecewise regular simple closed curve bounding a regular region not containing $(0,0)$ homeomorphic to a disc. However, this is impossible by Theorem 3.

Thus Γ and Γ' do not intersect. But then one of Q, Q' is contained in the inside of the other, so there exists an annular region P in the plane having boundary components $f(\Gamma)$ and $f(\Gamma')$. But then, since $K < 0$, the Gauss–Bonnet Theorem implies that

$$0 = 2\pi\chi(P) = 2\pi\chi(f^{-1}(P)) = \iint_{f^{-1}(P)} K\, dA < 0\,,$$

a contradiction. □

The example of the round cylinder shows that Theorem 6 is false if we only assume that $K \le 0$ everywhere. Also note that a surface diffeomorphic to a cylinder with $K < 0$ everywhere might have no simple closed geodesics (for example, the surface of revolution in $\mathbb{R}^3$ obtained by rotating the curve $v \mapsto (v, 0, 1/v)$, $v > 0$ around the z-axis), or one simple closed geodesic (for example, the catenoid).

We now discuss the final theorem of the chapter.

Theorem 7 (Jacobi) *Let $\boldsymbol{\alpha} : [a,b] \to \mathbb{R}^3$ be a closed curve with nowhere vanishing curvature κ, and suppose that the curve $\boldsymbol{n} : [a,b] \to S^2(1)$ defined by the principal normal to $\boldsymbol{\alpha}$ is simple. Then the trace of $\boldsymbol{n}$ divides $S^2(1)$ into two regular regions of equal area.*

Proof Assume that $\boldsymbol{\alpha}$ is parametrised by arc length s, and let $\boldsymbol{t} = d\boldsymbol{\alpha}/ds$, $\boldsymbol{b} = \boldsymbol{t} \times \boldsymbol{n}$. Let κ and τ denote the curvature and torsion of $\boldsymbol{\alpha}$.

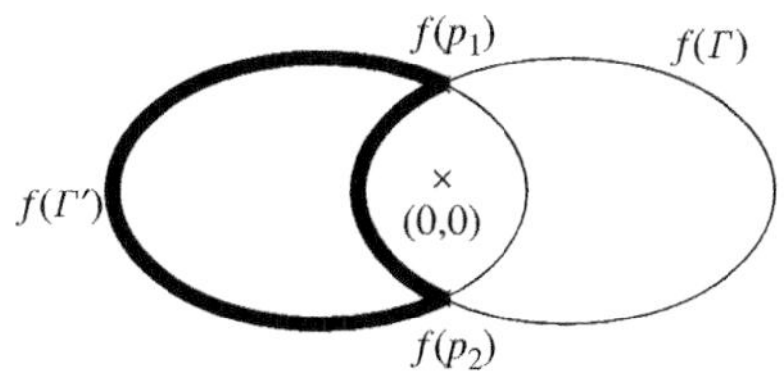

Figure 8.21 The image of intersecting closed geodesics

Then, letting $\tilde{s}$ be the arc length parameter of $\boldsymbol{n}$, we find that

$$\frac{d\boldsymbol{n}}{d\tilde{s}} = \frac{d\boldsymbol{n}}{ds}\frac{ds}{d\tilde{s}} = (-\kappa\boldsymbol{t} - \tau\boldsymbol{b})\frac{ds}{d\tilde{s}},$$

from which it follows that

$$\left(\frac{ds}{d\tilde{s}}\right)^2 = \frac{1}{\kappa^2 + \tau^2}.$$

If we let $'$ denote differentiation with respect to arc length s along $\boldsymbol{\alpha}$, then it follows from formula (7.2) for geodesic curvature that the geodesic curvature $\tilde{\kappa}_g$ of $\boldsymbol{n}$ on $S^2(1)$ is given by

$$\begin{aligned}
\tilde{\kappa}_g &= \left(\frac{ds}{d\tilde{s}}\right)^3 \boldsymbol{n}''.(\boldsymbol{n} \times \boldsymbol{n}') \\
&= \left(\frac{ds}{d\tilde{s}}\right)^3 (-\kappa'\boldsymbol{t} - \tau'\boldsymbol{b}).(\boldsymbol{n} \times (-\kappa\boldsymbol{t} - \tau\boldsymbol{b})) \\
&= \frac{\kappa'\tau - \kappa\tau'}{\kappa^2 + \tau^2}\frac{ds}{d\tilde{s}} \\
&= -\frac{d}{d\tilde{s}}\arctan\left(\frac{\tau}{\kappa}\right).
\end{aligned}$$

Thus, since $\boldsymbol{n}$ is a closed curve, we have

$$\int_a^b \tilde{\kappa}_g(\tilde{s})d\tilde{s} = 0\,.$$

We next note that, by the Jordan Curve Theorem, the trace of $\boldsymbol{n}$ divides $S^2(1)$ into two regular regions, each homeomorphic to a disc. Hence, applying the Gauss–Bonnet Theorem to one of the regions R into which $S^2(1)$ is divided by the trace of $\boldsymbol{n}$, we see that the area $A(R)$ of R is given by

$$A(R) = \iint_R dA = \iint_R K dA = 2\pi = \frac{1}{2}A\left(S^2(1)\right). \qquad \square$$

Exercises

8.1 Explain why each of the sets illustrated in Figure 8.4 is not a regular region.

8.2 The boundary of the shaded region of the plane illustrated in Figure 8.22 is made up of line segments meeting orthogonally. What are the values of the interior and exterior angles at each of the vertices v_1, v_2, v_3 and v_4? For one of the two possible orientations of the region, indicate the exterior angles from the incoming vector to the outgoing vector at each of v_1, v_2, v_3 and v_4.

8.3 Using the Theorem of Turning Tangents (Theorem 1 of §8.3), formulate and prove a similar theorem for the image under a local parametrisation of an n-gon in the plane.

8.4 Let b be a positive real number, and let

$$v \mapsto (f(v), 0, g(v))\,, \quad f(0) = 0\,, \quad f(v) > 0\ \forall v \in (0, b),$$

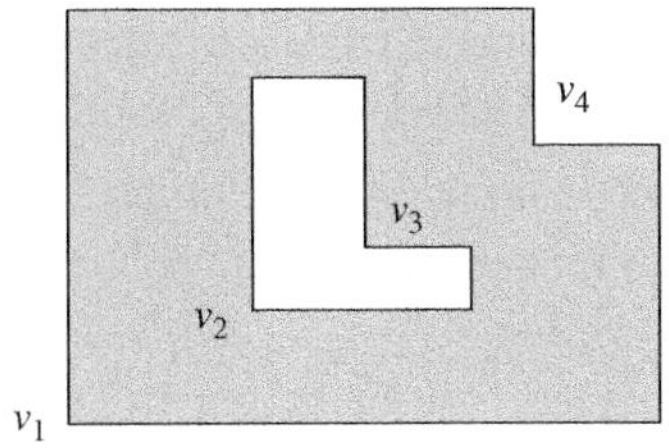

Figure 8.22 The region for Exercise 8.2

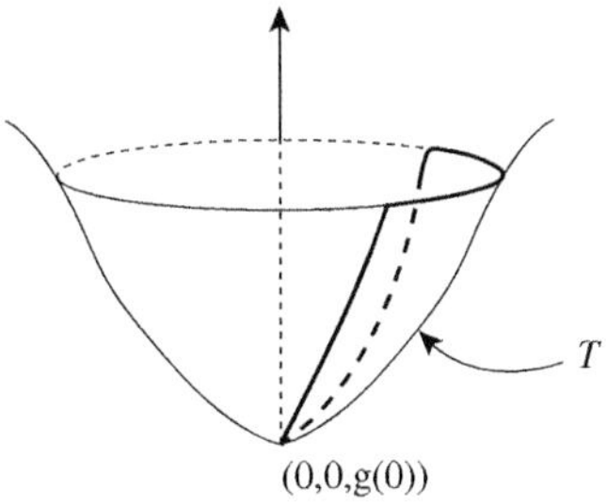

Figure 8.23 The triangle T in Exercise 8.4

be a regular curve parametrised by arc length. Let S be the surface of revolution obtained by rotating this curve about the z-axis (Figure 8.23) (note that S has a pole at $p = (0, 0, g(0))$), and let

$$\boldsymbol{x}(u, v) = (f(v)\cos u, f(v)\sin u, g(v)), \quad u \in (-\pi, \pi),$$

be the standard local parametrisation of S. Finally, if $u_0, u_1 \in (-\pi, \pi)$, with $u_0 < u_1$ and if $v_1 \in (0, b)$, let T denote the triangle which is the image under $\boldsymbol{x}$ of the set

$$\{(u, v) : u_0 \le u \le u_1,\ 0 \le v \le v_1\}$$

(you should remove the pole from T if this is a singular point of S).

(i) Show that

$$\int_{\partial T} \kappa_g ds = (u_1 - u_0) f'(v_1)\,.$$

(ii) If the pole is not a singular point of S, show that

$$\iint_T K dA = (u_1 - u_0)\left(1 - f'(v_1)\right).$$

(iii) Verify that if the pole is not a singular point of S then the Gauss–Bonnet Theorem for a triangle holds for T.

8.5 Let S be a compact connected surface in $\mathbb{R}^3$ which is not diffeomorphic to a sphere. Show that S has points where the Gaussian curvature is positive, points where it is negative, and points where it is zero.

8.6 Find the Gaussian curvature K at all points on the surface S with equation $x^2 + y^2 = z$. If a is a positive real number, evaluate $\iint_R K dA$, where R is the region of S

between the planes $z = 0$ and $z = a^2$. Find also $\int_{\partial R} \kappa_g ds$ and hence verify the Gauss–Bonnet Theorem directly for the region R.

8.7 Let S be the surface of revolution obtained by rotating the curve $(f(v), 0, g(v))$, $f(v) > 0\ \forall v$, about the z-axis, and let R be the region of S corresponding to taking $v_0 \leq v \leq v_1$. Evaluate $\iint_R K dA$ and $\int_{\partial R} \kappa_g ds$, and hence verify the Gauss–Bonnet Theorem directly for the region R. (The formulae are easier if you assume that the generating curve is parametrised by arc length.)

8.8 Find all the simple closed geodesics on the surface S in $\mathbb{R}^3$ defined by the equation

$$\frac{x^2}{a^2} + \frac{y^2}{b^2} = \cosh^2 z.$$

8.9 Show that any two simple closed geodesics on the surface S in $\mathbb{R}^3$ defined by the equation

$$x^4 + y^4 + z^4 = 1 \tag{8.7}$$

intersect. (*Hint:* this is similar to the case of a compact surface in $\mathbb{R}^3$ with $K > 0$, except that, as noted in Exercise 5.26, the surface given by (8.7) has points where $K = 0$.)

9 Minimal and CMC surfaces

We have now achieved our main aims in writing this book as described in the Preface. However, there are many other attractive and visual areas of the geometry of surfaces in $\mathbb{R}^3$ which we have not covered, and in this final chapter we explore our nomination for the best of these areas.

When a wire frame is dipped into soapy water and removed, the surface formed by the soap film which spans the frame minimises the area of all nearby surfaces which span the frame. These are examples of *minimal surfaces*, and, as our first example, it is clear from the physical description that (any bounded open subset of) a plane is a minimal surface.

For ease of study, minimal surfaces are defined to be surfaces which are **stationary** points of the area functional, not necessarily local minima. Loosely speaking, a minimal surface is one for which every bounded subset U is a stationary point of the area functional applied to all surfaces with the same boundary as U. In this regard, minimal surfaces are the 2-dimensional analogues of geodesics, since Theorem 3 of §7.4 characterises geodesics on a surface S as curves on S for which every bounded piece is a stationary point of the length functional applied to all curves on S with the same start point and end point as the piece. The question of whether a given minimal surface is actually area **minimising** is a difficult one, and is part of an active area of current research.

As well as occurring naturally as soap film surfaces, minimal surfaces also have several practical applications. For instance, they have been utilised in architecture; roofs made up of pieces modelled on soap films have been used to provide lightweight but very strong coverings over large areas.

The plane is an example of a minimal surface, and, as we shall see, so are catenoids and helicoids. However, a soap bubble (that is to say, a sphere) is not a minimal surface; although it minimises area, it does so subject to the constraint that it encloses a fixed volume of air. This constrained minimising problem leads to the study of surfaces of constant mean curvature, or *CMC surfaces*.

Minimal and CMC surfaces in $\mathbb{R}^3$ are often very beautiful, and there is an extensive visual library of these surfaces on the internet. The illustrations in this chapter are taken from images obtained by Katrin Leschke, using jReality, for her surface visualisation programme. We are grateful to Katrin for allowing us to use her images.

In this chapter we study the mathematics of minimal and CMC surfaces in $\mathbb{R}^3$. Surprisingly and excitingly, these surfaces may be studied using one of the most powerful and interesting areas of mathematics, namely complex analysis. A balanced course on the differential geometry of surfaces at the level we have aimed for could perhaps contain the material in the first five sections of this chapter; up to and including the material

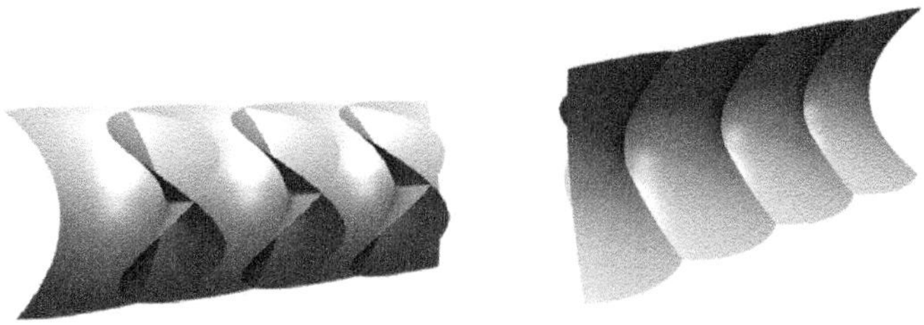

Figure 9.1 Two views from opposite sides of Catalan's surface

Figure 9.2 Two views of Enneper's surface

on associated families. In addition to the usual time constraints, it might be difficult to include the topics in the subsequent sections since they are rather more advanced and mathematically sophisticated than those previously considered in this book. These sections are intended to give an indication of the mathematical beauty that may be achieved with further study, and would be suitable for self-study by an interested student or could form the basis of a follow-on project for senior undergraduates or first year postgraduates.

In this chapter we shall encounter surfaces with self-intersections. Figures 9.1 and 9.2 show (finite parts of) two such minimal surfaces. One way that surfaces with self-intersections arise is as the image $S = \boldsymbol{x}(U)$ of an $\mathbb{R}^3$-valued smooth, but not necessarily injective, map $\boldsymbol{x}(u, v)$ defined on an open subset U of $\mathbb{R}^2$ with the property that $\boldsymbol{x}_u$ and $\boldsymbol{x}_v$ are linearly independent at each point of U. Theorem 1 of §2.5 on coordinate recognition shows that each point $(u, v) \in U$ has an open neighbourhood V in U such that $\boldsymbol{x}(V)$ is a surface parametrised by $\boldsymbol{x}|V$. This situation is the analogue of our study in Chapter 1 of regular (parametrised) curves in the plane, since these may also have self-intersections. In particular, for each $(u, v) \in U$, we have the corresponding *tangent plane* to S spanned by $\boldsymbol{x}_u(u, v)$ and $\boldsymbol{x}_v(u, v)$, and the corresponding *unit normal* $\boldsymbol{N} = \boldsymbol{x}_u \times \boldsymbol{x}_v / |\boldsymbol{x}_u \times \boldsymbol{x}_v|$. This means that we can discuss the first and second fundamental forms of S, and related concepts like area, Gaussian and mean curvatures, geodesics, and so on.

9.1 Normal variations

We wish to characterise surfaces which minimise area among all nearby surfaces (although, as mentioned above, we actually characterise those surfaces which are stationary points of the area functional, but not necessarily local **minimisers** of area). We do this by considering

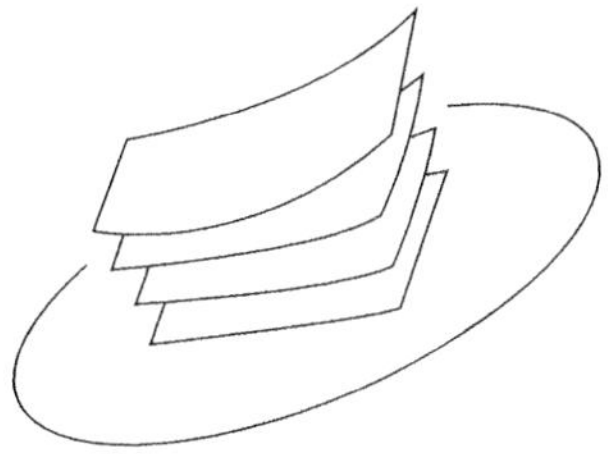

Figure 9.3 Normal variations

how the area of a piece of surface changes as we carry out appropriate deformations. This section may be regarded as a 2-dimensional analogue of §7.4, in which we considered variations of geodesics.

We begin by giving a mathematical description of the process of deforming a surface. To simplify the exposition while retaining the essential ideas, we restrict ourselves to considering certain deformations of the surface which may be described easily using local parametrisations.

Let $\boldsymbol{x} : U \to \mathbb{R}^3$ be a local parametrisation of a surface S in $\mathbb{R}^3$, and let $D \subset U \subseteq \mathbb{R}^2$ be a bounded open subset whose closure $\bar{D}$ is contained in U. If $h : \bar{D} \to \mathbb{R}$ is a smooth function, then the *normal variation of* $\boldsymbol{x}(\bar{D})$ *determined by* h is the family of maps $\boldsymbol{x}^r : \bar{D} \to \mathbb{R}^3$ given by (Figure 9.3)

$$\boldsymbol{x}^r(u, v) = \boldsymbol{x}(u, v) + rh(u, v)\boldsymbol{N}(u, v), \quad (u, v) \in \bar{D},\ r \in \mathbb{R}. \tag{9.1}$$

Here, as usual, $\boldsymbol{N}$ is the unit normal in the direction of $\boldsymbol{x}_u \times \boldsymbol{x}_v$.

We note that

$$\begin{aligned} (\boldsymbol{x}^r)_u &= \boldsymbol{x}_u + r(h\boldsymbol{N}_u + h_u\boldsymbol{N}), \\ (\boldsymbol{x}^r)_v &= \boldsymbol{x}_v + r(h\boldsymbol{N}_v + h_v\boldsymbol{N}), \end{aligned} \tag{9.2}$$

so that

$$(\boldsymbol{x}^r)_u \times (\boldsymbol{x}^r)_v = \boldsymbol{x}_u \times \boldsymbol{x}_v + R,$$

where $\lim_{r\to 0} R = 0$ for each u and v. It follows that there exists a sufficiently small $\epsilon > 0$ such that the vectors $(\boldsymbol{x}^r)_u$ and $(\boldsymbol{x}^r)_v$ are linearly independent for each $r \in (-\epsilon, \epsilon)$ and each $(u, v) \in D$, so, as discussed at the end of the introduction to this chapter, it makes sense to discuss the area $A(r)$ of $\boldsymbol{x}^r(\bar{D})$.

We say that S is a *minimal surface* if $A'(0) = 0$ for all normal variations of the surface determined by all local parametrisations of S.

The restriction that we consider only normal variations of the form given in (9.1) may be justified, at least intuitively, by noting that, firstly, any tangential variation of $\boldsymbol{x}(\bar{D})$ merely slides $\boldsymbol{x}(\bar{D})$ along itself, so if the boundary is fixed then the area doesn't change, and, secondly, **any** variation of $\boldsymbol{x}(\bar{D})$ in an orthogonal direction may, to first order, be written as in (9.1).

The following theorem gives a very satisfying mathematical characterisation of minimal surfaces in terms of the *mean curvature* H. We recall that mean curvature was defined in

§5.4, where we found that, in terms of the coefficients of the first and second fundamental forms,

$$H = \frac{1}{2}\frac{EN - 2FM + GL}{EG - F^2}. \tag{9.3}$$

Theorem 1 *A surface S in $\mathbb{R}^3$ is a minimal surface if and only if the mean curvature H of S is everywhere zero.*

Proof Integration on surfaces, and, in particular, area of surfaces, was discussed in §3.7. For the main part of the proof of the theorem, we shall use equation (3.20) to compute the area $A(r)$, and hence find $A'(0)$ for the normal variation $\boldsymbol{x}^r(u, v)$ determined by a function h.

Let E^r, F^r and G^r be the coefficients of the first fundamental form of $\boldsymbol{x}^r|\bar{D}$. It follows from (9.2) that

$$\begin{aligned}
E^r &= E + 2rh\,\boldsymbol{x}_u.\boldsymbol{N}_u + o(r),\\
F^r &= F + rh(\boldsymbol{x}_u.\boldsymbol{N}_v + \boldsymbol{x}_v.\boldsymbol{N}_u) + o(r),\\
G^r &= G + 2rh\,\boldsymbol{x}_v.\boldsymbol{N}_v + o(r),
\end{aligned}$$

where, in these and all the following equations in this proof, $o(r)$ stands for a remainder term R with the property that

$$\lim_{r\to 0}\frac{R}{r} = 0,$$

or, equivalently, both R and $\partial R/\partial r$ are zero when $r = 0$.

It now follows from the definitions (5.9) of L, M and N that

$$\begin{aligned}
E^r &= E - 2rhL + o(r),\\
F^r &= F - 2rhM + o(r),\\
G^r &= G - 2rhN + o(r),
\end{aligned}$$

so that, using (9.3) for the second equality,

$$\begin{aligned}
E^rG^r - (F^r)^2 &= EG - F^2 - 2rh(EN - 2FM + GL) + o(r)\\
&= (1 - 4rhH + o(r))\,(EG - F^2).
\end{aligned}$$

We therefore have

$$\sqrt{E^rG^r - (F^r)^2} = (1 - 2rhH + o(r))\,\sqrt{EG - F^2},$$

so, using equation (3.20), the area $A(r)$ of $\boldsymbol{x}^r(\bar{D})$ is given by

$$\begin{aligned}
A(r) &= \iint_{\bar{D}} \sqrt{E^rG^r - (F^r)^2}\,dudv\\
&= \iint_{\bar{D}} (1 - 2rhH + o(r))\,\sqrt{EG - F^2}\,dudv.
\end{aligned}$$

We now use the result (often called 'differentiating under the integral sign') which says that if $f(r,u,v)$ is smooth then

$$\frac{\partial}{\partial r}\left(\iint f\,dudv\right) = \iint \frac{\partial f}{\partial r}\,dudv.$$

Applying this result, we find that

$$A'(0) = -\iint_{\bar{D}} 2hH\sqrt{EG-F^2}\,dudv. \tag{9.4}$$

Our calculation of $A'(0)$ enables us to complete the proof of Theorem 1 quite easily. Indeed, if $H = 0$ then, from above, $A'(0) = 0$ for all normal variations, so that S is minimal. On the other hand, if $A'(0) = 0$ for all normal variations, consider the variation arising from taking $h(u,v) = H(u,v)$. Then the integrand in (9.4) becomes $2H^2\sqrt{EG-F^2}$, which must be identically zero, for otherwise $A'(0) < 0$. Hence $H = 0$. □

It follows from Theorem 1 that a minimal surface S is one for which the principal curvatures κ_1 and κ_2 satisfy $\kappa_1 + \kappa_2 = 0$. This means that the maximum and minimum values of the normal curvatures of curves on S through a point $p \in S$ are equal in magnitude but have opposite signs, and, in particular, the Gaussian curvature K is non-positive at all points. The saddle shape of a minimal surface around any of its points is apparent in the illustrations of minimal surfaces given in this chapter.

If we want to see whether a bounded piece of a minimal surface is actually locally area **minimising**, we would now need to compute $A''(0)$ (which is not too hard), and then work out conditions under which we could deduce that $A''(0) < 0$ for every local variation of the surface (and, if $A''(0) = 0$, look at higher derivatives). Even if we could do this, the question of whether the surface was globally (as opposed to locally) area minimising would still not be resolved. We mentioned in Chapter 7 that any sufficiently short piece of a geodesic on a surface S is a length **minimising** curve on S between its end points, and a similar result holds for minimal surfaces; it can be shown that a sufficiently small piece of a minimal surface is the surface of minimum area spanning its boundary.

The problem of **existence** of a surface of minimum area (or, more generally, a minimal surface) spanning any simple closed curve in $\mathbb{R}^3$ is known as *Plateau's problem*, named after a Belgian physicist who carried out experiments with soap films in the mid-nineteenth century. Jesse Douglas and Tibor Radó (independently, in the 1930s) were the first two people to obtain significant existence results for a large class of bounding curves. For instance, they proved that there exists a connected surface of minimum area spanning any given simple closed curve in $\mathbb{R}^3$.

9.2 Examples and first properties

Theorem 1 of §9.1 confirms that every plane is a minimal surface, while Example 5 of §5.6 shows that catenoids are also minimal. Similar calculations show that helicoids are minimal, so we now have three types of examples of minimal surfaces without self-intersections in $\mathbb{R}^3$. These surfaces are all closed unbounded subsets of $\mathbb{R}^3$.

Soap films provide our physical model and intuition for minimal surfaces, and it is interesting to explore whether intuition leads us in the right direction. Unlike soap bubbles, soap films must always be supported on wire frames, so we might expect that "free-standing" bounded minimal surfaces don't exist. This is indeed the case; the following proposition may be interpreted as saying that any bounded minimal surface must have a "supporting boundary".

Proposition 1 *There are no compact minimal surfaces in* $\mathbb{R}^3$.

Proof In Theorem 4 of §5.10, we showed that every compact surface in $\mathbb{R}^3$ has an elliptic point. At such a point the principal curvatures are both non-zero and have the same sign. The mean curvature, being the average of the principal curvatures, cannot be zero at an elliptic point. □

Catenoids and helicoids were found to be minimal by Euler and by Meusnier in the eighteenth century, but it proved very hard to find other examples of closed minimal surfaces without self-intersections in $\mathbb{R}^3$. In fact, the next ones were not found until 1835, when Scherk discovered a 1-parameter family of such surfaces. However, none of these surfaces have *finite topology*; that is to say, none of them are homeomorphic to a compact surface with a finite number of points removed (for instance, the plane is homeomorphic to a sphere with one point removed, while the catenoid is homeomorphic to a sphere with two points removed).

It was widely conjectured that planes, catenoids, and helicoids are the only closed minimal surfaces with no self-intersections and finite topology in $\mathbb{R}^3$. However, in 1982 a new closed minimal surface with finite topology was described mathematically by Celso Costa, and, helped by the then new science of computer visualisation, it was subsequently shown by David Hoffman, Jim Hoffman and Bill Meeks that Costa's minimal surface has no self-intersections. Of course, they could not tell this by simply looking at pictures on a computer screen; apart from inaccuracies of calculations, a computer can only display a bounded part of any surface, and hence, by Proposition 1, can never display the whole of a closed minimal surface. However, the pictures they obtained gave Hoffman and his colleagues the crucial clues they needed (concerning the symmetry of the surface) to enable them to provide a mathematical proof that Costa's surface had no self-intersections. See Figure 9.8.

The discovery of Costa's minimal surface and its properties proved to be a real breakthrough, and the impetus provided by these discoveries led to rapid and continuing progress in the global theory of minimal surfaces. In §9.8, we describe the *Weierstrass–Enneper representation*, which was used by Costa and Hoffman and co-workers, and is still used in current research to construct and investigate minimal surfaces.

We now give a characterisation of minimal surfaces in terms of the Gauss map $\boldsymbol{N}$. We begin by noting that $\boldsymbol{N}$ is conformal if and only if the principal curvatures κ_1 and κ_2 satisfy $\kappa_1 = \pm\kappa_2 \neq 0$. If $\kappa_1 = \kappa_2 \neq 0$ then the derivative $d\boldsymbol{N}$ preserves the orientation induced on S by $\boldsymbol{N}$, while if $\kappa_1 = -\kappa_2 \neq 0$ then $d\boldsymbol{N}$ reverses orientation.

A smooth map f between oriented surfaces is said to be *weakly conformal* if it preserves angles and orientation at those points where its derivative df is non-zero. Similarly, f is *weakly anti-conformal* if it preserves angles but reverses orientation at those points where its derivative is non-zero.

Proposition 2 *(i) A surface S is minimal if and only if the Gauss map $\boldsymbol{N}$ is weakly anti-conformal (when, as usual, the sphere is given the orientation determined by the outward-pointing unit normal).*

(ii) A connected surface S is an open subset of a sphere or a plane if and only if the Gauss map $\boldsymbol{N}$ is weakly conformal.

We note that (as is clear from the corresponding conditions on the principal curvatures) $\boldsymbol{N}$ is weakly anti-conformal (using the orientation on S determined by $\boldsymbol{N}$) if and only if $-\boldsymbol{N}$ is weakly anti-conformal (using the orientation on S determined by $-\boldsymbol{N}$), so the criteria given in the proposition are independent of the (local) orientation chosen on S. Indeed, S doesn't need to be an orientable surface in order to apply Proposition 2.

We conclude this section by giving another characterisation of minimal surfaces. This follows immediately from Exercise 5.22.

Proposition 3 *A surface S with $K < 0$ is minimal if and only if the asymptotic curves intersect orthogonally at all points.*

9.3 Bernstein's Theorem

The characterisation of minimal surfaces in $\mathbb{R}^3$ as those surfaces with zero mean curvature enables us to write down a differential equation which must be satisfied by a function $f(x, y)$ in order that its graph be a minimal surface. This equation, which we give in Lemma 1, is obtained by putting $EN - 2FM + GL = 0$ when we use the expressions obtained in Example 1 of §3.2 and Exercise 5.1 for E, F, G and L, M, N for the standard parametrisation of a graph. You are asked to prove this lemma in Exercise 9.4.

Lemma 1 *The graph $\Gamma(f)$ of a real-valued function $f(x, y)$ is a minimal surface if and only if*

$$(1 + {f_y}^2)f_{xx} - 2f_x f_y f_{xy} + (1 + {f_x}^2)f_{yy} = 0. \tag{9.5}$$

This is a highly non-linear elliptic partial differential equation, which indicates that, without some clever ideas, it is likely to be difficult to find solutions to the equation $H = 0$. Fortunately, as we shall see, there are some very clever ideas at hand.

Bernstein's Theorem, which he proved in 1917, concerns itself with looking for solutions of (9.5) defined over the **whole** of the xy-plane. It is clear that a function of the form $f(x, y) = ax + by + c$, with a, b, c being constant, is such a solution of (9.5), and, of course, the corresponding graph is a plane. Bernstein's Theorem states that these are the only solutions defined on the whole of the plane.

Theorem 2 (Bernstein) *let $\Gamma(f)$ be the graph of a function $f(x, y)$ defined on the whole of the xy-plane. If $\Gamma(f)$ is a minimal surface then $\Gamma(f)$ is a plane in $\mathbb{R}^3$.*

In Exercise 9.4, you are invited to find all solutions of (9.5) of the form $f(x, y) = g(x) + h(y)$. You will find that, modulo translations and rescaling, there are only two non-linear solutions, namely

$$f(x, y) = \pm \log \left(\frac{\cos x}{\cos y} \right). \tag{9.6}$$

The graph of each function is a non-planar minimal surface, but this does not contradict Bernstein's Theorem; the logarithm function is only defined for positive real numbers, so the function f is only defined over the "white squares" of an infinite square chess-board pattern on the plane. When completed by adding vertical lines over the corners of the squares of the chess-board (and taking the $+$ sign in (9.6) for definiteness), the resulting surface is *Scherk's first surface* (Figure 9.4), which has equation $e^z \cos y = \cos x$. This is a member of the 1-parameter family of closed minimal surfaces discovered by Scherk which we referred to in §9.2. It is clear that Scherk's first surface is doubly periodic; the left hand side of Figure 9.4 shows a basic piece, and the right hand side shows how the pieces fit together.

We give a proof of Bernstein's Theorem, but, for simplicity, we make the (correct!) assumption that if f is a function defined on the whole of the plane then its graph may be covered by a single isothermal parametrisation $\boldsymbol{x}(u, v)$ whose domain of definition is also the whole of the plane.

So, let $\boldsymbol{N}$ be the unit normal to $\Gamma(f)$ which has negative z-coordinate. Then the image of the Gauss map $\boldsymbol{N}$ lies in the lower hemisphere of $S^2(1)$, and, assuming that $\Gamma(f)$ is minimal, Proposition 2 of §9.2 shows that $\boldsymbol{N}$ is weakly anti-conformal. It follows from Example 2 of §3.4 that stereographic projection π from the north pole of $S^2(1)$ onto the equatorial plane is anti-conformal (when the sphere and the plane are given their standard orientations), so that $\pi \boldsymbol{N}$ is a weakly conformal map from $\Gamma(f)$ into the unit disc. It follows that $\pi \boldsymbol{N} \boldsymbol{x}$ is a weakly conformal map from the whole of $\mathbb{R}^2$ to the interior of the unit disc.

We now recall from complex analysis that, if we identify $\mathbb{R}^2$ with $\mathbb{C}$ in the usual way, then a smooth map of the plane is weakly conformal if and only if it is complex differentiable, and, by *Liouville's Theorem*, the only bounded complex differentiable maps defined

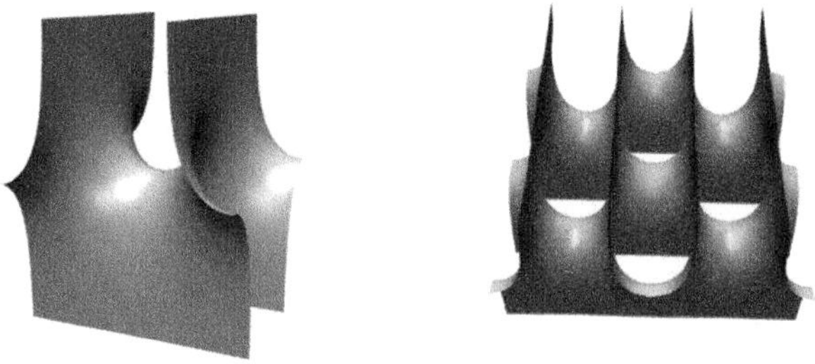

Figure 9.4 Two views of Scherk's first surface

on the whole of the complex plane are the constant maps. This implies that N is constant, and hence $\Gamma(f)$ is a plane.

Bernstein's Theorem has been improved by Osserman; he has shown that if the Gauss map of any closed minimal surface in $\mathbb{R}^3$ omits an open subset of the sphere then the surface must be a plane.

This use of the theory of complex differentiable functions leads us nicely to the next section.

9.4 Minimal surfaces and harmonic functions

The first indications of the crucial role played by complex variable theory in the study of minimal surfaces appeared at the end of §9.3. We push this idea further in this section when we investigate properties of isothermal parametrisations of minimal surfaces.

Isothermal parametrisations will prove to be of considerable importance in this chapter. As we have already remarked, any surface may be covered by a system of isothermal parametrisations, although actually doing this explicitly can be quite difficult or even impossible. However, if we do have an isothermal parametrisation of a surface S, computing the mean curvature H of S is surprisingly simple, as we now explain. For an isothermal parametrisation with $E = G = \lambda^2$, formula (9.3) for H simplifes to give

$$H = \frac{L+N}{2\lambda^2}, \tag{9.7}$$

but, in order to apply this, we would expect to have to find $\boldsymbol{x}_{uu} + \boldsymbol{x}_{vv}$ and the normal vector $\boldsymbol{N}$ in order to calculate $L + N$. However, as we now show, when we have an isothermal parametrisation there is no need to find $\boldsymbol{N}$ explicitly.

Lemma 1 *If $\boldsymbol{x} : U \to S$ is an isothermal local parametrisation of a surface S in $\mathbb{R}^3$, then $\boldsymbol{x}_{uu} + \boldsymbol{x}_{vv}$ is orthogonal to S.*

Proof Since $\boldsymbol{x}$ is isothermal, we have that

$$\boldsymbol{x}_u.\boldsymbol{x}_u = \boldsymbol{x}_v.\boldsymbol{x}_v, \quad \boldsymbol{x}_u.\boldsymbol{x}_v = 0.$$

Differentiating these expressions gives

$$\begin{aligned} \boldsymbol{x}_{uu}.\boldsymbol{x}_u &= \boldsymbol{x}_{uv}.\boldsymbol{x}_v, & \boldsymbol{x}_{uu}.\boldsymbol{x}_v + \boldsymbol{x}_u.\boldsymbol{x}_{uv} &= 0, \\ \boldsymbol{x}_{uv}.\boldsymbol{x}_u &= \boldsymbol{x}_{vv}.\boldsymbol{x}_v, & \boldsymbol{x}_{uv}.\boldsymbol{x}_v + \boldsymbol{x}_u.\boldsymbol{x}_{vv} &= 0. \end{aligned}$$

It follows that $(\boldsymbol{x}_{uu} + \boldsymbol{x}_{vv}).\boldsymbol{x}_u = 0 = (\boldsymbol{x}_{uu} + \boldsymbol{x}_{vv}).\boldsymbol{x}_v$, so that $(\boldsymbol{x}_{uu} + \boldsymbol{x}_{vv})$ is orthogonal to S as required. □

The following proposition is now immediate.

Proposition 2 *If $\boldsymbol{x} : U \to S$ is an isothermal local parametrisation of a surface S in $\mathbb{R}^3$ with $E = G = \lambda^2$, then the mean curvature H satisfies*

$$|H| = \frac{1}{2\lambda^2} |\boldsymbol{x}_{uu} + \boldsymbol{x}_{vv}| .$$

Example 3 (Catenoid) Up to rigid motions of $\mathbb{R}^3$, every catenoid has a parametrisation as a surface of revolution of the form

$$\boldsymbol{x}(u, v) = (a \cosh v \cos u, a \cosh v \sin u, av), \qquad -\pi < u < \pi,$$

where a is any positive real number. Then $E = G = a^2 \cosh^2 v$, $F = 0$, so the parametrisation is isothermal. A short calculation shows that

$$\boldsymbol{x}_{uu} + \boldsymbol{x}_{vv} = 0,$$

so that all catenoids are minimal.

It is a straightforward exercise to show that, apart from the plane, every connected minimal surface of revolution is (an open subset of) a catenoid (see Exercise 9.3).

Example 4 (Helicoid) For later convenience, we consider the helicoid parametrised by

$$\boldsymbol{x}(u, v) = (a \sinh v \sin u, -a \sinh v \cos u, au),$$

where a is any positive real number. Up to rigid motions (and possibly a reflection) of $\mathbb{R}^3$, all helicoids may be parametrised in this way. Again we have $E = G = a^2 \cosh^2 v$, $F = 0$, and

$$\boldsymbol{x}_{uu} + \boldsymbol{x}_{vv} = 0,$$

so that all helicoids are minimal.

Apart from the plane, every connected ruled minimal surface is (an open subset of) a helicoid. This is a theorem due to Catalan, and it is a rather tricky thing to show.

In order to exploit Proposition 2, we recall some material from complex analysis. Here, and for the rest of the chapter, we shall identify $\mathbb{R}^2$ with $\mathbb{C}$ in the usual way; $(u, v) \in \mathbb{R}^2$ being identified with $z = u + iv \in \mathbb{C}$.

Let $\psi(z) = x(u, v) + iy(u, v)$ be a complex valued function with real and imaginary parts $x(u, v)$ and $y(u, v)$ respectively. Then, by definition, $\psi(z)$ is complex differentiable at a point z_0 if

$$\psi' = \frac{d\psi}{dz} = \lim_{z \to z_0} \frac{\psi(z) - \psi(z_0)}{z - z_0}$$

exists, and so, in particular, is independent of the way that z approaches z_0. If we consider lines of approach parallel to the real axis and then parallel to the imaginary axis and equate the limits, we obtain

$$\psi' = x_u + iy_u = y_v - ix_v, \tag{9.8}$$

and comparing the real and imaginary parts we obtain the *Cauchy–Riemann equations*

$$x_u = y_v, \quad x_v = -y_u. \tag{9.9}$$

This shows that complex differentiability implies the Cauchy–Riemann equations, but, rather surprisingly, the converse is largely true. Specifically, a standard result in complex analysis says that if $x(u, v)$ and $y(u, v)$ are smooth functions then $\psi(z) = x(u, v) + iy(u, v)$ is complex differentiable at $z_0 \in \mathbb{C}$ if and only if the Cauchy–Riemann equations hold at z_0.

A complex valued function $\psi(z)$ is *holomorphic* on a subset D of $\mathbb{C}$ if it is complex differentiable on some open set containing D, and we recall that one of the most important properties (which has many applications) of holomorphic functions comes from the fact that their real and imaginary parts are both *harmonic*; that is to say they satisfy the 2-dimensional *Laplace equation*,

$$x_{uu} + x_{vv} = 0.$$

For example, we note that $\sin z = \sin(u + iv) = \sin u \cosh v + i \cos u \sinh v$, and a quick check shows that the real part $x(u, v) = \sin u \cosh v$ and the imaginary part $y(u, v) = \cos u \sinh v$ of $\sin z$ are both harmonic.

That the real part of a holomorphic function is harmonic is quite easy to see; we simply differentiate the first Cauchy–Riemann equation with respect to u, the second with respect to v, and then use the fact that $y_{uv} = y_{vu}$. That the imaginary part is also harmonic may be proved in a similar way.

The following result, which will be very useful for us, is an immediate consequence of Proposition 2.

Corollary 5 *Let $\mathbf{x} = (x_1, x_2, x_3)$ be an isothermal local parametrisation of a surface S in $\mathbb{R}^3$. Then that part of S covered by the image of $\mathbf{x}$ is minimal if and only if the coordinate functions x_1, x_2, x_3 of $\mathbf{x}$ are all harmonic.*

For convenience, we say that a $\mathbb{C}^3$-valued function is holomorphic if each of its coordinate functions is holomorphic. Similarly, an $\mathbb{R}^3$-valued function is said to be harmonic if each of its coordinate functions is harmonic.

9.5 Associated families

In this section we describe a remarkable feature of minimal surfaces; in a sense we make precise below, they come in 1-parameter families, each member of which has the same metric and the same Gauss map.

We begin with some topology. A *simple domain* is an open subset of $\mathbb{R}^2$ which is homeomorphic to the open unit disc. The idea is that a simple domain U is connected and has "no holes". This means that any two points p and q in U may be joined by a curve in U, and if γ_1 and γ_2 are curves in U joining p and q then γ_1 may be continuously deformed to γ_2 through curves in U joining p and q. It is clear that every point of an open subset V of the plane has an open neighbourhood in V which is a simple domain. In Figure 9.5 we give examples of sets which are not simple domains.

We have seen that the real part of a holomorphic function is harmonic, and we now recall that the converse holds on any simple domain. Specifically, we have the following result.

Figure 9.5 These are not simple domains

Proposition 1 *Let $x(u, v)$ be a smooth function defined on a simple domain U of $\mathbb{C}$. Then $x(u, v)$ is the real part of a holomorphic function $\psi(z) = x(u, v) + iy(u, v)$ defined on U if and only if x is harmonic.*

Given a harmonic function $x(u, v)$, the corresponding function $y(u, v)$, which is determined up to addition of a real constant, is called the *harmonic conjugate* of $x(u, v)$; it may be obtained by substituting for x_u and x_v in the Cauchy–Riemann equations (9.9) and then integrating.

Let U be a simple domain, and assume that $\boldsymbol{x} = (x_1, x_2, x_3) : U \to \mathbb{R}^3$ is an isothermal parametrisation of a minimal surface $S = \boldsymbol{x}(U)$. Then each of x_1, x_2 and x_3 is harmonic, and we let $\boldsymbol{y} = (y_1, y_2, y_3)$ be the harmonic conjugate of $\boldsymbol{x}$ (that is to say; y_j is the harmonic conjugate of x_j for $j = 1, 2, 3$) on U, so that (each coordinate function of) $\boldsymbol{\psi} = \boldsymbol{x} + i\boldsymbol{y}$ is holomorphic, and $\boldsymbol{x}$ and $\boldsymbol{y}$ satisfy the Cauchy–Riemann equations

$$\boldsymbol{x}_u = \boldsymbol{y}_v, \quad \boldsymbol{x}_v = -\boldsymbol{y}_u. \tag{9.10}$$

For each θ, let $\boldsymbol{x}_\theta = \cos\theta\, \boldsymbol{x} + \sin\theta\, \boldsymbol{y}$. Then $\boldsymbol{x}_\theta$ is the real part of $e^{-i\theta}\boldsymbol{\psi}$, and hence is harmonic (or just use the fact that the Laplace equation is linear). It follows from the Cauchy–Riemann equations that $\{(\boldsymbol{x}_\theta)_u, (\boldsymbol{x}_\theta)_v\}$ is obtained by rotating $\{\boldsymbol{x}_u, \boldsymbol{x}_v\}$ through angle $-\theta$ (using the orientation determined by the local parametrisation $\boldsymbol{x}$), so, in particular, $(\boldsymbol{x}_\theta)_u \times (\boldsymbol{x}_\theta)_v = \boldsymbol{x}_u \times \boldsymbol{x}_v$. Hence $\boldsymbol{x}_\theta$ is a local parametrisation of a surface S_θ (possibly with self-intersections), and each surface S_θ has the same Gauss map (or, more accurately, $\boldsymbol{N}\boldsymbol{x}_\theta$ doesn't change with θ). It also follows that the coefficients of the first fundamental forms E_θ, F_θ and G_θ are independent of θ, so that each S_θ has the same metric. In particular, each $\boldsymbol{x}_\theta$ is an isothermal parametrisation, so, since each coordinate function of $\boldsymbol{x}_\theta$ is harmonic, it follows from Corollary 5 of §9.4 that each surface S_θ is minimal.

The family $\{S_\theta\}$ of minimal surfaces is the *associated family* of S, and the minimal surface obtained for $\theta = \pi/2$ (that is to say, the surface parametrised by the harmonic conjugate $\boldsymbol{y}$), is often called the *conjugate* minimal surface – it is uniquely determined up to translation. It follows from the above working that the correspondences $f_\theta(\boldsymbol{x}(u, v)) = \boldsymbol{x}_\theta(u, v)$ give a 1-parameter family of isometries which deforms S through the associated family $\{S_\theta\}$ to the conjugate minimal surface $S_{\pi/2}$ in such a way that the Gauss map remains constant throughout the deformation.

Example 2 (Helicoid and catenoid) We consider

$$\boldsymbol{x}(u, v) = (a \sinh v \sin u, -a \sinh v \cos u, au),$$

which is the isothermal parametrisation of the helicoid given in Example 4 of §9.4. The conjugate minimal surface is

$$\boldsymbol{y}(u, v) = (a \cosh v \cos u, a \cosh v \sin u, av),$$

which (apart from its domain of definition) is the isothermal parametrisation of the catenoid given in Example 3 of §9.4. The map $\boldsymbol{x}(u, v) \to \boldsymbol{y}(u, v)$, which wraps the helicoid round the catenoid an infinite number of times, is essentially the same as that already discussed in Example 2 of §4.5 and illustrated in Figure 4.10. Animations of the corresponding 1-parameter family $\{f_\theta\}$ of deformations of the helicoid to the catenoid may be found on the internet.

The above example illustrates the proviso we mentioned earlier in the section; the surfaces S_θ may have self-intersections (or, as in the case of the catenoid, the surface may be covered more than once).

We recall that Scherk's first surface (which we discussed in §9.3) has equation $e^z \cos y = \cos x$. This minimal surface is clearly doubly periodic, and it turns out that the conjugate surface of a basic piece gives a basic piece of a complete singly periodic minimal surface, *Scherk's second surface*. Like his first surface, Scherk's second surface (Figure 9.6) has no self-intersections and has infinite topology. From afar, Scherk's second surface looks like two slightly deformed planes which intersect orthogonally in a series of holes.

You may wonder which metrics can occur as isothermal metrics of minimal surfaces. It is clear that not all isothermal metrics can occur; we have seen that minimal surfaces cannot have positive Gaussian curvature, and the Theorema Egregium states that the metric determines the Gaussian curvature. In particular, if $E = G = \lambda^2$, formula (6.11) for the Gaussian curvature in isothermal coordinates implies that, for a minimal surface, we must have that $\Delta \log \lambda \geq 0$, that is to say $\log \lambda$ is a *subharmonic function*. However, this is by no means the only restriction, as we shall see when we return to this question in §9.13.

It is also of interest to know, up to rigid motions, how many different minimal surfaces can have the same metric. We have seen in this section that all members of an associated family have the same metric, but, as we shall see at the end of §9.9, this is the only possibility; it turns out that any two minimal surfaces with the same metric are members of the same associated family (see Theorem 8 of §9.9).

9.6 Holomorphic isotropic functions

In the following sections we shall use some powerful theorems from complex analysis to study minimal surfaces. As mentioned at the beginning of the chapter, some of the material in the rest of the chapter is rather more advanced than the topics we have previously covered.

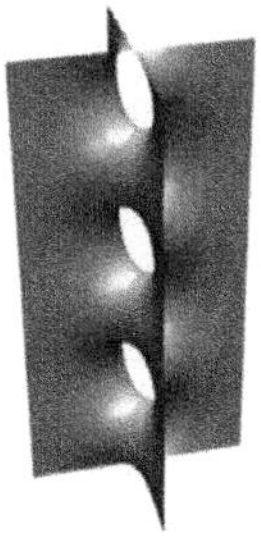

Figure 9.6 Scherk's second surface

In the next several sections we shall show how to construct (isothermal parametrisations of) minimal surfaces in $\mathbb{R}^3$ using holomorphic functions. This is achieved using the *Weierstrass–Enneper representation*, which is described in §9.8. Although this is a beautiful and remarkable aspect of the theory of minimal surfaces, it still leaves many interesting unanswered questions which belong to a vibrant area of current research.

Although it is easy to write down harmonic functions, the criterion for a minimal surface supplied by Corollary 5 of §9.4 is not straightforward to apply; a parametrisation $\boldsymbol{x} = (x_1, x_2, x_3)$ constructed from three arbitrary harmonic functions x_1, x_2, x_3 will not usually be isothermal, so that the resulting surface will not be minimal. In the next few sections we describe a method for dealing with this; rather than trying to construct $\boldsymbol{x}$ directly, it turns out to be easier to construct the partial derivatives of $\boldsymbol{x}$ and then integrate up to construct $\boldsymbol{x}$ itself.

We begin by extending the inner product of $\mathbb{R}^3$ to a symmetric complex-bilinear complex-valued form defined on $\mathbb{C}^3 \times \mathbb{C}^3$; specifically, if $\boldsymbol{z} = (z_1, z_2, z_3)$ and $\boldsymbol{w} = (w_1, w_2, w_3)$ then we let

$$\boldsymbol{z}.\boldsymbol{w} = z_1 w_1 + z_2 w_2 + z_3 w_3. \tag{9.11}$$

In particular,

$$\boldsymbol{z}.\boldsymbol{z} = {z_1}^2 + {z_2}^2 + {z_3}^2,$$

and we say that $\boldsymbol{z}$ is *isotropic* if

$$\boldsymbol{z}.\boldsymbol{z} = 0.$$

Lemma 1 *Let $\boldsymbol{z} \in \mathbb{C}^3$. Then $\boldsymbol{z}$ is isotropic if and only if the real and imaginary parts are orthogonal and have the same length.*

Proof Let $\boldsymbol{z} = \boldsymbol{a} + i\boldsymbol{b}$, where $\boldsymbol{a}$, $\boldsymbol{b}$ are the real and imaginary parts of $\boldsymbol{z}$. Then, using complex bilinearity,

$$\boldsymbol{z}.\boldsymbol{z} = (\boldsymbol{a} + i\boldsymbol{b}).(\boldsymbol{a} + i\boldsymbol{b}) = |\boldsymbol{a}|^2 - |\boldsymbol{b}|^2 + 2i\boldsymbol{a}.\boldsymbol{b},$$

and the lemma follows. □

We extend the definition of isotropic to cover functions also. A $\mathbb{C}^3$-valued function $\boldsymbol{\phi}$ is said to be *isotropic* if $\boldsymbol{\phi}(z)$ is isotropic for all z in the domain of $\boldsymbol{\phi}$.

Let U be a simple domain in $\mathbb{R}^2$, and let $\boldsymbol{x} : U \to S$ be an isothermal local parametrisation of a minimal surface S in $\mathbb{R}^3$. As we have seen, $\boldsymbol{x}$ is harmonic, and hence is the real part of a holomorphic function $\boldsymbol{\psi} : U \to \mathbb{C}^3$. Then, using (9.8) and the Cauchy–Riemann equations, we may write the derivative $\boldsymbol{\psi}'$ of $\boldsymbol{\psi}$ solely in terms of the partial derivatives of its real part, namely

$$\boldsymbol{\psi}' = \boldsymbol{x}_u - i\boldsymbol{x}_v. \tag{9.12}$$

That $\boldsymbol{\psi}'$ may be found from the real part $\boldsymbol{x}$, without having to know the corresponding imaginary part $\boldsymbol{y}$, will be very useful. Similarly, $\boldsymbol{\psi}'$ may also be written in terms of the partial derivatives of its imaginary part, namely

$$\boldsymbol{\psi}' = \boldsymbol{y}_v + i\boldsymbol{y}_u. \tag{9.13}$$

Lemma 2 *Let $\boldsymbol{\psi}$ be the holomorphic $\mathbb{C}^3$-valued function arising as above from an isothermal local parametrisation $\boldsymbol{x}(u, v)$ of a minimal surface, and let $\boldsymbol{\phi} = \boldsymbol{\psi}'$ be the derivative of $\boldsymbol{\psi}$. Then $\boldsymbol{\phi}$ is holomorphic and isotropic.*

Proof That $\boldsymbol{\phi}$ is holomorphic follows from a standard result in complex analysis, but the proof is easy so we give it. It follows from (9.12) that the Cauchy–Riemann equations for $\boldsymbol{\phi}$ are

$$\boldsymbol{x}_{uu} = -\boldsymbol{x}_{vv}, \quad \boldsymbol{x}_{uv} = \boldsymbol{x}_{vu}.$$

The first of these equations follows from the fact that $\boldsymbol{x}$ is harmonic, and the second from the commutativity of partial derivatives. Hence $\boldsymbol{\phi}$ is holomorphic, and Lemma 1, together with (9.12), shows that $\boldsymbol{\phi}$ is also isotropic. □

Example 3 (Helicoid) Let $\boldsymbol{x}(u, v)$ be the parametrisation of the helicoid given in Example 4 of §9.4. Then, taking $a = 1$ for simplicity,

$$\begin{aligned}\boldsymbol{\phi} = \boldsymbol{\psi}' &= \boldsymbol{x}_u - i\boldsymbol{x}_v \\ &= (\sinh v \cos u, \sinh v \sin u, 1) - i(\cosh v \sin u, -\cosh v \cos u, 0) \\ &= (-i \sin(iv) \cos u - i \cos(iv) \sin u, -i \sin(iv) \sin u + i \cos(iv) \cos u, 1) \\ &= (-i \sin(u + iv), i \cos(u + iv), 1).\end{aligned}$$

Hence, writing $z = u + iv$, we see that $\boldsymbol{\phi}(z) = (-i \sin z, i \cos z, 1)$, which is clearly holomorphic, and is also isotropic since $\boldsymbol{\phi}.\boldsymbol{\phi} = -\sin^2 z - \cos^2 z + 1 = 0$.

9.7 Finding minimal surfaces

We saw in §9.6 that we could associate a holomorphic isotropic $\mathbb{C}^3$-valued function $\boldsymbol{\phi}$ with any isothermal local parametrisation $\boldsymbol{x}$ of a minimal surface (as long as the domain of $\boldsymbol{x}$ is simple). This may well be interesting, but our main goal is to go in the opposite direction and construct minimal surfaces from suitable holomorphic data. In fact, even better, we shall construct isothermal parametrisations of minimal surfaces.

In this section we make progress towards this goal by reversing the process of the previous section and showing that, given a holomorphic isotropic $\mathbb{C}^3$-valued function $\boldsymbol{\phi}$ defined on a simple domain U, we may construct (an isothermal parametrisation of) a minimal surface. This will become particularly useful when we show in the next section how all such functions $\boldsymbol{\phi}$ may be constructed using suitable pairs of complex-valued holomorphic functions f and g.

In §9.6, we started with an isothermal parametrisation of a minimal surface. We then saw that this was the real part of a holomorphic $\mathbb{C}^3$-valued function $\boldsymbol{\psi}$ which we differentiated to give our holomorphic isotropic function $\boldsymbol{\phi}$. To reverse this process, we start with a holomorphic isotropic $\mathbb{C}^3$-valued function $\boldsymbol{\phi}$ defined on a simple domain U. We then use the theory of complex analysis to show that $\boldsymbol{\phi}$ has a primitive $\boldsymbol{\psi}$ defined on U (that is to say, a holomorphic function $\boldsymbol{\psi}$ with $\boldsymbol{\psi}' = \boldsymbol{\phi}$). Then, in Proposition 1, we show that the real part of $\boldsymbol{\psi}$ gives us an isothermal parametrisation of a minimal surface.

Using the usual notation of indefinite integrals, we let $\int \boldsymbol{\phi}(z)dz$ denote any primitive $\boldsymbol{\psi}$ of $\boldsymbol{\phi}$; this is well defined up to translations in $\mathbb{C}^3$. If $\boldsymbol{\phi} = (\phi_1, \phi_2, \phi_3)$ then $\int \boldsymbol{\phi}(z)dz$ is obtained by taking a primitive of each component, or, in symbols,

$$\int \boldsymbol{\phi}(z)dz = \left(\int \phi_1(z)dz, \int \phi_2(z)dz, \int \phi_3(z)dz\right).$$

That $\boldsymbol{\phi}$ has a primitive $\boldsymbol{\psi}$ defined on U is proved using the theory of contour integration. The idea is to pick a base point $z_0 \in U$ and note that, by Cauchy's Theorem, if $z \in U$ then the contour integral $\int_\gamma \boldsymbol{\phi}(t)dt$ along any contour $\gamma(t)$ in U joining z_0 to z is independent of the contour chosen. If we then define $\boldsymbol{\psi}(z)$ to be the common value of $\int_\gamma \boldsymbol{\phi}(t)dt$, the Converse of the Fundamental Theorem of Contour Integration says that $\boldsymbol{\psi}$ is a primitive of $\boldsymbol{\phi}$.

This is all very well, but, as we shall see, the primitive of a holomorphic function is usually found by applying the well-known process of 'anti-differentiation' (see Example 2, for instance).

Proposition 1 *Let $\boldsymbol{\phi} = (\phi_1, \phi_2, \phi_3)$ be a holomorphic isotropic $\mathbb{C}^3$-valued function defined on a simple domain U, and let $\boldsymbol{\psi}(z) = \int \boldsymbol{\phi}(z)dz$. If*

$$\boldsymbol{x}(u, v) = \mathrm{Re}\ \boldsymbol{\psi}(u + iv) = \mathrm{Re} \int \boldsymbol{\phi}(z)dz \tag{9.14}$$

is the real part of $\boldsymbol{\psi}(u + iv)$, then the partial derivatives of $\boldsymbol{x}$ are given by

$$\boldsymbol{x}_u - i\boldsymbol{x}_v = \boldsymbol{\phi}, \tag{9.15}$$

and, away from the zeros of $\boldsymbol{\phi}$, the map $\boldsymbol{x}$ is an isothermal parametrisation of a minimal surface S (possibly with self-intersections).

Proof Equation (9.15) follows immediately from (9.12), and it now follows from Lemma 1 of §9.6 that $\boldsymbol{x}_u$ and $\boldsymbol{x}_v$ are orthogonal and have the same length. It follows that, at those points where $\boldsymbol{\phi}$ is non-zero, $\boldsymbol{x}_u$ and $\boldsymbol{x}_v$ are also non-zero, and hence linearly independent, so that $\boldsymbol{x}$ is an isothermal parametrisation of a surface S (possibly with self-intersections). Finally, we note that $\boldsymbol{x}$ is harmonic, since it is the real part of the holomorphic function $\boldsymbol{\psi}$, so minimality of S follows from Corollary 5 of §9.4. □

Example 2 (Helicoid) We reverse the process carried out for the helicoid in Example 3 of §9.6. Let $\boldsymbol{\phi} : \mathbb{C} \to \mathbb{C}^3$ be defined by

$$\boldsymbol{\phi}(z) = (-i \sin z, i \cos z, 1).$$

We have seen that $\boldsymbol{\phi}$ is isotropic, and it is clear that

$$\boldsymbol{\psi}(z) = (i \cos z, i \sin z, z)$$

is a primitive of $\boldsymbol{\phi}$. Hence,

$$\begin{aligned} \boldsymbol{x}(u,v) &= \mathrm{Re} \int \boldsymbol{\phi}(z)dz = \mathrm{Re}\,(\boldsymbol{\psi}(u+iv)) \\ &= \mathrm{Re}\,(i\cos(u+iv), i\sin(u+iv), u+iv) \\ &= (\sinh v \sin u, -\sinh v \cos u, u), \end{aligned}$$

which gives the parametrisation of the helicoid used in Example 3 of §9.6.

Returning to our general discussion, we note that if we replace $\boldsymbol{\phi}$ by $e^{-i\theta}\boldsymbol{\phi}$ then $\boldsymbol{\psi}$ may be replaced by $e^{-i\theta}\boldsymbol{\psi}$ and we obtain the 1-parameter family S_θ of the corresponding associated minimal surface S (as discussed in §9.5). In particular, if we take $\theta = \pi/2$ then we obtain the conjugate minimal surface.

For instance, continuing with Example 2, in this case we find that

$$\begin{aligned} \mathrm{Re} \int -i\boldsymbol{\phi}(z)dz &= \mathrm{Re}\{-i\boldsymbol{\psi}(z)\} = \mathrm{Re}\,(\cos(u+iv), \sin(u+iv), -i(u+iv)) \\ &= (\cosh v \cos u, \cosh v \sin u, v), \end{aligned}$$

which is the parametrisation of the conjugate minimal surface, the catenoid, as discussed in Example 2 of §9.5.

9.8 The Weierstrass–Enneper representation

The construction described in the proof of Proposition 1 of §9.7 is potentially very useful because it shows how all minimal surfaces in $\mathbb{R}^3$ may (locally at least) be constructed from holomorphic isotropic $\mathbb{C}^3$-valued functions. However, we still have the question of how to actually **find** all such functions $\boldsymbol{\phi}$. We complete our description of the Weierstrass–Enneper representation by showing how to do this using a suitable pair of complex-valued functions.

Proposition 1 *Let f, g be complex-valued functions defined on an open set U in the complex plane. Assume that f is holomorphic on U, g is holomorphic on U except for poles, and the singularities of fg^2 are removable. Then the function $\boldsymbol{\phi} : U \to \mathbb{C}^3$ defined by*

$$\boldsymbol{\phi} = \frac{1}{2}\left(f(1-g^2), if(1+g^2), 2fg\right) \tag{9.16}$$

is holomorphic and isotropic in U. Conversely if $\boldsymbol{\phi} : U \to \mathbb{C}^3$ is holomorphic and isotropic then there exist unique functions f, g as above such that $\boldsymbol{\phi}$ is given by (9.16). *In fact if $\boldsymbol{\phi} = (\phi_1, \phi_2, \phi_3)$ then*

$$f = \phi_1 - i\phi_2, \quad g = \frac{\phi_3}{\phi_1 - i\phi_2}. \tag{9.17}$$

Finally, $\boldsymbol{\phi}$ is zero at some point if and only if f is zero at that point.

Remark 2 If g has a pole of order r at z_0, then fg^2 has a removable singularity at z_0 if and only if f has a zero of order at least $2r$ at z_0.

Proof of Proposition 1 If we are given functions f and g as in the first two sentences of the statement of the proposition, then some easy algebra shows that the holomorphic function $\boldsymbol{\phi}$ defined by (9.16) satisfies $\phi_1{}^2 + \phi_2{}^2 + \phi_3{}^2 = 0$.

Conversely, given a holomorphic isotropic $\mathbb{C}^3$-valued function $\boldsymbol{\phi}$, it is clear that (9.17) gives necessary conditions for (9.16) to hold, and more algebra shows that these conditions are also sufficient. Moreover, the singularities of fg^2 are removable since

$$fg^2 = \frac{\phi_3{}^2}{\phi_1 - i\phi_2} = -\frac{\phi_1{}^2 + \phi_2{}^2}{\phi_1 - i\phi_2} = -(\phi_1 + i\phi_2).$$

Finally, it is clear from (9.16) that $\boldsymbol{\phi}$ is zero if and only if f is zero. □

The formula (9.16) for ϕ in terms of f and g is called the *Weierstrass–Enneper formula*.

The following important theorem is the culmination of our description of the Weierstrass–Enneper representation; a method of constucting minimal surfaces using suitable pairs of complex functions.

Theorem 3 (The Weierstrass–Enneper representation) *Let $f(z), g(z)$ be complex-valued functions defined on a simple domain U. Assume that:*

(i) f is holomorphic on U,
(ii) g is holomorphic on U except for poles, and
(iii) if g has a pole of order r at z_0, then f has a zero of order at least $2r$ at z_0.

Let $\boldsymbol{x}$ be defined by

$$\boldsymbol{x}(u, v) = \mathrm{Re}\left\{\int \frac{1}{2}\left(f(1 - g^2), if(1 + g^2), 2fg\right) dz\right\}. \tag{9.18}$$

Then, away from the zeros of f, $\boldsymbol{x}$ is an isothermal parametrisation of a minimal surface (possibly with self-intersections) in $\mathbb{R}^3$.

Conversely, if U is a simple domain, and if $\boldsymbol{x} : U \to S$ is an isothermal local parametrisation of a minimal surface S in $\mathbb{R}^3$, then there exist functions f, g with properties (i), (ii) and (iii) such that (9.18) *holds.*

Finally, if (9.18) *holds, then the partial derivatives $\boldsymbol{x}_u$, $\boldsymbol{x}_v$ are related to f and g by*

$$\boldsymbol{x}_u - i\boldsymbol{x}_v = \boldsymbol{\phi},$$

where $\boldsymbol{\phi}$ is the integrand in (9.18).

Remark 4 The important direction in the above theorem is that, from a suitable choice of f and g, we may use (9.18) to construct a corresponding minimal surface. For this to work, we need that the integrand $\boldsymbol{\phi}$ in (9.18) should have an indefinite integral $\boldsymbol{\psi}$. This is assured by our assumption that U is a simple domain, but if we drop this latter assumption then the integrand in (9.18) given by a particular choice of f and g may well still have an indefinite integral $\boldsymbol{\psi}$ in which case (9.18) still defines an isothermal parametrisation $\boldsymbol{x} = \mathrm{Re}\,\boldsymbol{\psi}$ of a minimal surface. As an example of this, if we take $f(z) = z^2$ and $g(z) = 1/z^2$ then f

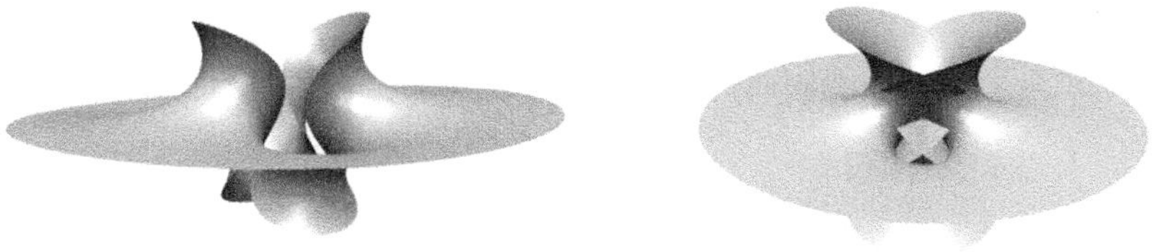

Figure 9.7 Two views of Richmond's surface

and g have the required properties on $U = \mathbb{C} \setminus \{0\}$. The corresponding function $\boldsymbol{\phi}$ (see Exercise 9.9) is given by

$$\boldsymbol{\phi}(z) = \frac{1}{2}\left(z^2 - \frac{1}{z^2}, iz^2 + \frac{i}{z^2}, 2\right),$$

which clearly has an indefinite integral $\boldsymbol{\psi}$ on U. Hence $\boldsymbol{x} : U \to \mathbb{R}^3$ given by taking $\boldsymbol{x} = \text{Re}\ \boldsymbol{\psi}$ gives an isothermal parametrisation of a minimal surface. This is *Richmond's surface* (Figure 9.7). You are asked to find an explicit formula for $\boldsymbol{x}(u, v)$ in Exercise 9.9.

Proof of Theorem 3 Assume that functions f, g satisfy (i), (ii) and (iii). Then Proposition 1 shows that the function $\boldsymbol{\phi}$ given by (9.16) is holomorphic and isotropic. Proposition 1 of §9.7 now shows that, if $\boldsymbol{x}$ is defined by (9.18) then, away from the zeros of f, $\boldsymbol{x}$ is an isothermal parametrisation of a minimal surface (possibly with self-intersections).

Conversely, assume that $\boldsymbol{x}$ is an isothermal local parametrisation of a minimal surface S. Then $\boldsymbol{x}$ is harmonic and hence is the real part of a holomorphic function $\boldsymbol{\psi}$. If we put $\boldsymbol{\psi}' = \boldsymbol{\phi}$ as usual, then Lemma 2 of §9.6 shows that $\boldsymbol{\phi}$ is holomorphic and isotropic, so, by Proposition 1 there exist functions f and g satisfying (i), (ii) and (iii) such that $\boldsymbol{\phi}$ is given by (9.16). Since $\boldsymbol{x} = \text{Re} \int \boldsymbol{\phi}(z)dz$ we see that $\boldsymbol{x}$ is given in terms of f and g by (9.18).

The final statement of the theorem follows immediately from Proposition 1 of §9.7. □

Example 5 (Helicoid) Taking $U = \mathbb{C}$ and $\boldsymbol{\phi} = (-i \sin z, i \cos z, 1)$ as in Example 2 of §9.7, we may use (9.17) to see that, in order to obtain the helicoid (as parametrised in Example 2 of §9.5), we should take

$$f(z) = -i \sin z + \cos z = e^{-iz}, \quad g(z) = e^{iz}.$$

Remark 6 As remarked earlier, if the holomorphic isotropic function $\boldsymbol{\phi}$ determines the minimal surface S, then $e^{-i\theta}\boldsymbol{\phi}$ determines the surface S_θ in the associated family. Hence, we may obtain S_θ by replacing f by $e^{-i\theta} f$ and leaving g fixed. In particular, to obtain the conjugate minimal surface, we replace f by $-if$. So, for example, the catenoid is obtained by taking $f(z) = -ie^{-iz}$, $g(z) = e^{iz}$.

We now give some more examples of how to use the Weierstrass–Enneper representation to construct minimal surfaces. We wish only to give a flavour of this vast area; much more information, often with excellent graphics, may be found on the internet.

Example 7 (Enneper's surface) Let $f(z) = 1$, $g(z) = z$. Then

$$\boldsymbol{\phi}(z) = \frac{1}{2}\left(1 - z^2, i(1 + z^2), 2z\right), \tag{9.19}$$

so, integrating, we see that

$$\boldsymbol{\psi}(z) = \frac{1}{2}\left(z - \frac{z^3}{3}, i(z + \frac{z^3}{3}), z^2\right) \tag{9.20}$$

is a primitive of $\boldsymbol{\phi}$. The resulting isothermal parametrisation of a minimal surface is

$$\boldsymbol{x}(u, v) = \frac{1}{2}\left(u - \frac{u^3}{3} + uv^2, -v + \frac{v^3}{3} - u^2 v, u^2 - v^2\right),$$

which is a parametrisation of *Enneper's surface* (Figure 9.2).

The associated family of Enneper's surface are simply rotations of that surface; no new minimal surfaces are produced. We now use properties of the function $\boldsymbol{\phi}$ given in (9.19) to prove this. In Exercise 9.10 you are invited to use similar ideas to give a slightly more computational proof using the formula (9.20) for $\boldsymbol{\psi}$.

If we rotate Enneper's surface about the vertical axis through angle μ, both $\boldsymbol{x}_u$ and $\boldsymbol{x}_v$ are also rotated about the vertical axis through angle μ, and it follows that the holomorphic isotropic function $\tilde{\boldsymbol{\phi}}$ for the rotated surface is obtained by extending this rotation to a complex linear map from $\mathbb{C}^3$ to itself. Specifically,

$$2\tilde{\boldsymbol{\phi}}(z) = \left((1 - z^2)\cos\mu - i(1 + z^2)\sin\mu, (1 - z^2)\sin\mu + i(1 + z^2)\cos\mu, 2z\right). \tag{9.21}$$

We note that the map $\tilde{\boldsymbol{\psi}}$ for the rotated surface is obtained from $\boldsymbol{\psi}$ in the same way.

A short calculation using (9.21) shows that

$$\begin{aligned} 2\tilde{\boldsymbol{\phi}}(z) &= \left(e^{-i\mu} - e^{i\mu}z^2, i\left(e^{-i\mu} + e^{i\mu}z^2\right), 2z\right) \\ &= e^{-i\mu}\left(1 - e^{2i\mu}z^2, i\left(1 + e^{2i\mu}z^2\right), 2e^{i\mu}z\right), \end{aligned}$$

so that

$$\tilde{\boldsymbol{\phi}}(e^{-i\mu}z) = e^{-i\mu}\boldsymbol{\phi}(z).$$

Hence $\tilde{\boldsymbol{\phi}}$ is obtained from $e^{-i\mu}\boldsymbol{\phi}$ by the change of variable $w = e^{-i\mu}z$. Then

$$\begin{aligned} \tilde{\boldsymbol{\psi}}(w) &= \int \tilde{\boldsymbol{\phi}}(w)\, dw = e^{-i\mu}\int \boldsymbol{\phi}(z)\frac{dw}{dz}dz \\ &= e^{-2i\mu}\int \boldsymbol{\phi}(z)\, dz = e^{-2i\mu}\boldsymbol{\psi}(z), \end{aligned} \tag{9.22}$$

so, if $\tilde{\boldsymbol{x}}(w) = \operatorname{Re}\tilde{\boldsymbol{\psi}}(w)$, then $\tilde{\boldsymbol{x}}(w)$ is obtained from $\operatorname{Re}\{e^{-2i\mu}\boldsymbol{\psi}(z)\}$ by the change of variable $w = e^{-i\mu}z$. Hence, for each θ, the surface obtained by rotating Enneper's surface about the vertical axis through angle $\mu = \theta/2$ is obtained from the member S_θ of the associated family of Enneper's surface by a change of variable $w = e^{-i\theta/2}z$. In particular, the conjugate surface is obtained by rotation through $\pi/4$. You are asked to give a direct proof of this in Exercise 9.5.

Although Enneper's surface is closed and has finite topology, it also has self-intersections.

Example 8 (Scherk's second surface) Let $f(z) = 4/(1 - z^4)$, $g(z) = iz$, and let U be the open unit disc. The resulting minimal surface is a basic piece of *Scherk's second surface* (Figure 9.6); the singly periodic closed minimal surface which, as we mentioned in §9.5, may be obtained by considering the conjugate surface to a basic piece of the (doubly periodic) first surface of Scherk. This surface, although having no self-intersections, has infinite topology.

The Weierstrass–Enneper representation has played and continues to play a crucial role in research into the local and global properties of minimal surfaces in $\mathbb{R}^3$. The usefulness of the Weierstrass–Enneper representation is clear – it allows us to construct many examples of minimal surfaces in $\mathbb{R}^3$. The drawbacks are also clear; the shape and properties of the minimal surfaces (9.18) described by the representation are not at all apparent – do they have self-intersections, for instance?

The rapid advances in computer graphics have enabled mathematicians to see on their screens models of finite pieces of the minimal surfaces which they construct using the Weierstrass–Enneper representation. This sometimes gives them insights into essential properties of the surfaces (for instance, possible symmetries) which they have then been able to establish mathematically. This, in turn, has enabled them to decide some of the difficult global properties such as the existence or otherwise of self-intersections. This type of investigation was carried out by Hoffman, Hoffman and Meeks in their pioneering work (establishing the properties of Costa's minimal surface (Figure 9.8), for example) mentioned in §9.2.

9.9 Finding I, II, N and K

Let S be the minimal surface resulting from applying the Weierstrass–Enneper representation to suitable functions f and g. Although the actual shape of the surface may not be at all clear (without using computer graphics), many of the geometrical quantities of S may be determined directly from f and g. In this section we obtain expressions in terms of f and g for the first fundamental form, the Gauss map, the second fundamental form, and the Gaussian curvature of S.

Figure 9.8 Costa's minimal surface

Proposition 1 *Let $\boldsymbol{x}(u, v)$ be the isothermal parametrisation of a minimal surface constructed from functions f and g via the Weierstrass–Enneper representation* (9.18). *Then the coefficients E, F and G of the first fundamental form are given by*

$$E = G = \frac{1}{4}|f|^2(1 + |g|^2)^2, \quad F = 0. \tag{9.23}$$

Proof That $E = G$ and $F = 0$ follows since $\boldsymbol{x}$ is an isothermal parametrisation.

Since $\boldsymbol{\phi} = \boldsymbol{x}_u - i\boldsymbol{x}_v$, if we put $E = G = \lambda^2$ then, denoting complex conjugation as usual by $\bar{\ }$,

$$\boldsymbol{\phi}.\bar{\boldsymbol{\phi}} = (\boldsymbol{x}_u - i\boldsymbol{x}_v).(\boldsymbol{x}_u + i\boldsymbol{x}_v) = 2\lambda^2,$$

so that, using (9.16),

$$\begin{aligned} 2\lambda^2 = \boldsymbol{\phi}.\bar{\boldsymbol{\phi}} &= \frac{1}{4}|f|^2(|1 - g^2|^2 + |1 + g^2|^2 + 4|g|^2) \\ &= \frac{1}{2}|f|^2(1 + |g|^2)^2, \end{aligned}$$

and the result follows. □

Continuing our theme of capturing the geometry of the minimal surface obtained from f and g in the Weierstrass–Enneper representation, we now show that the map g is essentially the Gauss map $\boldsymbol{N}$ of the corresponding minimal surface. To do this we shall use *stereographic projection* of $S^2(1)$ from the north pole $(0, 0, 1)$ onto the xy-plane (identified with $\mathbb{C}$ as usual). This is the map $\pi : S^2(1) \setminus \{(0, 0, 1)\} \to \mathbb{C}$ which maps a point $(x, y, z) \in S^2(1) \setminus \{(0, 0, 1)\}$ to the intersection with the xy-plane of the line in $\mathbb{R}^3$ through $(0, 0, 1)$ and (x, y, z). A short calculation (see, for instance, Exercise 2.1) shows that

$$\pi(x, y, z) = \frac{x + iy}{1 - z}. \tag{9.24}$$

Theorem 2 *Let S be the minimal surface constructed from functions f and g via the Weierstrass–Enneper representation* (9.18). *Then*

$$\pi \boldsymbol{N} = g,$$

where $\boldsymbol{N}$ is the Gauss map of S and π is stereographic projection from the north pole of the unit sphere.

Proof Let $\boldsymbol{x}(u, v)$ be the isothermal parametrisation of S constructed from f and g. Then

$$\begin{aligned} \boldsymbol{x}_u \times \boldsymbol{x}_v &= \frac{1}{2i}(\boldsymbol{x}_u - i\boldsymbol{x}_v) \times (\boldsymbol{x}_u + i\boldsymbol{x}_v) \\ &= \frac{1}{2i}\boldsymbol{\phi} \times \bar{\boldsymbol{\phi}}, \end{aligned}$$

so that

$$\boldsymbol{N} = \frac{\boldsymbol{x}_u \times \boldsymbol{x}_v}{|\boldsymbol{x}_u \times \boldsymbol{x}_v|} = -i\frac{\boldsymbol{\phi} \times \bar{\boldsymbol{\phi}}}{|\boldsymbol{\phi} \times \bar{\boldsymbol{\phi}}|}. \tag{9.25}$$

However, a little calculation using (9.16) shows that

$$\boldsymbol{\phi} \times \bar{\boldsymbol{\phi}} = \frac{f\bar{f}(1+g\bar{g})}{2}\left(i(g+\bar{g}), g-\bar{g}, i(g\bar{g}-1)\right), \tag{9.26}$$

so that, from (9.25),

$$\begin{aligned} \boldsymbol{N} &= \frac{(g+\bar{g}, -i(g-\bar{g}), g\bar{g}-1)}{|g|^2+1} \\ &= \frac{(2\operatorname{Re} g, 2\operatorname{Im} g, |g|^2-1)}{|g|^2+1}. \end{aligned} \tag{9.27}$$

The result now follows from the formula (9.24) for stereographic projection. □

Example 3 (Catenoid) Using the standard parametrisation of the catenoid, we saw in Example 1 of §4.1 that

$$\boldsymbol{N} = \frac{(\cos u, \sin u, -\sinh v)}{\cosh v},$$

so that

$$\pi \boldsymbol{N} = \frac{\cos u + i\sin u}{\cosh v}\frac{1}{1+\tanh v} = \frac{\cos u + i\sin u}{e^v} = e^{-v+iu},$$

which is the expression for g we found in Remark 6 of §9.8 for the standard parametrisation of the catenoid using the Weierstrass–Enneper representation.

Theorem 2 fits in nicely with some of our earlier results. We showed in Proposition 2 of §9.2 that the Gauss map $\boldsymbol{N}$ of a minimal surface is weakly anti-conformal, while, as noted in §9.3, stereographic projection from the north pole of $S^2(1)$ is anti-conformal at all points of $S^2(1)$ (except the north pole, where it is not defined). Thus we obtain the map g (which is holomorphic, and hence weakly conformal, away from its poles) as the composite of two weakly anti-conformal maps.

Having obtained an expression for the Gauss map $\boldsymbol{N}$ in terms of g, we may differentiate this to obtain the coefficients L, M and N of the second fundamental form in terms of f and g.

Proposition 4 *Let L, M and N be the coefficients of the second fundamental form of the minimal surface constructed from f and g via the Weierstrass–Enneper representation. Then $L = -N$, and*

$$L - iM = -fg'.$$

In particular, the function $L - iM$ is holomorphic and

$$L = -N = -\operatorname{Re}(fg'), \quad M = Im(fg').$$

Proof Since we are dealing with an isothermal local parametrisation of a minimal surface, (9.7) gives that $L + N = 0$, so that, using the definition of L, M and N, and recalling that we have extended the inner product to a complex-valued bilinear form on $\mathbb{C}^3 \times \mathbb{C}^3$,

$$\left(\frac{\partial}{\partial u} - i\frac{\partial}{\partial v}\right)\boldsymbol{N}.\boldsymbol{\phi} = (\boldsymbol{N}_u - i\boldsymbol{N}_v).(\boldsymbol{x}_u - i\boldsymbol{x}_v)$$
$$= 2(-L + iM). \tag{9.28}$$

We now use the expression (9.27) for $\boldsymbol{N}$ obtained in the proof of Theorem 2 to find an expression for the left hand side of the above equation in terms of f and g.

We first note that (9.12) and (9.13) applied to g give that

$$\left(\frac{\partial}{\partial u} - i\frac{\partial}{\partial v}\right)\operatorname{Re} g = i\left(\frac{\partial}{\partial u} - i\frac{\partial}{\partial v}\right)\operatorname{Im} g = g',$$

so, using the fact that $\boldsymbol{N}.\boldsymbol{\phi} = \boldsymbol{N}.(\boldsymbol{x}_u - i\boldsymbol{x}_v) = 0$, we may use (9.27) to find that

$$\left(\frac{\partial}{\partial u} - i\frac{\partial}{\partial v}\right)\boldsymbol{N}.\boldsymbol{\phi} = \frac{2}{1+|g|^2}(g', -ig', g'\bar{g}).\boldsymbol{\phi}$$
$$= \frac{fg'}{1+|g|^2}(1 - g^2 + 1 + g^2 + 2g\bar{g})$$
$$= 2fg',$$

and the required expressions for L, M and N follow from (9.28). □

Example 5 (Catenoid) As noted in Remark 6 of §9.8, if we take $f(z) = -ie^{-iz}$ and $g(z) = e^{iz}$ then the Weierstrass–Enneper representation gives the standard parametrisation of the catenoid. Since $fg' = 1$, we see immediately that $L = -N = -1$ and $M = 0$.

We may use the expressions for L, M and N obtained in Proposition 4 to find a formula for the Gaussian curvature K in terms of f and g.

Theorem 6 *The Gaussian curvature K of the minimal surface constructed from f and g via the Weierstrass–Enneper representation* (9.18) *is given by*

$$K = -\frac{16|g'|^2}{|f|^2(1+|g|^2)^4}. \tag{9.29}$$

Proof From Proposition 4,

$$LN - M^2 = -\left(\operatorname{Re}(fg')\right)^2 - \left(\operatorname{Im}(fg')\right)^2 = -|f|^2|g'|^2,$$

so, using Proposition 1 and Proposition 4,

$$K = \frac{LN - M^2}{EG - F^2} = -\frac{|L - iM|^2}{EG - F^2} = -\frac{16|f|^2|g'|^2}{|f|^4(1+|g|^2)^4} = -\frac{16|g'|^2}{|f|^2(1+|g|^2)^4},$$

as required. □

The above result confirms the fact that, for a minimal surface, $K \leq 0$.

Example 7 For the standard parametrisation of the catenoid, we have $K = -\cosh^{-4} v$. A quick check shows that we obtain the same expression for K if we take $f(z) = -ie^{-iz}$, $g(z) = e^{iz}$ in (9.29).

We have seen that all members of the associated family of a minimal surface have the same metric. We now show that, conversely, if two minimal surfaces have the same metric then, up to rigid motions of $\mathbb{R}^3$, the two surfaces are in the same associated family.

Theorem 8 *Let $\mathbf{x}(u, v)$, $\tilde{\mathbf{x}}(u, v)$ be isothermal parametrisations (with connected domain) of minimal surfaces S, $\tilde{S}$ with $E = \tilde{E}$. Then $\mathbf{x}$ and $\tilde{\mathbf{x}}$ are in the same associated family.*

Proof It follows from the Theorema Egregium that $K = \tilde{K}$, and so $LN - M^2 = \tilde{L}\tilde{N} - \tilde{M}^2$. However, $L + N = 0 = \tilde{L} + \tilde{N}$ so that

$$L^2 + M^2 = \tilde{L}^2 + \tilde{M}^2,$$

which means that the functions $L - iM$ and $\tilde{L} - i\tilde{M}$ have the same modulus. Proposition 4 says that both these functions are holomorphic, which implies that $L - iM = e^{i\theta}(\tilde{L} - i\tilde{M})$ for some constant θ. We may now use Proposition 4 again, together with Remark 6 of §9.8, to show that $\tilde{S}$ has the same coefficients of first and second fundamental forms as the element S_θ of the associated family of S. Hence, by Bonnet's Theorem (Theorem 1 of §6.3), $\tilde{S}$ may be obtained by applying a rigid motion of $\mathbb{R}^3$ to S_θ. □

9.10 Surfaces of constant mean curvature

We conclude this chapter by briefly considering surfaces of constant mean curvature or *CMC surfaces* in $\mathbb{R}^3$. As would be expected, they are somewhat more difficult to deal with than the special case of minimal surfaces, but they have some beautiful properties.

As mentioned earlier, CMC surfaces arise from the variational problem of finding compact surfaces of stationary area bounding a given volume. Simple examples of non-minimal CMC surfaces are given by the sphere and the cylinder, each of which has constant principal curvatures. The sphere is the shape assumed by a soap bubble, and is the solution to the problem of finding a surface of minimal area which encloses a given volume or, equivalently, maximising the volume enclosed by a surface of a given area. In other words it is the solution of the 2-dimensional version of the *isoperimetric problem* for curves in the plane (which says that the circle is the closed plane curve of shortest length enclosing a given area).

9.11 CMC surfaces of revolution

In this section we give an account of the description due to Delaunay of all CMC surfaces of revolution. As usual with such problems, we seek to reduce the problem to that of solving an ordinary differential equation for the generating curve.

To save on calculation, we consider that part of the surface of revolution S for which the generating curve is not orthogonal to the axis of rotation. We may thus assume a parametrisation of the form

$$\mathbf{x}(u, v) = (f(v)\cos u, f(v)\sin u, v), \qquad f(v) > 0\ \forall v, \tag{9.30}$$

and we have seen in Proposition 4 of §5.6 that the mean curvature of S is given by

$$H = \frac{ff'' - f'^2 - 1}{2f(1+f'^2)^{3/2}}.$$

It follows from this that

$$\frac{d}{dv}\left(\frac{f}{(1+f'^2)^{1/2}}\right) = -2ff'H,$$

so that

$$\frac{d}{dv}\left(Hf^2 + \frac{f}{(1+f'^2)^{1/2}}\right) = \frac{dH}{dv}f^2 + 2ff'H - 2ff'H = \frac{dH}{dv}f^2.$$

Hence S has constant mean curvature H if and only if $f(v)$ satisfies

$$Hf^2 + \frac{f}{(1+f'^2)^{1/2}} = c, \tag{9.31}$$

for some constant c.

If $H = 0$, that is to say, S is minimal, then (9.31) integrates up to give the equation of a catenary, the generating curve of the catenoid (which confirms our earlier remark that the catenoid is the only minimal surface of revolution). If $H \neq 0$ but $c = 0$, then $H < 0$ and the solutions of (9.31) are $f(v) = (H^{-2} - v^2)^{1/2}$, so that the corresponding surface is a sphere of radius $1/|H|$. If neither H nor c are zero then it turns out that the generating curve is the locus of the focus of a conic, as the conic is rolled along a straight line. If H and c have the same sign, the conic is an ellipse; if they have opposite signs, the conic is a hyperbola. These *roulettes* (Figure 9.9) are called *undularies* (for the ellipse) and *nodaries* (for the hyperbola); the corresponding surfaces of revolution are *unduloids* and *nodoids*. The roulette of a parabola is a catenary, so the corresponding surface of revolution is the (minimal) catenoid.

In all the above cases, the axis of rotation is the line along which the conic is rolled (which we call the *axis* of the roulette) The corresponding surfaces of revolution are the *Delaunay surfaces* (Figure 9.10); these (together with planes and spheres, of course) are the only CMC surfaces of revolution.

Figure 9.9 Roulettes of ellipse and hyperbola

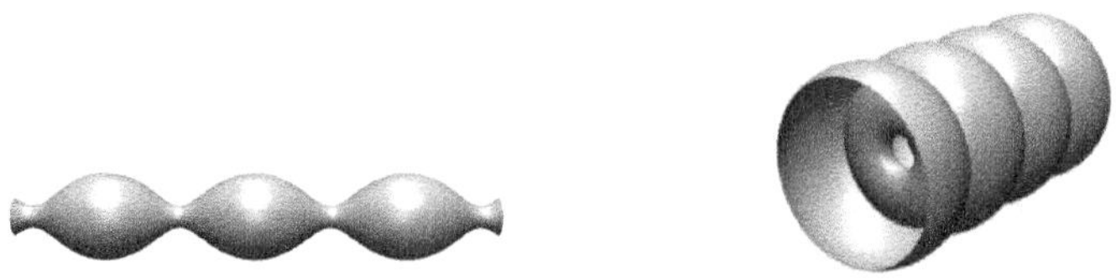

Figure 9.10 Delaunay CMC surfaces of revolution

Theorem 1 (Delaunay's Theorem) *Other than planes and spheres, the connected surfaces of revolution in $\mathbb{R}^3$ with constant mean curvature are precisely those obtained by rotating about their axes the roulettes of the conic sections.*

9.12 CMC surfaces and complex analysis

As in the case of minimal surfaces, the theory of holomorphic functions plays a key role in the description and analysis of CMC surfaces in $\mathbb{R}^3$. As we did for minimal surfaces, we shall consider isothermal local parametrisations of our surfaces, and in this section it will again be convenient to put $E = G = \lambda^2$.

The following lemma, involving the coefficients L, M and N of the second fundamental form, is a crucial first step. It is a generalisation of part of Proposition 4 of §9.9.

Lemma 1 *Let S be a connected surface in $\mathbb{R}^3$. Then S has constant mean curvature if and only if, for every isothermal local parametrisation $\boldsymbol{x}(u, v)$, the complex valued function $L - N - 2iM$ is a holomorphic function of $u + iv$.*

Proof We have seen that, for an isothermal local parametrisation, equation (9.3) for the mean curvature may be written as

$$2H\lambda^2 = L + N, \tag{9.32}$$

and that the Codazzi–Mainardi equations obtained in Chapter 6 simplify to give

$$L_v - M_u = 2H\lambda\lambda_v, \quad N_u - M_v = 2H\lambda\lambda_u. \tag{9.33}$$

Differentiating (9.32), we obtain

$$2H\lambda\lambda_u = L_u + N_u - 2\lambda^2 H_u, \quad 2H\lambda\lambda_v = L_v + N_v - 2\lambda^2 H_v,$$

and, substituting for $H\lambda\lambda_u$ and $H\lambda\lambda_v$ in (9.33), the Codazzi–Mainardi equations (9.33) may be written as

$$(L - N)_v - 2M_u = -2EH_v,$$
$$(L - N)_u + 2M_v = 2EH_u.$$

Hence $H_u = H_v = 0$ if and only if $(L - N)_u = -2M_v$ and $(L - N)_v = 2M_u$. But these last two equations are the Cauchy–Riemann equations for the complex-valued function $L - N - 2iM$. □

We note that, when using an isothermal local parametrisation, $L = N$ and $M = 0$ if and only if $\boldsymbol{x}_u$ and $\boldsymbol{x}_v$ are eigenvectors of the Weingarten map with equal eigenvalues, that is to say, if and only if we are at an umbilic. Using this, we can prove the following corollary of Lemma 1; it follows from the fact the zeros of a non-zero holomorphic function are isolated.

Corollary 2 *Let S be a connected CMC surface which is not part of a sphere or a plane. Then the umbilics of S occur at isolated points.*

We now generalise results we have already obtained for minimal surfaces by showing that CMC surfaces having the same metric occur in 1-parameter families. More precisely, we show the following.

Theorem 3 *Let $\boldsymbol{x}(u,v)$ be an isothermal local parametrisation (with connected domain) of a surface S in $\mathbb{R}^3$ with constant mean curvature c.*

(i) There is a 1-parameter family $\{\boldsymbol{x}_\theta(u,v)\}$ of isothermal local parametrisations of surfaces $\{S_\theta\}$ in $\mathbb{R}^3$ which have the same metric and the same constant mean curvature c as S.

(ii) Let $\tilde{\boldsymbol{x}}(u,v)$ be an isothermal local parametrisation of a surface $\tilde{S}$ in $\mathbb{R}^3$ having the same metric and same constant mean curvature c as S. Then, up to rigid motions of $\mathbb{R}^3$, $\tilde{\boldsymbol{x}}$ is a member of the family $\{\boldsymbol{x}_\theta\}$.

Proof (i) Let L, M and N be the coefficients of the second fundamental form of $\boldsymbol{x}$. For a real number θ, define functions $\tilde{L}$, $\tilde{M}$ and $\tilde{N}$ in terms of L, M and N by the following two equations:

$$\tilde{L}+\tilde{N}=L+N, \tag{9.34}$$

and

$$\tilde{L}-\tilde{N}-2i\tilde{M}=e^{i\theta}(L-N-2iM). \tag{9.35}$$

We show that the Gauss formula and the Codazzi–Mainardi equations hold for $\tilde{L}$, $\tilde{M}$ and $\tilde{N}$ (keeping the metric fixed), and the result will then follow from (9.34) and Bonnet's Theorem (Theorem 1 of §6.3).

To check that the Gauss formula holds, we need only check that $\tilde{L}\tilde{N}-\tilde{M}^2=LN-M^2$. To do this we note that

$$|\tilde{L}-\tilde{N}-2i\tilde{M}|^2=(\tilde{L}-\tilde{N})^2+4\tilde{M}^2=(\tilde{L}+\tilde{N})^2-4(\tilde{L}\tilde{N}-\tilde{M}^2), \tag{9.36}$$

with a similar expression for $|L-N-2iM|^2$. Equations (9.34) and (9.35) now imply that the Gauss formula holds.

We now check that the Codazzi–Mainardi equations in the form of (6.16) and (6.17) hold for $\tilde{L}$, $\tilde{M}$ and $\tilde{N}$. Using (9.34), this is equivalent to showing that

$$\tilde{L}_v-\tilde{M}_u=L_v-M_u \quad \text{and} \quad \tilde{M}_v-\tilde{N}_u=M_v-N_u.$$

However, (9.35) implies that $\tilde{L}-\tilde{N}-2i\tilde{M}$ is holomorphic, so using the Cauchy–Riemann equations we find

$$2(\tilde{L}_v-\tilde{M}_u)=2\tilde{L}_v-(\tilde{L}_v-\tilde{N}_v)=\tilde{L}_v+\tilde{N}_v=L_v+N_v=2(L_v-M_u).$$

The second equation is proved in a similar manner.

(ii) We show that the coefficients of the second fundamental form of $\tilde{\boldsymbol{x}}(u,v)$ satisfy (9.34) and (9.35) for some real number θ. The result will then follow from Bonnet's Theorem.

That (9.34) holds is immediate from the assumptions. For (9.35), we note that similar working to that employed in the proof of (i) shows that since $K=\tilde{K}$,

$$|\tilde{L}-\tilde{N}-2i\tilde{M}|^2=|L-N-2iM|^2, \tag{9.37}$$

so Lemma 1 now implies that (9.35) holds for some real number θ. □

Remark 4 In Theorem 3, if S is a minimal surface then the corresponding 1-parameter family $\{S_\theta\}$ is the associated family of S.

9.13 Link with Liouville and sinh-Gordon equations

In this section we pick up on a comment made in §9.5 and discuss the question of which isothermal metrics can occur as the metric of a minimal, or, more generally CMC, surface in $\mathbb{R}^3$. In doing so, we discover an interesting and useful relationship between CMC surfaces and two very important equations, namely *Liouville's equation* and the *sinh-Gordon equation*.

We first show, in Proposition 2, that we may pick a particularly useful isothermal local parametrisation of a CMC surface, namely one for which $M = 0$ and $L - N$ is constant. For convenience later in this section, we take this constant to be -2. As we have seen in Example 5 of §9.9, the standard parametrisation of the catenoid has this property, as has the parametrisation of Enneper's surface given in Example 7 of §9.8.

We begin by noting that if we extend the second fundamental form to a complex quadratic form (we did similar in §9.6 for the inner product) then the function $L-N-2iM$ can be written as

$$L - N - 2iM = II(\boldsymbol{x}_u - i\boldsymbol{x}_v). \tag{9.38}$$

We use this to show how the complex function $L - N - 2iM$ transforms under change of isothermal local parametrisations. This will be useful in the next two sections, but we shall require Lemma 1 from the optional §3.9 for the proof.

Lemma 1 *Let $\boldsymbol{x}(u, v)$ and $\tilde{\boldsymbol{x}}(\tilde{u}, \tilde{v})$ be two isothermal local parametrisations of a surface S in $\mathbb{R}^3$ with transition function $h(u, v) = (\tilde{u}(u, v), \tilde{v}(u, v))$ (so that $\tilde{\boldsymbol{x}}h = \boldsymbol{x}$). If $\boldsymbol{x}$ and $\tilde{\boldsymbol{x}}$ induce the same orientation on the overlap of their images, then h is a non-singular holomorphic function and the components L, M, N and $\tilde{L}, \tilde{M}, \tilde{N}$ of the second fundamental form with respect to the parametrisations $\boldsymbol{x}(u, v)$ and $\tilde{\boldsymbol{x}}(\tilde{u}, \tilde{v})$ are related by*

$$(\tilde{L} - \tilde{N} - 2i\tilde{M})h'^2 = L - N - 2iM. \tag{9.39}$$

Proof We note that $h = \tilde{\boldsymbol{x}}^{-1}\boldsymbol{x}$, so that h is a conformal orientation-preserving diffeomorphism between open sets of the Euclidean plane. Hence h is non-singular and holomorphic.

A short calculation using the change of variable formula (3.31), and (9.12) and (9.13), shows that

$$\boldsymbol{x}_u - i\boldsymbol{x}_v = (\tilde{\boldsymbol{x}}_{\tilde{u}} - i\tilde{\boldsymbol{x}}_{\tilde{v}})h',$$

and the result now follows from (9.38). □

Proposition 2 *Let p be a non-umbilic point on a CMC surface S in $\mathbb{R}^3$. Then there exists an isothermal local parametrisation of an open neighbourhood of p such that*

$$L - N = -2, \quad M = 0. \tag{9.40}$$

Proof Let $\boldsymbol{x}(u, v)$ be an isothermal local parametrisation of an open neighbourhood of p in S which contains no umbilics. It follows from Lemma 1 of §9.12 that $L - N - 2iM$ is a nowhere zero holomorphic function of $z = u + iv$, so, by restricting to a sufficiently small open neighbourhood of p, we may write $L - N - 2iM = \mu^2$ for some nowhere zero holomorphic function μ. Equation (9.39) then shows that the required change of variables may be found on a suitable open neighbourhood of p by taking h to be a primitive of $i\mu/\sqrt{2}$ (constructed as described in §9.7). □

We note from (9.39) that the isothermal parametrisation for which (9.40) holds is unique up to translations and change of sign. The geometrical significance of this local parametrisation is that (see Lemma 1 of §5.6) the coordinate curves are also lines of curvature.

In the following theorems, Δ denotes the Laplacian $\partial^2/\partial^2 u + \partial^2/\partial^2 v$. Also, to get the equations in a nice form, we take $E = G = e^{\omega}$.

Theorem 3 *(i) Each non-umbilic point on a minimal surface S determines a solution ω of the* Liouville equation

$$\Delta\omega - 2e^{-\omega} = 0.$$

(ii) For each solution ω of the Liouville equation, there is a 1-parameter family $\{\boldsymbol{x}_\theta\}$ of isothermal parametrisations of minimal surfaces $\{S_\theta\}$ which have $E = G = e^{\omega}$, $F = 0$.

Proof (i) Let $\boldsymbol{x}(u, v)$ be an isothermal parametrisation of an open neighbourhood of p in S as in Proposition 2. Then $L = -1$, $M = 0$ and $N = 1$, so the expression (6.11) for K in terms of an isothermal parametrisation with $E = G = e^{\omega}$ gives

$$-e^{-2\omega} = -(1/2)e^{-\omega}\Delta\omega,$$

which leads directly to the Liouville equation.

(ii) Conversely, if ω is a solution of the Liouville equation, then for each θ the functions $E = G = e^{\omega}$, $F = 0$, $L_\theta = -\cos\theta$, $M_\theta = \sin\theta$, $N_\theta = \cos\theta$, satisfy the conditions of Bonnet's Theorem, so there exists a corresponding parametrisation $\boldsymbol{x}_\theta$ of a surface, which is clearly minimal. □

We now consider the corresponding result for CMC surfaces which are not minimal. We note that, by rescaling and changing the direction of the unit normal if necessary, we may assume that such a surface has constant mean curvature $c = 1$.

The proof of the following theorem is similar to that of Theorem 3. In this case, though, we take $L_\theta = -\cos\theta + e^{\omega}$, $M_\theta = \sin\theta$, $N_\theta = \cos\theta + e^{\omega}$.

Theorem 4 *(i) Each non-umbilic point on a surface S of constant mean curvature $c = 1$ determines a solution ω of the* sinh-Gordon equation

$$\Delta\omega + 4\sinh\omega = 0. \tag{9.41}$$

(*ii*) *For each solution ω of the sinh-Gordon equation, there is a 1-parameter family $\{\boldsymbol{x}_\theta\}$ of isothermal parametrisation of surfaces $\{S_\theta\}$ of constant mean curvature $c = 1$ which have $E = G = e^\omega$, $F = 0$,*

The Liouville equation and the sinh-Gordon equation are both elliptic partial differential equations, for which there is a well-developed theory of existence and uniqueness of solutions. For instance, it is easily checked that $\omega = 2\log(\cosh v)$ is a solution of Liouville's equation, and, from Example 5 of §9.9, the corresponding 1-parameter family of minimal surfaces is the associated family of the catenoid.

The above ideas have been of crucial importance in many of the strides which have been taken in the understanding of CMC surfaces. An example of this is given towards the end of §9.14, when we give a very brief description of some work of Wente.

9.14 CMC spheres

We begin this section by explaining the section heading. In this section, by a "sphere" we shall mean a surface which is **diffeomorphic** to the sphere $S^2(1)$. We shall distinguish our standard model by referring to $S^2(r)$ as a **round** sphere.

The round spheres are our first, and so far our only, examples of compact CMC surfaces in $\mathbb{R}^3$. As we shall see in this section, there are very good reasons for this lack of compact examples. In order to prove Theorem 2, we need a deep result in complex analysis called the Uniformization Theorem. The proof of this is beyond the scope of this book, so we content ourself with a statement in the following form.

Theorem 1 (Uniformization Theorem) *If there is a diffeomorphism from a surface S onto the round sphere $S^2(1)$, then there is a conformal diffeomorphism from S onto $S^2(1)$.*

Using this, we can now prove the following.

Theorem 2 (Hopf) *Let S be a CMC sphere in $\mathbb{R}^3$ (possibly with self-intersections). Then S is a round sphere.*

Proof We shall show that every point of S is an umbilic and hence, by Theorem 1 of §5.8, S is a round sphere.

Using the Uniformization Theorem, we let $f : S^2(1) \to S$ be a conformal diffeomorphism. We use this to define two isothermal local parametrisations whose images cover S. Specifically, we let $\boldsymbol{x} : \mathbb{C} \to S^2(1)$ be the inverse of stereographic projection from the north pole, and let $\tilde{\boldsymbol{x}} : \mathbb{C} \to S^2(1)$ be complex conjugation followed by the inverse of stereographic projection from the south pole. If we now let $\boldsymbol{y} = f\boldsymbol{x}$ and $\tilde{\boldsymbol{y}} = f\tilde{\boldsymbol{x}}$, then $\boldsymbol{y}$ and $\tilde{\boldsymbol{y}}$ are our required isothermal parametrisations of S. Moreover, equation (4.18) gives that

$$\tilde{\boldsymbol{y}}(\tilde{z}) = \boldsymbol{y}(z) \quad \text{where} \quad \tilde{z} = h(z) = 1/z, \tag{9.42}$$

Figure 9.11 Wente torus

so, from (9.39), we now have that

$$(\tilde{L} - \tilde{N} - 2i\tilde{M})\left(-\frac{1}{z^2}\right)^2 = L - N - 2iM. \quad (9.43)$$

Lemma 1 of §9.12 says that $L - N - 2iM$ is a holomorphic function on the whole of $\mathbb{C}$, so we may write this as a Taylor series in z,

$$L - N - 2iM = \sum_{k=0}^{\infty} a_k z^k,$$

for suitable constants a_k, $k = 0, 1, \ldots$. Similarly,

$$\tilde{L} - \tilde{N} - 2i\tilde{M} = \sum_{k=0}^{\infty} \tilde{a}_k \tilde{z}^k,$$

for suitable constants $\tilde{a}_k$, $k = 0, 1, \ldots$, so that (9.43) gives

$$\sum_{k=0}^{\infty} \tilde{a}_k \left(\frac{1}{z}\right)^k = z^4 \sum_{k=0}^{\infty} a_k z^k,$$

which is possible if and only if $a_k = \tilde{a}_k = 0$ for all k. This means that $L = N$ and $M = 0$, so that every point is an umbilic, as required. □

A related result, due to Alexandrov, states that round spheres are the only compact connected CMC surfaces without self-intersections in $\mathbb{R}^3$. Indeed, Alexandrov proved a higher dimensional version of this result. As a result of his own theorem and that of Alexandrov, Hopf conjectured that, even if we allow the possibility of self-intersections, spheres are the only compact connected CMC surfaces in $\mathbb{R}^3$. However, a counterexample to this conjecture was found by Wente in 1984. He constructed an example (Figure 9.11) of a CMC torus, necessarily with self-intersections, in $\mathbb{R}^3$ using the method described at the end of §9.13. Specifically, he found certain solutions ω of the sinh-Gordon equation (9.41) which are doubly periodic, and showed that, for a suitable choice of the parameters involved, a member of the corresponding family $\{\boldsymbol{x}_\theta\}$ found in Theorem 4 of §9.13 is also doubly periodic, and hence the corresponding surface S_θ is a CMC torus.

Exercises

9.1 Show that, by applying a suitable re-scaling of Euclidean space, and by changing the direction of the unit normal if necessary, we may always assume that any CMC surface in $\mathbb{R}^3$ is either minimal or has constant mean curvature $H = 1$.

9.2 As noted in Chapter 5, although the mean curvature H of a surface S in $\mathbb{R}^3$ changes sign when $\boldsymbol{N}$ is replaced by $-\boldsymbol{N}$, the *mean curvature vector* $\boldsymbol{H} = H\boldsymbol{N}$ is independent of choice of sign of $\boldsymbol{N}$. Show that

$$\boldsymbol{x}^r = \boldsymbol{x} + r\boldsymbol{H}$$

defines a normal variation of the whole of S which reduces the area of every open neighbourhood of any point of S at which $H \neq 0$.

9.3 In this exercise we give a method of showing that the catenoid is the only connected minimal surface of revolution. To save on calculation, we will consider the (slightly special) case of a minimal surface of revolution which may be parametrised in the form

$$\boldsymbol{x}(u, v) = (f(v)\cos u,\ f(v)\sin u,\ v), \qquad v \in I,\ 0 < u < 2\pi.$$

(i) Use Proposition 4 of §5.6 to show that $ff'' - f'^2 - 1 = 0$.
(ii) By re-writing the above equation in the form

$$\frac{2f'f''}{f'^2 + 1} = \frac{2f'}{f},$$

show that $f'^2 + 1 = (kf)^2$ for some positive constant k.
(iii) Show that the graph of f is a catenary.

9.4 Let $f(x, y)$ be a smooth real-valued function defined on an open subset U of $\mathbb{R}^2$. Show that:

(i) the graph S of $f(x, y)$ is a minimal surface in $\mathbb{R}^3$ if and only if $f(x, y)$ satisfies the minimal graph equation (9.5)

$$(1 + {f_y}^2)f_{xx} - 2f_x f_y f_{xy} + (1 + {f_x}^2)f_{yy} = 0\,;$$

(ii) if $f(x, y)$ is a function of x only, then S is minimal if and only if S is a plane whose normal vector is orthogonal to the y-axis and not parallel to the x-axis;
(iii) if $f(x, y) = g(x) + h(y)$, for smooth non-zero functions $g(x)$ and $h(y)$, then S is minimal if and only if either S is a plane or f has the form

$$f(x, y) = c\,\log\left(\frac{\cos\left((x - k)/c\right)}{\cos\left((y - \tilde{k})/c\right)}\right) + d,$$

where $c, k, \tilde{k}, d$ are constants with $c \neq 0$.

Deduce that, modulo translations and re-scaling of $\mathbb{R}^3$, there are only two non-linear solutions of the minimal graph equation of the form $f(x, y) = g(x) + h(y)$, namely

$$f(x, y) = \pm\log\left(\frac{\cos x}{\cos y}\right).$$

Taking the $+$ sign gives **Scherk's first minimal surface**, as described in §9.3.

9.5 **(The conjugate to Enneper's surface)** Let

$$\boldsymbol{x}(u, v) = \frac{1}{2}\left(u - \frac{u^3}{3} + uv^2, -v + \frac{v^3}{3} - u^2 v, u^2 - v^2\right)$$

be the parametrisation of Enneper's surface given in Example 7 of §9.8. Find the conjugate minimal surface $\boldsymbol{y}(u, v)$ by finding the harmonic conjugates to each of the coordinate functions of $\boldsymbol{x}$. Let $\tilde{\boldsymbol{x}}$ be the composite of $\boldsymbol{x}$ followed by a rotation through angle $\pi/4$ about the vertical axis. Show that, if

$$\tilde{u} = (u + v)/\sqrt{2}, \quad \tilde{v} = (-u + v)/\sqrt{2},$$

then

$$\tilde{\boldsymbol{x}}(\tilde{u}, \tilde{v}) = \boldsymbol{y}(u, v),$$

so that $\tilde{\boldsymbol{x}}$ is obtained from $\boldsymbol{y}$ by the change of variable given above.

9.6 Let $\boldsymbol{x}$ be an isothermal local parametrisation of a minimal surface S in $\mathbb{R}^3$, and let $\{\boldsymbol{x}_\theta\}$ be the family of local parametrisations of the associated family $\{S_\theta\}$ of minimal surfaces of S as described in §9.5. Show that the coefficients L_θ, M_θ, N_θ of the second fundamental form of S_θ are given by

$$L_\theta = -N_\theta = L\cos\theta - M\sin\theta, \quad M_\theta = L\sin\theta + M\cos\theta.$$

The following exercises use material in §9.6 to §9.14.

9.7 Let $\boldsymbol{x}(u, v)$ be a local parametrisation of a (not necessarily minimal) surface S in $\mathbb{R}^3$, and define a nowhere zero $\mathbb{C}^3$-valued function $\boldsymbol{\phi}(u + iv)$ by setting

$$\boldsymbol{\phi} = \boldsymbol{x}_u - i\boldsymbol{x}_v. \tag{9.44}$$

Show that:

(i) $\boldsymbol{x}$ is isothermal if and only if $\boldsymbol{\phi}$ is isotropic, in which case if $E = G = \lambda^2$ then $\boldsymbol{\phi}.\overline{\boldsymbol{\phi}} = 2\lambda^2$;

(ii) $\boldsymbol{x}$ is harmonic if and only if $\boldsymbol{\phi}$ is holomorphic, in which case if $\boldsymbol{\psi}$ is a holomorphic $\mathbb{C}^3$-valued function whose real part is $\boldsymbol{x}$, then $\boldsymbol{\phi} = \boldsymbol{\psi}'$.

9.8 Let $\boldsymbol{x}(u, v)$ be the isothermal parametrisation of a minimal surface obtained from the Weierstrass–Enneper representation by taking $f(z) = (1 - e^{iz})/2$ and $g(z) = e^{-iz/2}$. Show that $\boldsymbol{x}(u, v)$ may be written as

$$\boldsymbol{x}(u, v) = \frac{1}{2}\left(u - \sin u \cosh v, 1 - \cos u \cosh v, 4\sin(u/2)\sinh(v/2)\right),$$

which is the isothermal parametrisation of **Catalan's surface** (Figure 9.1) that we investigated in Exercise 7.12.

9.9 Find the isothermal parametrisation $\boldsymbol{x}(u, v)$ of a minimal surface obtained from the Weierstrass–Enneper representation by taking $f(z) = z^2$ and $g(z) = 1/z^2$. This is **Richmond's surface**, which is illustrated in Figure 9.7. (In fact, the function $\boldsymbol{\phi}$ constructed from this choice of f and g has a singularity at $z = 0$, but as you will see, $\boldsymbol{\phi}$ still has a primitive $\boldsymbol{\psi}$ on $\mathbb{C} \setminus \{0\}$.)

9.10 **(The associated family of Enneper's surface)** This question generalises the result of Exercise 9.5, and gives an alternative treatment of some of the material in Example 7 of §9.8. Let ψ be as in (9.20), namely

$$\psi(z) = \frac{1}{2}\left(z - \frac{z^3}{3}, i(z + \frac{z^3}{3}), z^2\right),$$

and let $\boldsymbol{x}$ be the real part of ψ. Show that if $\tilde{\boldsymbol{x}}$ is the composite of $\boldsymbol{x}$ followed by a rotation through angle μ about the vertical axis then $\tilde{\boldsymbol{x}}$ is the real part of $\tilde{\psi}$, where

$$2\tilde{\psi}(z) = \left(e^{-i\mu}z - e^{i\mu}\frac{z^3}{3}, i\left(e^{-i\mu}z + e^{i\mu}\frac{z^3}{3}\right), z^2\right).$$

Now make the change of variable $w = e^{-i\mu}z$, and show that

$$\tilde{\psi}(w) = e^{-2i\mu}\psi(z),$$

which gives a slightly different way of deriving the relation between $\tilde{\psi}(w)$ and $\psi(z)$ obtained in equation (9.22), thus showing that the members of the associated family of Enneper's surface are (re-parametrisations of) rotations of the surface about the vertical axis.

9.11 We saw in Example 7 of §9.8 that the parametrisation of Enneper's surface given in Exercise 9.5 arises from the Weierstrass–Enneper representation by taking $f(z) = 1$ and $g(z) = z$. By performing a calculation similar to that outlined in Example 3 of §9.9, verify Theorem 2 of §9.9 for Enneper's surface by showing that, in standard notation, $\pi \boldsymbol{N} = g$ in this case.

9.12 Find the solution of the Liouville equation (as discussed in the proof of Theorem 3(i) of §9.13) which corresponds to the parametrisation of Enneper's surface given in Example 7 of §9.8.

9.13 Use the parametrisation

$$\boldsymbol{x}(u, v) = (\sinh v \sin u, -\sinh v \cos u, u)$$

of the helicoid given in Example 3 of §9.6 to find an isothermal parametrisation of the helicoid with $\tilde{L} = -1$, $\tilde{M} = 0$ and $\tilde{N} = 1$. Which solution of the Liouville equation does this correspond to under the correspondence given in the proof of Theorem 3(i) of §9.13?

9.14 Let ω be a solution of the Liouville equation, and let $\boldsymbol{x}(u, v)$ be the corresponding isothermal parametrisation of a minimal surface with $L = -1$, $M = 0$ and $N = 1$. If $\boldsymbol{x}_\theta$ is a member of the associated family of $\boldsymbol{x}$, find the corresponding solution of Liouville's equation as described in the proof of Theorem 3(i) of §9.13.

10 Hints or answers to some exercises

Chapter 1

1.1 It is clear that all points in the image of $\boldsymbol{\alpha}$ satisfy the equation of the astroid. Conversely, if $x^{2/3} + y^{2/3} = 1$, then there exists $u \in \mathbb{R}$ such that $(x^{1/3}, y^{1/3}) = (\cos u, \sin u)$. Thus every point of the astroid is in the image of $\boldsymbol{\alpha}$.

Trigonometric identities may be used to show that $\boldsymbol{\alpha}' = (3/2)\sin 2u(-\cos u, \sin u)$, which is zero only when u is an integer multiple of $\pi/2$. The corresponding points of the astroid are the cusps in Figure 10.1.

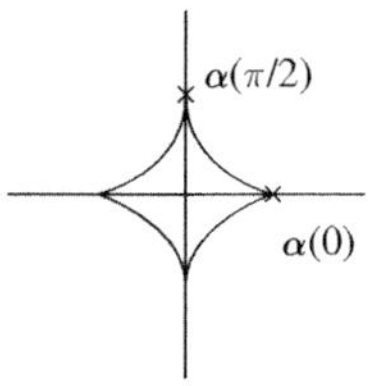

Figure 10.1 Astroid (Exercise 1.1)

The required length is

$$\frac{3}{2}\int_0^{\pi/2} \sin 2u \, du = \frac{3}{2}\,.$$

1.3 When $r = 1$, a calculation shows that $\boldsymbol{\alpha}' = \tanh u \operatorname{sech} u(\sinh u, -1)$, so that, for $u \geq 0$, $\boldsymbol{t} = \operatorname{sech} u(\sinh u, -1)$. It follows that $\boldsymbol{\alpha} + \boldsymbol{t} = (u, 0)$. A tractrix is sketched in Figure 10.2(a).

1.5 For $u \geq 0$, $|\boldsymbol{\alpha}'| = \tanh u$ and $\boldsymbol{t} = (\tanh u, -\operatorname{sech} u)$. It follows that $d\boldsymbol{t}/ds = (|\boldsymbol{\alpha}'|)^{-1}\boldsymbol{t}' = \boldsymbol{n}/\sinh u$. Hence $\kappa = \operatorname{cosech} u$.

1.6 Either: use Exercise 1.4 to show that the curvature of the catenary $\boldsymbol{\alpha}(u) = (u, \cosh u)$ is given by $\kappa = \operatorname{sech}^2 u$,
or: proceed as follows:
$\boldsymbol{\alpha}' = (1, \sinh u)$, so that $|\boldsymbol{\alpha}'| = \cosh u$ and $\boldsymbol{t} = (\operatorname{sech} u, \tanh u)$. Hence $\boldsymbol{n} = (-\tanh u, \operatorname{sech} u)$, and

$$\frac{d\boldsymbol{t}}{ds} = \frac{1}{|\boldsymbol{\alpha}'|}\boldsymbol{t}' = \frac{1}{\cosh^2 u}(-\tanh u, \operatorname{sech} u) = \frac{1}{\cosh^2 u}\boldsymbol{n}\,.$$

Hence $\kappa = \operatorname{sech}^2 u$. Three catenaries are sketched in Figure 10.2(b).

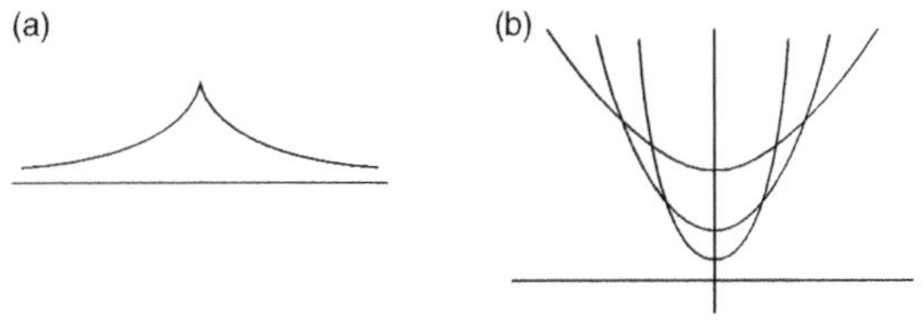

Figure 10.2 (a) A tractrix (Exercise 1.3); (b) Three catenaries (Exercise 1.6)

1.9 (i) Let s_α be arc length along $\boldsymbol{\alpha}$ measured from $u = 0$. Since $\boldsymbol{\alpha}' = (1, \sinh u)$ we see that $ds_\alpha/du = |\boldsymbol{\alpha}'| = \cosh u$. Hence $s_\alpha(u) = \sinh u$ and $\boldsymbol{t}_\alpha = (\text{sech}\, u, \tanh u)$. The result follows from formula (1.9) for the involute.

(ii) The evolute of $\boldsymbol{\alpha}$ is given by

$$\boldsymbol{\beta} = \boldsymbol{\alpha} + \frac{1}{\kappa_\alpha}\boldsymbol{n}_\alpha.$$

Here, we have (from Exercise 1.6) that $\kappa_\alpha = \text{sech}^2 u$ and $\boldsymbol{n}_\alpha = (-\tanh u, \text{sech}\, u)$. A direct substitution gives the result.

A short calculation shows that $\boldsymbol{\beta}' = 0$ if and only if $u = 0$, so this gives the only singular point of $\boldsymbol{\beta}$ (where the curve $\boldsymbol{\beta}$ has a cusp). A catenary and its evolute are sketched in Figure 10.3.

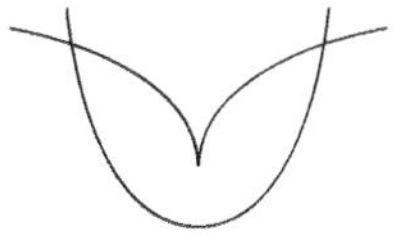

Figure 10.3 A catenary and its evolute (Exercise 1.9)

1.13 Calculations similar to those of Example 2 of §1.5 show that $|\boldsymbol{\alpha}'| = \sqrt{2}\cosh u$, $\boldsymbol{t} = (\tanh u, 1, \text{sech}\, u)/\sqrt{2}$, $\kappa = (1/2)\text{sech}^2 u$, $\boldsymbol{n} = (\text{sech}\, u, 0, -\tanh u)$, and $\boldsymbol{b} = (-\tanh u, 1, -\text{sech}\, u)/\sqrt{2}$. Differentiating one more time, we find that $d\boldsymbol{b}/ds = (1/2)\text{sech}^2 u(-\text{sech}\, u, 0, \tanh u)$, so that $\tau = -(1/2)\text{sech}^2 u$.

1.17 Assume there is a unit vector $\boldsymbol{X}_0$ such that $\boldsymbol{t}.\boldsymbol{X}_0 = c$, a constant. Then $\boldsymbol{n}.\boldsymbol{X}_0 = 0$, so that $\boldsymbol{X}_0 = c\boldsymbol{t} + c_1\boldsymbol{b}$ for some constant c_1. Then $0 = \boldsymbol{X}_0{}' = |\boldsymbol{\alpha}'|(c\kappa + c_1\tau)\boldsymbol{n}$, so that $\kappa/\tau = -c_1/c$ which is constant.

Conversely, if $\kappa/\tau = k$, a constant, the Serret–Frenet formulae may be used to show that $\boldsymbol{t} - k\boldsymbol{b}$ is constant. The result follows since $\boldsymbol{t}.(\boldsymbol{t} - k\boldsymbol{b}) = 1$.

1.18 The assumption on $\boldsymbol{\alpha}$ implies the existence of a smooth function $r(u)$ such that $\boldsymbol{\alpha} + r\boldsymbol{n} = \boldsymbol{p}$. Now differentiate and use Serret–Frenet to find r, κ and τ.

1.19 The given information implies that $\boldsymbol{n}_\alpha = \pm\boldsymbol{n}_\beta$.

(i) Differentiating $\boldsymbol{t}_\alpha.\boldsymbol{t}_\beta$ (with respect to u), and using Serret–Frenet, gives that $\boldsymbol{t}_\alpha.\boldsymbol{t}_\beta$ is constant.

(ii) The given information implies the existence of a smooth function $r(u)$ such that $\boldsymbol{\beta} = \boldsymbol{\alpha} + r\boldsymbol{n}_\alpha$. Now differentiate and use Serret–Frenet.

Chapter 2

2.1 The line through $(u, v, 0)$ and $(0, 0, 1)$ may be parametrised by $\boldsymbol{\alpha}(t) = t(u, v, 0) + (1 - t)(0, 0, 1)$. This line intersects $S^2(1)$ when $|(tu, tv, (1 - t))|^2 = 1$, and a short calculation gives that $t = 0$ or $t = 2/(u^2 + v^2 + 1)$. Since $t = 0$ corresponds to $(0, 0, 1)$, we quickly see that $\boldsymbol{x}(u, v)$ is as claimed.

The formula for $\boldsymbol{F}$ follows from consideration of similar triangles, and that $\boldsymbol{F}\boldsymbol{x}(u, v) = (u, v)$ is a routine calculation. That $\boldsymbol{x}$ is a local parametrisation as claimed follows quickly by checking conditions (S1) and (S2).

2.3(a) (i) There are many ways. The one which perhaps is closest to that given in Example 4 of §2.1 is to cover the cylinder by four local parametrisations as follows. Firstly, let $U = \{(u, v) \in \mathbb{R}^2 : -1 < u < 1,\ v \in \mathbb{R}\}$ and let $\boldsymbol{x} : U \to \mathbb{R}^3$ be given by $\boldsymbol{x}(u, v) = (u, \sqrt{1 - u^2}, v)$. Then find W and $\boldsymbol{F}$ such that conditions (S1) and (S2) are satisfied. Now cover the cylinder using three other similar local parametrisations.

(ii) Use the fact that the image of (x, y, z) under rotation about the z-axis through angle u is $(x \cos u - y \sin u, x \sin u + y \cos u, z)$.

(iii) Use $f(x, y, z) = x^2 + y^2 - 1$.

2.5(a) It is clear that grad $f = 0$ exactly when $x + y + z = 1$, and these are the points of $\mathbb{R}^3$ which are mapped to zero under f. It follows that the equation $f(x, y, z) = k$ defines a surface for any $k > 0$, while $f(x, y, z) = k$ is the empty set (and so not a surface) if $k < 0$. Note that, although grad f vanishes at *every* point satisfying $f(x, y, z) = 0$, this set is still a surface, namely the plane $x + y + z = 1$.

This example shows that Theorem 1 of §2.4 is not an "if and only if" theorem.

2.7 A routine check shows that if $f(x, y, z) = (x^2/a^2) + (y^2/b^2) + (z^2/c^2)$ then grad f is never zero on S. Hence S is a surface.

(i) One suitable choice of $\boldsymbol{F} : W \to \mathbb{R}^2$ is obtained by taking

$$W = \{(x, y, z) \in \mathbb{R}^3 : |z| < c, \text{ and if } y = 0 \text{ then } x > 0\},$$

and

$$\boldsymbol{F}(x, y, z) = \left(\operatorname{Arg}\left(\frac{x}{a} + i\frac{y}{b}\right), \arcsin\left(\frac{z}{c}\right)\right).$$

(ii) We check conditions (1), (2) and (3) of that theorem. Firstly, $f\boldsymbol{x}(u, v) = \cos^2 v + \sin^2 v = 1$, so (1) holds. Secondly, if $\boldsymbol{x}(u_1, v_1) = \boldsymbol{x}(u_2, v_2)$ then comparing the third component gives that $v_1 = v_2$. The first two components then show that $u_1 = u_2$. Finally, it is clear that $\boldsymbol{x}_u$ and $\boldsymbol{x}_v$ are linearly independent unless $\cos v = 0$, which never happens for the given range of values of v.

2.9 Show S is a surface by applying Theorem 1 of §2.4 to the function $f(x, y, z) = x \sin z - y \cos z$. It is clear that $f\boldsymbol{x}(u, v) = 0$ for all $(u, v) \in \mathbb{R}^2$, so that the image of $\boldsymbol{x}$ is a subset in S. To show that the image of $\boldsymbol{x}$ is the whole of S, we note that if $(x, y, z) \in S$ then $\boldsymbol{x}(z, x \cos z + y \sin z) = (x, y, z)$.

Part (i) is straightforward, and the working above shows that, for (ii), we can take $\boldsymbol{F}(x, y, z) = (z, x \cos z + y \sin z)$.

Chapter 3

3.1 Here, $\boldsymbol{x}_u = (1, 0, 2u)$ and $\boldsymbol{x}_v = (0, 1, 3v^2)$. Since these vectors span the tangent plane, we would need both $(1, 0, 2u).(-1, 1, 0)$ and $(0, 1, 3v^2).(-1, 1, 0)$ to be zero. Clearly, neither expression can be zero!

3.3 Let $f(x, y, z) = 2x^2 - xy + 4y^2$. Then $(\text{grad } f)(0, 1/2, 2) = (-1/2, 4, 0)$, so the unit normal there is $(-1, 8, 0)/\sqrt{65}$. A basis for the tangent plane at $(0, 1/2, 2)$ is provided by any pair of linearly independent vectors orthogonal to the normal, for instance $(0, 0, 1)$ and $(8, 1, 0)$.

3.7 Let $f(x, y, z) = (x^2/a^2) + (y^2/b^2) + (z^2/c^2)$. Then $\text{grad } f = 2(x/a^2, y/b^2, z/c^2)$, and it follows that the equation of the tangent plane based at $(a/2, b/2, c/\sqrt{2})$ is $x/a + y/b + \sqrt{2}z/c = 2$.

3.8 Here, $\boldsymbol{x}_u = (\cos v, \sin v, 1)$ and $\boldsymbol{x}_v = (-u \sin v, u \cos v, -\tan v)$. Hence $E = \boldsymbol{x}_u.\boldsymbol{x}_u = 2$, and, similarly, $F = -\tan v$, and $G = u^2 + \tan^2 v$.

3.10 Routine calculations show that the surface S has $E = G = \cosh^2 v$, $F = 0$. That $\boldsymbol{x}_u = \tilde{\boldsymbol{x}}_v$ and $\boldsymbol{x}_v = -\tilde{\boldsymbol{x}}_u$ is straightforward, and the rest of the solution is now quick.

3.12 (i) A short calculation shows that the required length is $2\int_0^1 \cosh t \, dt = 2 \sinh 1$.
(ii) The two given curves intersect when $t = r = 0$, which gives $u = v = 0$. The result follows since $\boldsymbol{\alpha}'(0).\boldsymbol{\beta}'(0) = (\boldsymbol{x}_u + \boldsymbol{x}_v).(\boldsymbol{x}_u - \boldsymbol{x}_v)(0, 0) = (E - G)(0, 0) = 0$.

3.15 Here

$$\boldsymbol{x}_u = \frac{2}{(u^2 + v^2 + 1)^2}(-u^2 + v^2 + 1, -2uv, 2u),$$

from which it follows that

$$E = \frac{4}{(u^2 + v^2 + 1)^2}.$$

That $F = 0$ and $G = 4(u^2 + v^2 + 1)^{-2}$ may be shown in a similar manner.

3.17 Let $f(u+iv) = x(u, v)+iy(u, v)$, where $x(u, v)$ and $y(u, v)$ are the real and imaginary parts of f. Then $\boldsymbol{x}(u, v) = (u, v, x(u, v), y(u, v))$ so that $\boldsymbol{x}_u = (1, 0, x_u, y_u)$ and $\boldsymbol{x}_v = (0, 1, x_v, y_v)$. That $E = G$ and $F = 0$ follows from the Cauchy–Riemann equations, $x_u = y_v$ and $x_v = -y_u$.

3.19 Here, $E = G = 1$ and $F = 0$. The family $\mathcal{F}$ is given by $\phi(u, v) = \text{constant}$, where $\phi(u, v) = v \sin u/(1 - \cos u)$, and it follows that the tangent vectors to the family $\mathcal{F}$ are scalar multiples of $u'\boldsymbol{x}_u + v'\boldsymbol{x}_v$, where $u'v = v' \sin u$. This shows that the tangent vectors to $\mathcal{F}$ are scalar multiples of $\sin u \boldsymbol{x}_u + v\boldsymbol{x}_v$, so that $\boldsymbol{\beta}(r) = \boldsymbol{x}\,(u(r), v(r))$ is an orthogonal trajectory of $\mathcal{F}$ if and only if

$$\frac{du}{dr} \sin u = -v\frac{dv}{dr}.$$

Integrating, we see that the orthogonal trajectories of $\mathcal{F}$ are given by $2 \cos u - v^2 =$ constant. Now check that $|\boldsymbol{x}(u, v)|^2$ is constant on each of these trajectories.

3.24 The given line lies on S if and only if, for all $\lambda \in \mathbb{R}$,

$$(p_1 + \lambda v_1)(p_2 + \lambda v_2) = (p_3 + \lambda v_3)\,.$$

Since $(p_1, p_2, p_3) \in S$, we see that the above holds if and only if $v_1 p_2 + p_1 v_2 = v_3$ and $v_1 v_2 = 0$. This implies that (v_1, v_2, v_3) is a scalar multiple of either $(0, 1, p_1)$ or

$(1, 0, p_2)$. Hence S is a doubly ruled surface, and the rulings intersect orthogonally if and only if $p_1 p_2 = 0$.

3.26 Let S be a ruled surface of revolution. Let ℓ be a line of the ruling, let $a \geq 0$ be the perpendicular distance of ℓ from the axis of rotation, and let θ be the angle between ℓ and the axis of rotation.

By applying a rigid motion of $\mathbb{R}^3$, we may assume that the axis of rotation is the z-axis, and that ℓ is the line parametrised by $\boldsymbol{\alpha}(t) = (a, 0, 0) + t(0, \sin\theta, \cos\theta)$.

If $a = 0$ we have a cone or a plane, so we now assume $a > 0$. When the point $\boldsymbol{\alpha}(t)$ is rotated about the z-axis to be in the half-plane $y = 0$, $x > 0$, we obtain the point $\boldsymbol{\beta}(t) = \left(\sqrt{a^2 + t^2 \sin^2\theta}, 0, t\cos\theta\right)$, so that $\boldsymbol{\beta}$ is a parametrisation of the curve with equation $x^2 - z^2 \tan^2\theta = a^2$. Considering the various values of θ gives that the ruled surfaces of revolution are planes, cones, cylinders, and those hyperboloids of one sheet discussed in Example 2 of §3.6 for which $a = b$.

3.27 If we parametrise $S^2(1)$ by rotating $(\cos v, 0, \sin v)$ about the z-axis, then $E = \cos^2 v$, $F = 0$ and $G = 1$. Hence the area of the southern hemisphere is (as expected!)

$$\int_{-\pi/2}^{0} \int_{-\pi}^{\pi} \cos v \, du \, dv = 2\pi.$$

3.29 We parametrise $S^2(1)$ as in the solution to Exercise 3.27, and let E, F and G be the coefficients of the first fundamental form.

If we define a local parametrisation of the cylinder by letting $\tilde{\boldsymbol{x}}(u, v) = f\boldsymbol{x}(u, v) = (\cos u, \sin u, \sin v)$, then routine calculations show that $\tilde{E}\tilde{G} - \tilde{F}^2 = EG - F^2$, and the result follows from (3.20).

Chapter 4

4.3 (i) Grad $f = (2x, 2y, -1)$, from which it follows that $\boldsymbol{N} = (2x, 2y, -1)/\sqrt{4z + 1}$. Hence the image of $\boldsymbol{N}$ is contained in the lower hemisphere of $S^2(1)$. However, if $X^2 + Y^2 + Z^2 = 1$ with $-1 \leq Z < 0$, then

$$\boldsymbol{N}\left(-\frac{X}{2Z}, -\frac{Y}{2Z}, \frac{1}{4Z^2}(1 - Z^2)\right) = (X, Y, Z),$$

so it follows that the image of $\boldsymbol{N}$ is the lower hemisphere (excluding the equator).

Alternatively: since S is a surface of revolution, the image of $\boldsymbol{N}$ will be obtained by rotating the image under $\boldsymbol{N}$ of the generating curve $(v, 0, v^2)$, $v \geq 0$ of S. If $(X, Z) = (\cos\theta, \sin\theta)$ for $-\pi/2 \leq \theta < 0$, then, when $v = (-1/2)\cot\theta \geq 0$, $\boldsymbol{N}(v, 0, v^2) = (X, 0, Z)$. The result follows.

4.5 Here, h is the restriction to S of the map $g : \mathbb{R}^3 \to \mathbb{R}$ given by the same formula. Since g is linear, it is smooth and is equal to its own derivative at each point. It follows that $h : S \to \mathbb{R}$ is smooth and if $\boldsymbol{X} \in T_pS$ then $dh_p(\boldsymbol{X}) = \boldsymbol{X}.\boldsymbol{v}$. This map is identically zero if and only if $\boldsymbol{v}$ is orthogonal to S at p.

4.7 The hypotheses imply the existence of a point $q_0 \in \mathbb{R}^3$ and a function $f : S \to \mathbb{R}$, smooth (except at q_0 if this latter point is in S) such that, if $p \in S$, then $p + f(p)\boldsymbol{N}(p) = q_0$. Now differentiate this.

4.8 Let $(X, Y, Z) \in T_{(x,y,z)}S^2(1)$. Then $xX + yY + zZ = 0$, and $df_{(x,y,z)}(X, Y, Z) = (aX, bY, cZ)$. The vector $(x/a, y/b, z/c)$ is orthogonal to $\tilde{S}$ at (ax, by, cz), and the inner product of $(x/a, y/b, z/c)$ with (aX, bY, cZ) is zero.

4.10 We first note that f is well-defined since, if $(r\cos\theta, r\sin\theta) = (\tilde{r}\cos\phi, \tilde{r}\sin\phi)$ then $r = \tilde{r}$ and ϕ and θ differ by an integer multiple of 2π, so that $(r\cos n\theta, r\sin n\theta) = (\tilde{r}\cos n\phi, \tilde{r}\sin n\phi)$. It is clear that the image of f is on $\tilde{S}$, while f is surjective since every point of $\tilde{S}$ may be written in the form $(r\cos\mu, r\sin\mu, br)$, $r > 0$. We parametrise the plane S (minus the non-positive real axis) by using polar coordinates; $\boldsymbol{x}(r, \theta) = (r\cos\theta, r\sin\theta, 0)$, $r > 0$, $-\pi < \theta < \pi$. Then $f(r, \theta) = (1/n)(r\cos n\theta, r\sin n\theta, br)$, and it follows that if $b = \sqrt{n^2 - 1}$ then $f_r.f_r = (1 + b^2)/n^2 = 1 = E$, $f_r.f_\theta = 0 = F$, and $f_\theta.f_\theta = r^2 = G$. The result follows from Proposition 2 of §4.3.

To see the geometry of f, note that $f(\tilde{r}e^{i\tilde{\theta}}) = f(re^{i\theta})$ if and only if $\tilde{r} = r$ and $\tilde{\theta} = \theta + (2k/n)\pi$ for some integer k. So, to model the map, cut a line from the edge of the paper to the centre, then make a cone with vertex at the centre of the paper by sliding the edges of the cut past each other until they line up again after $n - 1$ circuits.

4.13 If we parametrise the helicoid as in Example 2 of §4.5, then $E = G = \cosh^2 v$, $F = 0$. A calculation gives that $\boldsymbol{N}_u.\boldsymbol{N}_u = \boldsymbol{N}_v.\boldsymbol{N}_v = \cosh^{-2} v$ and $\boldsymbol{N}_u.\boldsymbol{N}_v = 0$, so Proposition 3 of §4.4 shows that $\boldsymbol{N}$ is conformal with conformal factor $\lambda = \cosh^{-2} v$.

4.16 Figure 10.4 shows the curves of intersection of $S^2(1)$ with (i) $x = 0$, (ii) $z = 0$, and also shows their images under $\tilde{f}$. The circle $x^2 + z^2 = 1$, $y = 0$, is left setwise fixed.

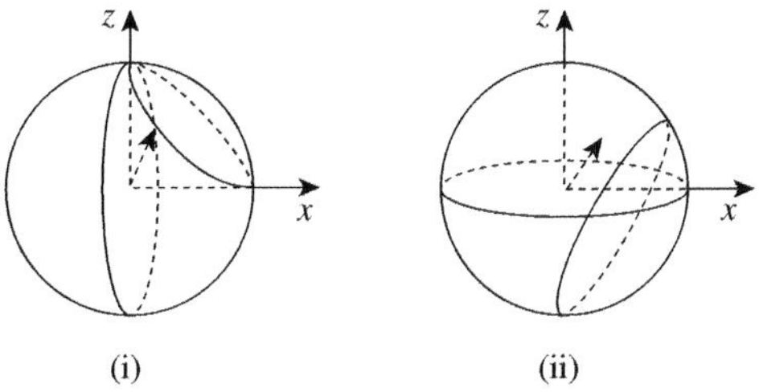

Figure 10.4 For the solution to Exercise 4.16

4.18 First show that $\boldsymbol{\alpha}'(v) = (-v^{-2}, 0, v^{-2}\sqrt{v^2 - 1})$.

It is clear that the image of f is equal to S, and calculations similar to those used to compute the coefficients of the first fundamental form of a surface of revolution show that $f_u.f_u = f_v.f_v = 1/v^2$, and $f_u.f_v = 0$. The result now follows from Proposition 3 of §4.4.

4.20 We compute the conformal factor of the map f given in Example 4 of Appendix 2. We have that

$$g(f_u, f_u) = \frac{(ad - bc)^2}{(cz + d)^2(c\bar{z} + d)^2(\text{im } f(z))^2}.$$

A short calculation shows that if $z = u + iv$ then the imaginary part of $f(z)$ is equal to $(ad - bc)v(cz + d)^{-1}(c\bar{z} + d)^{-1}$. It then follows that $g(f_u, f_u) = 1/v^2$, so that the conformal factor of f is equal to 1.

Chapter 5

5.3 Here, $N = (-\sin u, \cos u, -\sinh v)/\cosh v$. Also,

$$\begin{aligned} x_{uu} &= (-\sinh v \cos u, -\sinh v \sin u, 0), \\ x_{uv} &= (-\cosh v \sin u, -\cosh v \cos u, 0), \\ x_{vv} &= (\sinh v \cos u, \sinh v \sin u, 0). \end{aligned}$$

Hence $L = x_{uu}.N = 0$, and similarly $M = 1$ and $N = 0$. Hence $K = (LN - M^2)/(EG - F^2) = -\cosh^{-4} v$, and $H = 0$. It follows that the principal curvatures are $\pm\cosh^{-2} v$.

5.5 (a) Here, $x_u = \alpha' + v\alpha''$, $x_v = \alpha'$, so that N is the unit vector in direction $\alpha'' \times \alpha'$. This is independent of v.

5.8 Corollary 5 of §5.5 says that $K = -f''/f$.

(i) Here, $f'' + f = 0$, so, replacing v by $v + c$ for a suitable constant c, we can take $f(v) = A \cos v$ for a positive constant A and $-\pi/2 < v < \pi/2$. However, since $|\alpha'| = 1$ we need $|f'(v)| < 1$ (since $K = 0$ if $g'(v) = 0$). Hence the domain of the generating curve is as stated in the exercise.

Having determined $f(v)$ then $g(v)$ is given by the indefinite integral

$$g(v) = \int \sqrt{1 - (f'(v))^2}\, dv.$$

Figure 10.5 gives sketches of the requested generating curves. Note that the case $A = 1$ gives the unit sphere.

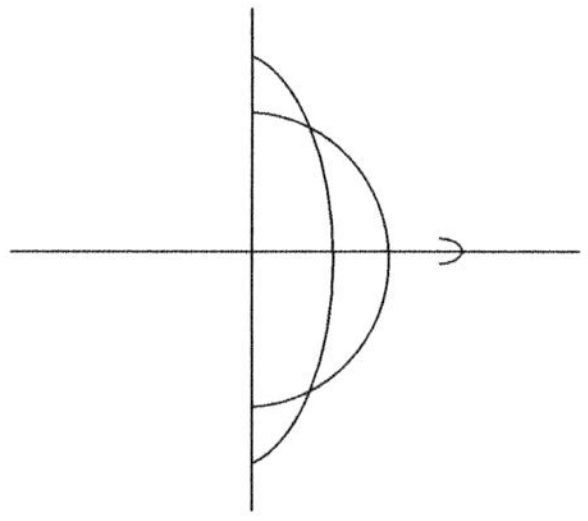

Figure 10.5 For the solution to Exercise 5.8(i)

(ii) Here $f'' = 0$, so that $f(v) = Av + B$ for suitable constants A and B, and we need $-1 \le A \le 1$. Considering various values of A and B gives the result.

(iii) Here, $f'' - f = 0$, so that $f(v) = \lambda e^v + \mu e^{-v}$.

If λ and μ are both non-zero and have the same sign (which must be positive), then, replacing v by $v + c$ for a suitable constant c, we may write $f(v) = A \cosh v$ for some positive constant A. Then the domain of the generating curve is $(-v_0, v_0)$, where $A \sinh v_0 = 1$.

If one of λ and μ is zero, we may assume that $f(v) = e^{-v}$, $v > 0$. The corresponding surface is the pseudosphere.

If λ and μ are both non-zero and have opposite signs, then, replacing v by $\pm v + c$ for a suitable constant c, we may write $f(v) = B \sinh v$, $v > 0$, for some positive constant B. This time the domain of the generating curve is $(0, v_0)$, where $B \cosh v_0 = 1$. In particular, $0 < B < 1$.

Figure 10.6 gives sketches of the requested generating curves.

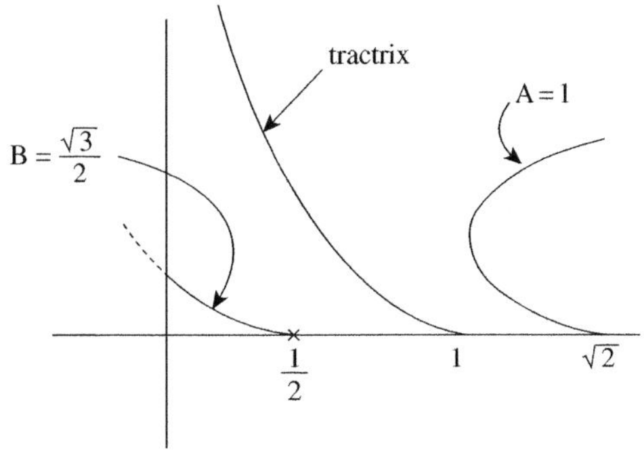

Figure 10.6 For the solution to Exercise 5.8(iii)

5.11 It follows from the formulae in Example 6 of §5.6 that the torus of revolution has points where $H = 0$ if and only if we can find a value of v such that $2b \cos v = -a$. Such a v exists if and only if $a \leq 2b$.

5.13 (i) If we parametrise the (upper half of the) cone as usual by taking

$$\boldsymbol{x}(u, v) = (v \cos u, v \sin u, v), \qquad v > 0,$$

then calculations give that $E = v^2$, $F = 0$, $G = 2$, while $L = -v/\sqrt{2}$, $M = N = 0$. At $\boldsymbol{\alpha}(t)$ we have $u = t$ and $v = e^t$, so that, at $\boldsymbol{\alpha}(t)$, we have $E = e^{2t}$, $F = 0$ and $G = 2$, while $L = -e^t/\sqrt{2}$, $M = N = 0$. Formula (5.33) now shows that $\kappa_n = -(3\sqrt{2}e^t)^{-1}$.

5.14 Let $\boldsymbol{x}(u, v)$ be a local parametrisation whose image is V. It follows that $\boldsymbol{x}.\boldsymbol{N}$ is constant, so that V is contained in a plane.

5.19 Routine calculations show that $E = \operatorname{sech}^2 v$, $F = 0$ and $G = \tanh^2 v$. We may use (5.22) to show that $L = -N = -\operatorname{sech} v \tanh v$, and $M = 0$. It now follows from (5.43) that the asymptotic curves are $u \pm v = \text{constant}$. The tangent vectors to the asymptotic curves are $\boldsymbol{x}_u \mp \boldsymbol{x}_v$, and the angle θ between these is given by

$$\cos\theta = \frac{(\boldsymbol{x}_u + \boldsymbol{x}_v).(\boldsymbol{x}_u - \boldsymbol{x}_v)}{|\boldsymbol{x}_u + \boldsymbol{x}_v|\,|\boldsymbol{x}_u - \boldsymbol{x}_v|}\,.$$

The answer now follows from a short calculation.

5.21 Let $\boldsymbol{\alpha}$ be an asymptotic curve on a surface S, and assume that $\boldsymbol{\alpha}$ is parametrised by arc length. Then $\boldsymbol{\alpha}''.\boldsymbol{N} = 0$, so that the principal normal $\boldsymbol{n}$ of $\boldsymbol{\alpha}$ is also orthogonal to $\boldsymbol{N}$. Hence, if $\boldsymbol{b}$ is the binormal of $\boldsymbol{\alpha}$ then (by choosing the correct sign for $\boldsymbol{N}$), we have that $\boldsymbol{b} = \boldsymbol{N}$. Hence $\boldsymbol{b}' = \boldsymbol{N}' = \tau\boldsymbol{n}$, so, if θ is the angle between $\boldsymbol{\alpha}'$ and the principal direction with principal curvature κ_1, then, by (5.35), $\tau^2 = |\boldsymbol{N}'|^2 = {\kappa_1}^2 \cos^2\theta + {\kappa_2}^2 \sin^2\theta$. Now use the fact that $\boldsymbol{\alpha}$ is an asymptotic curve to obtain the given formula.

5.25 We may use Exercise 5.1 to show that, if $D = 1 + g_u{}^2 + g_v{}^2$, then $L = 6u/\sqrt{D}$, $M = -6v/\sqrt{D}$, and $N = -6u/\sqrt{D}$. Hence $LN - M^2 < 0$ except at $u = v = 0$, in which case $LN - M^2 = 0$.

5.29 We use the argument and notation of the proof of Theorem 4 of §5.12. If $\boldsymbol{\alpha}(t)$ is a curve on S with $\boldsymbol{\alpha}(0) = p_0$ we have that $\boldsymbol{\alpha}'(0).q_0 = 0$ and $\boldsymbol{\alpha}''(0).q_0 \leq 0$. It follows that $\boldsymbol{N}(p_0) = q_0$, and all the normal curvatures of S at p_0 are non-positive. Hence p_0 isn't a hyperbolic point.

Chapter 6

6.2 The expressions for the Christoffel symbols follow immediately from equations (6.4). The Gauss formula (6.9) now shows that $GK = -G_{uu}/2 + G_u{}^2/4G$, from which the result follows. The formula for the Gaussian curvature may also be obtained using equation (6.10) for K in orthogonal coordinates.

6.6 The given parametrisation $\boldsymbol{x}$ of the helicoid has $E = 1 + v^2$, $F = 0$, $G = 1$, $L = N = 0$, $M = 1/\sqrt{1+v^2}$. Hence $K = -(1+v^2)^{-2}$. The formula given in Exercise 5.9 may be used to show that if $\tilde{K}$ is the Gaussian curvature of $\tilde{S}$ then $\tilde{K} = -(1+v^2)^{-2}$ also. However, the given map isn't an isometry since $\tilde{E} = v^2 + v^{-2} \neq E$.

6.8 The coordinate curves are lines of curvature if and only if $F = M = 0$, and, in this case, $\kappa_1 = L/E$ and $\kappa_2 = N/G$. The first Codazzi–Mainardi equation (6.16) becomes $2L_v = E_v(\kappa_1 + \kappa_2)$, which holds if and only if $2L_v - 2E_v\kappa_1 + (\kappa_1 - \kappa_2)E_v = 0$, which holds if and only if

$$2\frac{L_v E - L E_v}{E^2} + \frac{E_v}{E}(\kappa_1 - \kappa_2) = 0\,,$$

which quickly leads to the first of the given equations. The equivalence of the second pair of equations is proved similarly (or simply interchange u and v).

6.9 Using the equations from the previous question,

$$\begin{aligned}(L-N)_v = \{E(\kappa_1 - \kappa_2)\}_v &= E_v(\kappa_1 - \kappa_2) + E(\kappa_1 - \kappa_2)_v \\ &= -2E(\kappa_1)_v + E(\kappa_1 - \kappa_2)_v = -E(\kappa_1 + \kappa_2)_v\,.\end{aligned}$$

Hence, if $\kappa_1 + \kappa_2$ is constant then $(L - N)_v = 0$. Interchanging u and v shows that $(L - N)_u = 0$ also, so that $L - N$ is constant.

Chapter 7

7.1 We note that $d\boldsymbol{\alpha}/ds = \boldsymbol{\alpha}'/|\boldsymbol{\alpha}'|$, so that

$$\frac{d^2\boldsymbol{\alpha}}{ds^2} = \frac{\boldsymbol{\alpha}''}{|\boldsymbol{\alpha}'|^2} + \frac{1}{|\boldsymbol{\alpha}'|}\frac{d}{dt}\left(\frac{1}{|\boldsymbol{\alpha}'|}\right)\boldsymbol{\alpha}'\,.$$

It follows from (7.1) that $\kappa_g = (d^2\boldsymbol{\alpha}/ds^2).(\boldsymbol{N} \times d\boldsymbol{\alpha}/ds)$, and the desired formula follows if we substitute the expressions obtained above for $d\boldsymbol{\alpha}/ds$ and $d^2\boldsymbol{\alpha}/ds^2$.

7.6 That $\boldsymbol{\alpha}$ lies on the given cone is easy to check. Also,

$$\boldsymbol{\alpha}' = e^t(\cos t - \sin t, \sin t + \cos t, 1)\,, \text{so } |\boldsymbol{\alpha}'| = \sqrt{3}e^t\,,$$

while

$$\boldsymbol{\alpha}'' = e^t(-2\sin t, 2\cos t, 1) \text{ and } \boldsymbol{N} = \frac{1}{\sqrt{2}}(\cos t, \sin t, -1)\,.$$

A calculation using formula (7.2) for geodesic curvature now shows that $\kappa_g = (1/\sqrt{6})e^{-t}$.

7.8 (a) Let $\boldsymbol{\alpha}(s) = \boldsymbol{x}(u(s), v_0)$ be a parametrisation of $v = v_0$ by arc length. Then $du/ds = 1/\sqrt{E}$, and, since $\boldsymbol{x}_u$ and $\boldsymbol{x}_v$ are orthogonal, $\boldsymbol{N} \times d\boldsymbol{\alpha}/ds = \boldsymbol{x}_v/|\boldsymbol{x}_v|$. Hence,

$$\kappa_g = \frac{d^2\boldsymbol{\alpha}}{ds^2}.\frac{\boldsymbol{x}_v}{|\boldsymbol{x}_v|} = \frac{1}{\sqrt{E}}\frac{d}{du}\left(\frac{1}{\sqrt{E}}\boldsymbol{x}_u\right).\frac{\boldsymbol{x}_v}{\sqrt{G}} = -\frac{E_v}{2E\sqrt{G}}\,,$$

which leads to the given answer.

(b) Substitute the expressions for E, F and G in terms of f and g.

(c) Using (a), we see that the geodesic curvature of the coordinate curve $u =$ constant is $(1/2\sqrt{E})(\log G)_u$, and the result follows.

(d) Clear.

7.11 (a) Using the standard notation, the assumptions imply that $\boldsymbol{N}' = \lambda\boldsymbol{\alpha}'$ and $\boldsymbol{\alpha}'' = \mu\boldsymbol{N}$ for some scalar functions λ and μ. It follows that $(\boldsymbol{N} \times \boldsymbol{\alpha}')' = 0$ so that $\boldsymbol{N} \times \boldsymbol{\alpha}'$ is constant. It now follows that $\boldsymbol{\alpha}.(\boldsymbol{N} \times \boldsymbol{\alpha}')$ is also constant, so that $\boldsymbol{\alpha}$ lies on a plane with normal $\boldsymbol{N} \times \boldsymbol{\alpha}'$ and the result follows.

7.15 (a) The segment of the unit circle which lies in the sector has length ϕ, and when we bend the sector to form the cone, this segment maps to the parallel on the cone which is a circle of radius r where $r^2 + \beta^2 r^2 = 1$ (Figure 10.7). Hence $r = 1/\sqrt{1+\beta^2}$ so that $\phi = 2\pi/\sqrt{1+\beta^2}$, and the result follows.

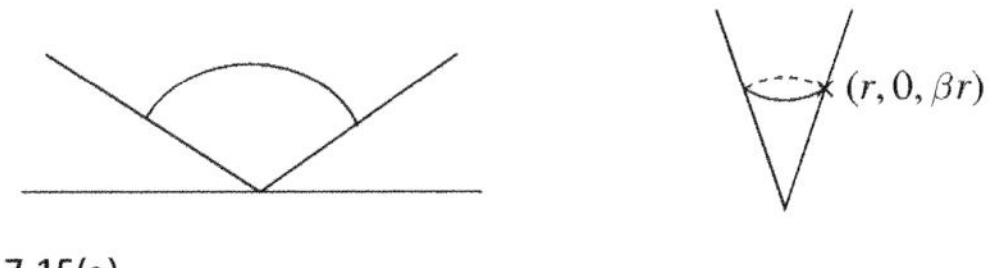

Figure 10.7 For the solution to Exercise 7.15(a)

(b), (c) and (d) Cut the plane along a half-line to the origin, and then form a cone by sliding the edges of the cut past each other until the line segments forming an angle ϕ at the origin line up to give a generator of the cone. A line ℓ in the plane not intersecting the cut gives a maximal geodesic on the cone, and two points p_1 and p_2 on ℓ give the same point on the geodesic if and only if they are equidistant from the origin 0 and the angle $\angle p_1 0 p_2$ is an integer multiple of ϕ.

If $\phi \geq \pi$, then no geodesic has self-intersections. If $\pi > \phi \geq \pi/2$ then every geodesic which is not a meridian has exactly one self-intersection. If $\pi/n > \phi \geq \pi/(n+1)$ for some integer n then every geodesic which is not a meridian has exactly n self-intersections. Figure 10.8 illustrates this for $n = 2$. In this figure, the rays from the origin drawn in a solid line map to the same meridian of the cone, and those drawn in a dashed line map to the opposite meridian. The points p_1 and p_2 (resp. q_1 and q_2) give a point where the geodesic has a self-intersection.

The result now follows from (a).

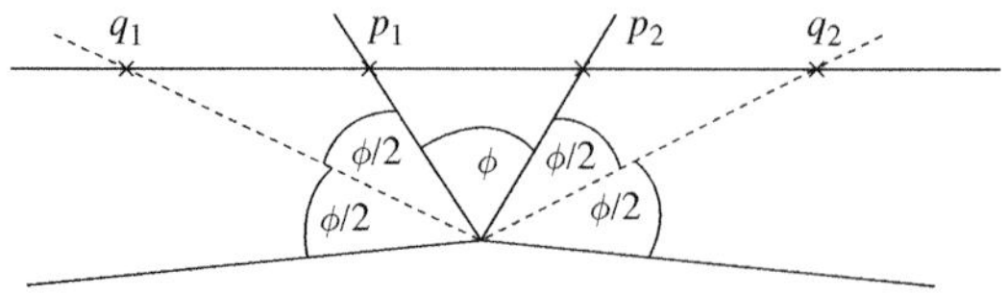

Figure 10.8 The horizontal line gives a geodesic on the cone (Exercise 7.15)

7.22 We may obtain this surface by rotating $(\cosh v, 0, \sinh v)$ about the z-axis. Let $\boldsymbol{x}(u, v)$ be the standard parametrisation of the surface, and let $\boldsymbol{\alpha}(t) = \boldsymbol{x}(u(t), v(t))$ be a closed geodesic. If $\cosh v(t)$ attains its maximum at $t = t_0$, then $|v(t)|$ also attains its maximum at $t = t_0$. Hence $v'(t_0) = 0$ so that $\boldsymbol{\alpha}$ is tangential to the parallel at $\boldsymbol{\alpha}(t_0)$. Hence from Corollary 3 of §7.6, $\cosh v(t)$ also attains its minimum at t_0, so that $\cosh v(t)$, and hence $v(t)$, is constant and $\boldsymbol{\alpha}$ is a parallel. However, for a parallel to be a geodesic, we need $\sinh v = 0$, so that the only closed geodesic is the parallel $v = 0$ (when parametrised proportional to arc length).

Chapter 8

8.2 The interior angles at v_1, v_2, v_3 and v_4 are $\pi/2$, $3\pi/2$, $\pi/2$ and $3\pi/2$ respectively. The exterior angles are $\pi/2$, $-\pi/2$, $\pi/2$ and $-\pi/2$ respectively. Figure 10.9 shows the exterior angles at v_1, v_2 and v_4 for the indicated orientation.

8.6 Parametrise the surface S as a surface of revolution,

$$\boldsymbol{x}(u, v) = (v\cos u, v\sin u, v^2), \quad -\pi < u < \pi, \ v > 0.$$

Routine calculations show that $K = (LN - M^2)/(EG - F^2) = 4(1 + 4v^2)^{-2}$ (or use (5.25)), from which it follows that

$$\iint_R K\,dA = 2\pi\left(1 - \frac{1}{\sqrt{1 + 4a^2}}\right).$$

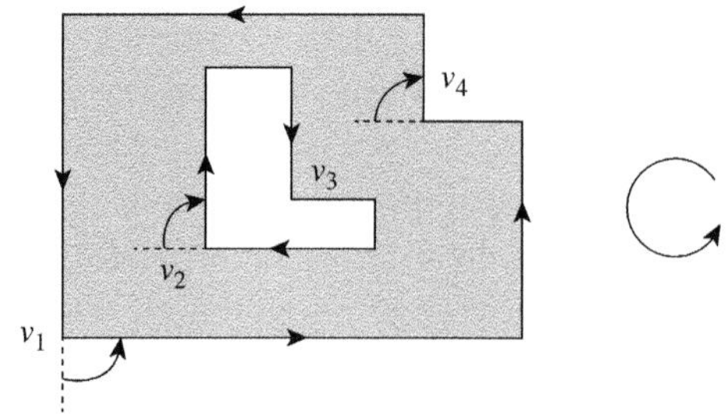

Figure 10.9 For the solution to Exercise 8.2

The boundary ∂R is the parallel $v = a$, and $\boldsymbol{n}_{\text{in}} = -\boldsymbol{x}_v/|\boldsymbol{x}_v|$. A calculation using (8.3) now shows that

$$\kappa_g = \frac{1}{a\sqrt{1+4a^2}}.$$

The length of the parallel is $2\pi a$ so that $\int_{\partial R} \kappa_g ds = 2\pi/\sqrt{1+4a^2}$. Since $\chi(R) = 1$, we now see that the Gauss–Bonnet Theorem holds for R.

8.8 This surface is diffeomorphic to a cylinder, and routine calculations show that $K < 0$ everywhere on S. It now follows from Theorem 6 of §8.6 that S has at most one simple closed geodesic. The surface intersects the plane $z = 0$ orthogonally (with the curve of intersection being an ellipse) so it follows from Exercise 7.10 that this ellipse is a geodesic on S.

Index